Symmetry and Pairing in Superconductors

NATO Science Series

*A Series presenting the results of activities sponsored by the NATO Science Committee.
The Series is published by IOS Press and Kluwer Academic Publishers, in conjunction
with the NATO Scientific Affairs Division.*

General Sub-Series

A. **Life Sciences**	IOS Press
B. **Physics**	Kluwer Academic Publishers
C. **Mathematical and Physical Sciences**	Kluwer Academic Publishers
D. **Behavioural and Social Sciences**	Kluwer Academic Publishers
E. **Applied Sciences**	Kluwer Academic Publishers
F. **Computer and Systems Sciences**	IOS Press

Partnership Sub-Series

1. **Disarmament Technologies**	Kluwer Academic Publishers
2. **Environmental Security**	Kluwer Academic Publishers
3. **High Technology**	Kluwer Academic Publishers
4. **Science and Technology Policy**	IOS Press
5. **Computer Networking**	IOS Press

*The Partnership Sub-Series incorporates activities undertaken in collaboration with NATO's
Partners in the Euro-Atlantic Partnership Council – countries of the CIS and Central and Eastern
Europe – in Priority Areas of concern to those countries.*

NATO-PCO-DATA BASE

The NATO Science Series continues the series of books published formerly in the NATO ASI
Series. An electronic index to the NATO ASI Series provides full bibliographical references (with
keywords and/or abstracts) to more than 50000 contributions from international scientists published
in all sections of the NATO ASI Series.
Access to the NATO-PCO-DATA BASE is possible via CD-ROM "NATO-PCO-DATA BASE" with
user-friendly retrieval software in English, French and German (© WTV GmbH and DATAWARE
Technologies Inc. 1989).

The CD-ROM of the NATO ASI Series can be ordered from: PCO, Overijse, Belgium.

3. High Technology – Vol. 63

Symmetry and Pairing in Superconductors

edited by

Marcel Ausloos

SUPRAS, Institute of Physics,
University of Liège,
Belgium

and

Sergei Kruchinin

Bogolyubov Institute for
Theoretical Physics,
Kiev, Ukraine

Kluwer Academic Publishers

Dordrecht / Boston / London

Published in cooperation with NATO Scientific Affairs Division

Proceedings of the NATO Advanced Research Workshop on
Symmetry and Pairing in Superconductors
Yalta, Ukraine
April 29 - May 2, 1998

Library of Congress Cataloging-in-Publication Data

```
Symmetry and pairing in superconductors : proceedings of a NATO
  Advanced Research Workshop held at Oreanda Hotel, Yalta, Ukraine on
  April 29-May 2, 1998 / edited by Marcel Ausloos, Sergei Kruchinin.
     p.   cm. -- (NATO ASI series. Partnership sub-series 3, High
  technology ; vol. 63)
    Includes index.
    ISBN 0-7923-5520-2 (hb : alk. paper)
    1. High temperature superconductors--Congresses. 2. Symmetry
  (Physics)--Congresses. 3. Pairing correlations (Nuclear physics)-
  -Congresses.  I. Ausloos, M. (Marcel), 1943-   . II. Kruchinin,
  Sergei. III. Series.
  QC611.98.H54S9  1999
  537.6'23--dc21                              98-43888
```

ISBN 0-7923-5520-2 (HB)

ISBN 0-7923-5521-0 (PB)

Published by Kluwer Academic Publishers,
P.O. Box 17, 3300 AA Dordrecht, The Netherlands.

Sold and distributed in North, Central and South America
by Kluwer Academic Publishers,
101 Philip Drive, Norwell, MA 02061, U.S.A.

In all other countries, sold and distributed
by Kluwer Academic Publishers,
P.O. Box 322, 3300 AH Dordrecht, The Netherlands.

Printed on acid-free paper

TABLE OF CONTENTS

PART II.   .... than experimental   269

viii

PART III. .... with respect to other reports.

══

late submission

FOREWORD

These Proceedings of a NATO-ARW (HTECH ARW 97 18 43) held at the Oreanda Hotel, Yalta, Ukraine from April 29 till May 2 , 1998 resulted from many discussions between various workers, concerning the need for a gathering of all (if possible) who were concerned with the subject of symmetry of the order parameter and pairing states for superconductivity. We applied ourselves in particular to High critical Temperature Superconductors (HTS), but also studied other unconventional superconductors. The study of HTS is one of the most prominent research subjects in solid state sciences. The understanding of the role of symmetry and pairing conditions are also thought to be necessary before technological applications since these features may be influenced by external fields.

The workshop discussions have touched upon theoretical and experimental aspects, but also on related topics. These served as initiators for a very great amount of discussions with many comments from the audience. More than thirty "long lectures" and one on-going "poster session" were held. Private discussions went unrecorded but obviously took place at many locations : lecture halls, staircases, cafes, bedrooms, bars, beach,bus, plane ... Arguments openly reported for the first time were often quite sharp ones, - and this is an understatement.

The ARW was held starting from 09:00 every day; the morning session ending at 01:30 p.m.; lunch till 02:00 p.m.; free afternoon for discussion or short excursions allowing for exchange of ideas; after each supper an evening session was held, one of them lasting till 11: 30 p.m. Due to the need of a charter plane to be taken between Kiev and Semfiropol/Yalta, **almost all** participants stayed (had to stay !) during the whole ARW.

The Proceedings are more or less arbitrarily divided into three parts : one is reserved for papers mainly considering theoretical aspects, though not neglecting experimental data.

The second part rather contains papers where the data analysis seems predominant, without neglecting theoretical insight and calculations. The third part contains papers which could also have found their place in Part I or Part II... but a ternary structure is always more appealing. The various presentations were allowed some space for these proceedings. Due to various committments not all could be inserted in due time. It is worth noting that there were many participants for which the main activity is considered to be "more experimental" than theoretical.

The editors had to choose some order of presentation. This was very hard to do. We decided to choose an "anti-alphabetical" order in each Part. In so-doing, the paper by one of the co-organizer comes first, and nobody should be upset; that is a privilege. The first Part has thus nice boundaries. The same is fortunately true for the second and third Part. Moreover in choosing that order and the three part structure, the "last contributed paper in print" serves as a fine conclusion paper, even though the content was much more elaborate than what appears from these proceedings.

The editors and their co-organizers are very pleased to acknowledge the financial support of NATO for this ARW. The financial matters were dealt with extremely well. We have a deep appreciation for the help and cooperation of the NATO Scientific Affair Division staff and leaders, who allowed us much freedom in the organizational and scientific matters (within NATO rules of course). This NATO-ARW was very successful.

Moreover, we emphasize here our expression of the deepest appreciation for the help and cooperation of our collaborators. Special thanks go to our family and to our scientific coworkers as well for their understanding now and then. Dr. E. Thomas is kindly thanked for editorial help. No need to say that the editorial work was quite heavy due to TEX conditions and the various ways of reporting the manuscripts.

Marcel AUSLOOS & Sergei KRUCHININ

PART I.

Apparently considered to be more theoretical …

DAY OF UNCONVENTIONAL SUPERCONDUCTORS

H. WON
Department of Physics and IRC, Hallym University
Chunchon 200-702, South Korea

AND

K. MAKI
Department of Physics and Astronomy,
University of Southern California,
Los Angeles, CA 90089-0484, USA

Abstract. First, we shall review a variety of unconventional superconductors, so far identified and unidentified. It is very remarkable there are 3 nonunitary superconducting states in the list. Second, the peak in the out of plane magnetoresistance in organic superconductors, κ-(ET)$_2$X with X=Cu(NCS)$_2$ is analyzed in terms of the standard superconducting fluctuation theory. A similar analysis has been done by Varlamov and his coworkers on the c-axis magnetoresistance in high T_c cuprate superconductors. However, unlike high T_c cuprate the organic superconductors are in the clean limit. Then the peak in the out of plane magnetoresistance suggests strongly d-wave pairing in organic superconductors.

Key words: unconventional superconductor, *d*-wave superconductor, organic superconductor, superconducting fluctuation.

1. Introduction

Since the discovery of high T_c cuprate superconductors by Bednorz and Müller[1] in 1986, the most momentous event is the realization that *d*-wave symmetry is involved in hole-doped high T_c cuprate superconductors [2, 3]. Both the presence of nodes in the superconducting order parameter $\Delta(\vec{k})$ as seen by ARPES in Bi2212 [4, 5], the T-linear dependence of the magnetic penetration depth [6], the T^2-dependence of the electronic specific

3

M. Ausloos and S. Kruchinin (eds.), Symmetry and Pairing in Superconductors, 3–10.
© *1999 Kluwer Academic Publishers. Printed in the Netherlands.*

heat [7, 8], the electronic Raman scattering [9], and the phase sensitive experiment using the Josephson coupling between YBCO and an s-wave superconductor like Pb [10] and Nb [11] and the generation of $\frac{1}{2}$ quantum flux at the center of the tricrystal samples [2] provide the definitive evidences of d-wave pairing. Indeed the tricrystal geometry method appears to be not only elegant but also versatile. In this way Tsuei, Kirtley and their collaborators are able to establish the d-wave symmetry in YBCO, Tl2201, Bi2212, GdBCO, and LSCO. The realization of d-wave symmetry in high T_c cuprate superconductors has naturally an enormous impact on the theory of superconductivity and on the theory of strongly correlated electron systems. First of all, the superconductivity in the Coulomb dominated system should be unconventional in a sharp contrast to classical s-wave superconductors generated by the electron-phonon interaction. Perhaps the unconventional superconductivity is the most effective way to take care of the strong correlation energy. In the following we shall list up some of unconventional superconductors.

a)d-wave superconductors

This is the simplest of unconventional superconductivity. Also we have seen the hole doped high T_c cuprate superconductors belong to this group. However, it is important to realize that the superconductivity in the electron doped cuprates appears to belong to s-wave superconductors [12, 13, 14]. Also from both the Pauli limiting of the upper critical field and the temperature dependence of the Knight shift some of heavy fermion superconductors, CeCu$_2$Si$_2$, URu$_2$Si$_2$, and UPd$_2$Al$_3$ appears to belong to d-wave supercondcutors [15]. However a further study on these systems is highly desirable. In the following, we present the case that the organic superconductors in κ-(ET)$_2$X with X=Cu(NCS)$_2$, Cu[N(CN)$_2$]Br, Cu[N(CN)$_2$]Cl, and CuCN[N(CN)$_2$] belong to this group.

b)p-wave superconductors

Absence of the Pauli limiting in the upper critical field of Bechgard salts (TMTSF)$_2$ClO$_4$ and (TMTSF)$_2$PF$_6$ under pressure [16] suggests strongly the equal-spin pairing(i.e. the triplet pairing) is involved here though further study is clearly needed.

The recently discovered superconductivity in Sr$_2$RuO$_4$ [17] is most likely not only p-wave but also of nonunitary state [15]. The simplest definition of the nonunitary state is $E_\uparrow(\vec{k}) \neq E_\downarrow(\vec{k})$. The quasi-particle energy of the up-spin state is different from the one in the down-spin state. In other words the ground state breaks not only the usual gauge symmetry but also the time reversal symmetry. Also the simplest possibility for UBe$_{13}$ is of p-wave. At least the superconductivity of UBe$_{13}$ belong to the triplet state. Further in the compound U$_{1-x}$Th$_x$Be$_{13}$ the C phase [19] is most likely to be the nonunitary state [20].

c) f-wave superconductors

Recently the f-wave superconductivity in UPt$_3$ is established from the anisotropy in the thermal conductivity at low temperatures [21, 22] and from the temperature independence of the spin susceptibility [23]. More recently the Knight shift study of UPt$_3$ in a small magnetic field [24] revealed that the B-phase is the nonunitary state. Therefore, not only the most of novel supercondcutors are unconventional but also some of them are nonunitary. Therefore it is clear that our concept of supercondcutivity will be completely changed when nature and properties of these unconventional superconductors are understood. In the following we shall consider a particular problem of the organic superconductor κ-(ET)$_2$X salts.

2. Out of plane magnetoresistance

Since the discovery of a new class of organic κ-(ET)$_2$X superconductors, superconductors with X=Cu(NCS)$_2$, Cu[N(CN)$_2$]Br, Cu[N(CN)$_2$]Cl, and CuCN[N(CN)$_2$] in 1988 [25], their normal and superconducting properties have been rather extensively studied [26, 27]. These systems have the superconducting transition temperature T$_c$ ranging 10-13 K, the quasi two dimensional electronic properties very similar to high T$_c$ cuprates superconductors, and the superconductivity appears in the vicinity of the antiferromagnetic state suggesting the strong electron correlation or the Coulomb dominance [28]. As to the symmetry of the superconducting order parameter, it is still very controversial at the present moment [27]. However, the recent acceptance of the d-wave pairing in the hole doped high T$_c$ cuprate superconductors will have a great influence on this question. Indeed, both recent T_1^{-1} measurement of κ-(ET)$_2$Cu[N(CN)$_2$]Br [29, 30] and low temperature specific heat [31] indicate the presence of nodes in $\Delta(\vec{k})$ consistent with d-wave superconductivity. Further a STM study of κ-(ET)$_2$Cu(NCS)$_2$ indicates the presence of nodes in $\Delta(\vec{k})$ [32]. On the other hand, the temperature dependence of the superfluid density $\rho_s(T)$ which should give the crucial information on this question is unfortunately very controversial up to now [27].

It is known for sometimes that the out of plane magnetoresistance in κ-(ET)$_2$X with X=Cu(NCS)$_2$ exhibits a peak before the superconducting transition [33]. Confronted with a similar behavior in the c-axis resistance in the high T$_c$ cuprate, Anderson and Zou even proposed that the semiconducting behavior for the out of plane resistance [34]. We believe that the peak in the out of plane resistance in high T$_c$ cuprate superconductors like Bi2212 and YBCO are now well understood in terms of density of states(DOS) effect in the superconducting fluctuation as studied in greater detail by Varlamov and his collaborators [35, 36]. In the standard fluctuation theory there are three distinct classes of terms contributing to the magnetoconductance; the Aslamazov-Larkin(AL) term [37], the so called Maki-Thompson(MT) term [38], and the density of states(DOS) term. In the usual situation, both the AL and MT terms dominate, while the DOS term can be neglected, since the AL and MT terms are more diverging in the vicinity of the superconducting transition. However, in the layered compounds like high T_c cuprates or organic superconductors we are now considering, the AL term is suppressed except in the immediate vicinity of the superconducting transition, since the AL term involves the fourth power of t_z, the interlayer transfer integral, while all other terms are proportional to $(t_z)^2$. Finally, the MT term is suppressed if there is strong pair-breaking or the underlying superconductivity is of non-s-wave [39]. In these circumstances DOS term which is negative in signe dominates the fluctuation conductivity. We believe this is the situation exactly realized in κ-$(ET)_2X$ superconductors as well as in high T_c cuprate superconductors. However, since organic superconductors are in the clean limit (i.e. $\tau T_c \gg 1$, where τ is the transport lifetime and T_c is the superconducting transition temperature), the pair-breaking scattering cannot be so strong. Therefore the peak in the out of plane magnetoresistance in κ-$(ET)_2X$ superconductors indicates that the underlying superconductivity is not of s-wave but d-wave.

3. Formulation

In the following, we shall work out the theoretical expression of the magnetoconductivity in the presence of superconducting fluctuation. We follow the analysis given in [35, 36] but we are concerned with the clean limit and d-wave superconductors. On the other hand, we take the fluctuation propagator à la Lawrence-Doniach [40] as in [35, 36]. Then the fluctuation propagator in a magnetic field perpendicular to the plane is given by

$$\mathcal{D}^{-1}(B, q_z, \omega_\nu) = N_0(\epsilon + 4e\xi^2 B(n + \frac{1}{2}) + r\sin^2(\frac{sq_z}{2}) + \frac{\pi}{8T}|\omega_\nu|) \quad (1)$$

where $\epsilon = \ln(\frac{T}{T_c})$, $\xi^2 = \frac{7\zeta(3)v^2}{2(4\pi T)^2}$, $r = \frac{7\zeta(3)(4t_z)^2}{2(4\pi T)^2}$, and N_o and ω_ν are the electron density of states at the Fermi-energy and the Matsubara frequency and $\zeta(3) = 1.202\ldots$.

Then, making use of the standard formalism, we obtain;

$$\frac{\sigma_{AL}}{\sigma_n} = \frac{\pi}{32} \frac{1}{\tau E_F} \frac{r}{b^2} \sum_{n=0}^{\infty} \frac{1}{[(n + \frac{1}{2} + \frac{\epsilon}{b})(n + \frac{1}{2} + \frac{\epsilon+r}{b})]^{\frac{3}{2}}} \quad (2)$$

$$\frac{\sigma_{MT}}{\sigma_n} = \frac{C}{(\tau E_F)} \frac{2b}{r} \sum_{n=0}^{\infty} \left\{ \frac{n + \frac{1}{2} + \frac{\epsilon + r/2}{b}}{\sqrt{(n + \frac{1}{2} + \frac{\epsilon}{b})(n + \frac{1}{2} + \frac{\epsilon + r}{b})}} - 1 \right\} \tag{3}$$

$$\frac{\sigma_{DOS}}{\sigma_n} = -\frac{C}{(\tau E_F)} \left\{ \sum_{n=0}^{\infty} \left(\frac{1}{\sqrt{(n + \frac{1}{2} + \frac{\epsilon}{b})(n + \frac{1}{2} + \frac{\epsilon + r}{b})}} - \frac{1}{n + \frac{1}{2} + \frac{\epsilon + r/2}{b}} \right) \right.$$
$$\left. + \psi\left(\frac{1 + \epsilon + r/2}{b} + \frac{1}{2} \right) - \psi\left(\frac{1}{2} + \frac{\epsilon + r/2}{b} \right) \right\} \tag{4}$$

where

$$C = (T\tau)^2 \frac{\pi^3}{7\zeta(3)} - T\tau \ , \qquad b = 4e\xi^2 B$$
$$\sigma_n = \frac{sm}{2\pi} (2et_z)^2 \tau \tag{5}$$

and $\psi(z)$ is the di-gamma function.

Finally, the magnetoresistance is given by

$$R(B,T)/R_n = \left(1 + \frac{\sigma_{AL}}{\sigma_n} + \frac{\sigma_{MT}}{\sigma_n} + \frac{\sigma_{DOS}}{\sigma_n} \right)^{-1}$$

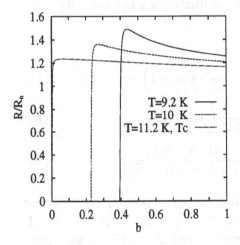

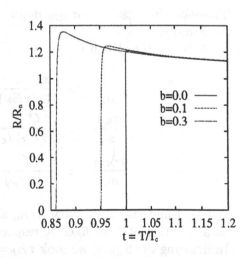

Fig.1 $R(B,T)/R_n$ versus b for a few t.

Fig.2 $R(B,T)/R_n$ versus t for a few b.

It is very easy to see that only σ_{DOS} is negative. So in order to have that σ_{DOS} dominance, it is necessary to have the suppression of both σ_{AL} and σ_{MT}. From Eq.(2)-(4), it is readily seen that σ_{AL} is suppressed when $r \ll 1$. Also, in a d-wave superconductor as in the present analysis $\sigma_{MT} < |\sigma_{DOS}|$

except in the immediate vicinity of the superconducting transition. Perhaps some asymptotic behaviors are useful;

1) $B \simeq H_{c2}$, or, $b = -2\epsilon$

$$\frac{\sigma_{AL}}{\sigma_n} \simeq \frac{\pi}{32} \frac{1}{(\tau E_F)} \frac{b}{\sqrt{r}} \frac{1}{(\frac{1}{2}b + \epsilon)^{3/2}} \tag{6}$$

$$\frac{\sigma_{MT}}{\sigma_n} = -\frac{\sigma_{DOS}}{\sigma_n} \simeq \frac{C}{(\tau E_F)\sqrt{r}\sqrt{\frac{1}{2}b + \epsilon}} \tag{7}$$

In this region, clearly σ_{AL} dominates, while σ_{MT} and σ_{DOS} cancel each other.

2) $B \gg H_{c2}(t)$ ($b \gg -2\epsilon$)

$$\frac{\sigma_{AL}}{\sigma_n} \simeq \frac{\pi}{32} \frac{1}{(\tau E_F)} \frac{r}{b^2} (7\zeta(3) - \frac{\pi^4}{2} \frac{(\epsilon + \frac{r}{2})}{b}) \tag{8}$$

$$\frac{\sigma_{MT}}{\sigma_n} \simeq \frac{C}{(\tau E_F)} \frac{r}{4b} (\frac{\pi^2}{2} - 14\zeta(3) \frac{(\epsilon + \frac{r}{2})}{b}) \tag{9}$$

$$\frac{\sigma_{DOS}}{\sigma_n} \simeq -\frac{C}{(\tau E_F)} (\psi(\frac{1}{b} + \frac{1}{2}) - \psi(\frac{1}{2})) \tag{10}$$

Therefore for $b \gg -2\epsilon$, σ_{DOS} dominates the fluctuation conductivity.

Finally,

3) $B = 0$ (i.e. $b = 0$)

$$\frac{\sigma_{AL}}{\sigma_n} \simeq \frac{\pi}{32} \frac{1}{(\tau E_F)} \frac{4}{r} \left(\frac{\epsilon + \frac{r}{2}}{\sqrt{\epsilon(\epsilon + r)}} - 1 \right) \tag{11}$$

$$\frac{\sigma_{MT}}{\sigma_n} \simeq \frac{C}{(\tau E_F)} \frac{\frac{r}{2}}{(\epsilon + \frac{r}{2} + \sqrt{\epsilon(\epsilon + r)})} \tag{12}$$

$$\frac{\sigma_{DOS}}{\sigma_n} \simeq -\frac{C}{(\tau E_F)} \ln \left(\frac{4(1 + \epsilon + \frac{r}{2})}{\epsilon + \frac{r}{2} + \sqrt{\epsilon(\epsilon + r)}} \right) \tag{13}$$

In Fig. 1 and Fig. 2, we show R/R_n as a function of b for fixed $t = T/T_c$ and as a function of t for fixed b, respectively. Note $b = 2B/H_{c2}(t)$ for $t < 1$. In drawing the figure, we took $\tau E_F = 5$, $r = 0.2$, and $C = 87.13$. These figures describe semi-quantitatively the observed out of plane magnetoresistance from κ-(ET)$_2$Cu[N(CN)$_2$]Br [41].

4. Concluding Remarks

We have shown that the peak in the out of plane magnetoresistance in κ-(ET)$_2$X superconductors is described in terms of the standard fluctuation theory. On the other hand, in order that the theory works, the superconductivity should be non-s-wave, suggesting strongly a d-wave superconductor.

We have shown elsewhere that the pseudogap phenomenon is nothing but the DOS effect due to the superconducting fluctuation [42]. Therefore, the dominance of the DOS term in the electronic conductivity implies the related pseudogap phenomenon in κ-$(ET)_2X$ superconductors. Although we are not aware of the pseudogap phenomenon in organic superconductors, this will be another fruitful direction for exploring the organic superconductors. Also in order to confirm the d-wave symmetry, the phase-sensitive experiments in organic superconductors, as done in high Tc cuprate superconductors [2, 3] are highly desirable.

5. Acknowledgements

We acknowledge the hospitality of Peter Wyder at LCMI-MPI Grenoble where this work started. We have benefited from discussions with Mark Kartsovnik on his more recent experimental data. H. Won acknowledges support in part from the Korea Science and Engineering Foundation under grant number 96-0702-02-01-3 and in part from the Korea Ministry of Education through the Basic Science Research Institute program. The present work is also supported by the National Science Foundation under grant number DMR95-31720.

References

1. Bednorz J. G. and Müller, K. A. (1986) Z. Phys. B **64**, 189.
2. Tsuei, C. C. and Kirtley, J. R. (1997) Physica C **282-287**, 4.
3. Van Harlingen, D. J. (1997) Physica C **282-287**, 128.
4. Shen, Z. X., et al. (1993) Phys. Rev. Lett. **70**, 1553.
5. Ding, H. et al. (1994) Phys. Rev. Lett. **72**, 1333.
6. Hardy, W. H. et al. (1993) Phys. Rev. Lett. **70**, 3999; Bonn, D. A. et al. (1993) Phys. Rev. B **47**, 11314.
7. Moler, K. A. et al. (1994) Phys. Rev. Lett. **73**, 2744.
8. Momono, N. et al. (1994) Physica C **233**, 395.
9. Devereaux, T. P. et al. (1994) Phys. Rev. Lett. **72**, 396.
10. Wollman, D. A. et al. (1993) Phys. Rev. Lett. **71**, 2134; (1995) Phys. Rev. Lett. **74**, 797.
11. Brawner, D. A. and Ott, H. R. (1996) Phys. Rev. B **50**, 6530; (1996) B **53**, 8249.
12. Wu, P. H. et al. (1993) Phys. Rev. Lett. **70**, 85; Anlage, S. M. et al. (1996) Phys. Rev. B **50**, 523.
13. Stadlober, B. et al. (1995) Phys. Rev. Lett. **74**, 4911.
14. Maki, K. and Puchkaryov, E. (1998) Europhys. Lett. **42**, 209.
15. Sigrist, M. and Ueda, K. (1991) Rev. Mod. Phys. **63**, 239.
16. Naughton, M. J., Lee, I. J. , Chaikin, P. M. and Danner, G. M. (1997) Synth. Metal **85**, 1481; Lee, I. J., Naughton, M. J., Darmer, G. M., and Chaikin, P. M. (1997) Phys. Rev. Lett. **78**, 3555.
17. Maeno, Y. et al. (1994) nature (London) **372** 532; Nishigaki, S., Maeno, Y., Famer, S., Ikeda, S., Fujita, T. (1998) J. Phys. Soc. Jpn. **67**, 560.

10

18. Sigrist, M., Zhitomirsky, M. E. (1996) J. Phys. Soc. Jpn. **65**, 3720.
19. Heffner, R. H. and Norman, M. R. (1996) Comments in Condensed Matt. Phys. **17**, 325.
20. Sigrist, M and Rice, T. M. (1989) Phys. Rev. B **31**, 2200.
21. Lussier, B., Ellman, B., and Taillefer, L. (1996) Phys. Rev. B **53**, 5145.
22. Norman, M. R. and Hirschfeld, P. J. (1996) Phys. Rev. B **53**, 5706; Graf, M. J., Yip, S.-K., and Sauls, J. A. (1996) J. Low Temp. Phys. **102**, 307.
23. Tou, H. Kitaoka, Y., Asayama, K., Kimura, N., Ohnuki, Y., Yamamoto, E., and Maezawa, K. (1996) Phys. Rev. Lett. **77**, 1374.
24. Tou, H., Kitaoka, Y., Ishida, K., Asayama, K., Kimura, N., Onuki, Y., Yamamoto, E., Haga, Y., and Maezawa, K. (1998) Phys. Rev. Lett. **80**, 3129.
25. Urayama, K., Yamochi, H., Saito, G., Nozawa, K., Sugano, T., Kinoshita, M., Sato, S., Oshima, K., Kawamoto, A., Tanaka (1988) J., Chem. Lett. 55.
26. Ishiguro, T. and Yamaji, K., "Organic Superconductors", (1990) Springer Verlag, Berlin.
27. Lang, M.,(1996) Superconductivity Review **2**, 1-115.
28. Maki, K. and Won, H.J. (1996) Physique I (Paris) **6**, 2317.
29. Mayaffre, H., Wzietek, P., Lenoir, C., Jerome, D., and Batail, P., Phys. (1995) Rev.Lett. **75**, 4122.
30. Kanoda, K., Miyagawa, K., Kawamoto, A., and Nakazawa, Y. (1996) Phys. Rev. B **54**, 76.
31. Nakazawa, Y., and Kanoda, K. (1997) Phys. Rev. B **55**, R8670.
32. Ichimaru, K., Arai, T., Nomura, K., Takasaki, S., Yamada, J., Nakatsuji, S., and Anzai, H. (1997) Physica C **282-287**, 1895.
33. Ito, H., Ishiguro, T., Komatsu, T., Saito, G., Anzai, H. (1994) Physica B **201**, 470; (1994) Journal of Superconductivity **7**, 667.
34. Anderson, P.W., and Zou, Z. (1988) Phys. Rev. Lett. **60**, 132.
35. Ioffe, L., Larkin, A., Varlamov, A. A., Yu, L. (1993) Phys.Rev. B **47**, 3936.
36. Dorin, V., Klemm, R., Varlamov, A. A., Livanov, D., Buzdin, A. (1993) Phys. Rev. B **48**, 12951.
37. Aslamazov, L. G., and Larkin, A. I. (1968) Physics Lett. A **26**, 238; (1968) Soviet Phys. Solid State Physics **10**, 1104.
38. Maki, K. (1968) Prog.Theor.Phys. (Kyoto) **39**, 897; (1968) **40**, 193; Thompson, R. S. (1970) Phys.Rev. B **1**, 327.
39. Yip, S. K. (1990) Phys.Rev. B **41**, 2612.
40. Lawrence, W. E., and Doniach, S. (1971) Proc. Twelfth Int. Conf. Low Temp. Phys., Kanda, E., edit., Keygaku, Tokyo, p.361.
41. Kartsovnik, M., private communication.
42. Maki, K. and Won, H. (1997) Physica C **282-287**, 1839.

TUNNELING SPECTROSCOPY OF THE HEAVY ELECTRON SUPERCONDUCTOR UBe_{13}

CH. WÄLTI AND H. R. OTT
Laboratorium für Festkörperphysik, ETH Zürich,
CH-8093 Zürich, Switzerland

1. Introduction

Among other indications, the anomalous behavior of the specific heat [1] and the NMR-relaxation rate [2] in the heavy-electron superconductor UBe_{13} have been interpreted as evidence that superconductivity in this compound is of unconventional nature. Inspired by the analogy of the behavior of these properties to those of superfluid He^3, it has been proposed that the superconducting state is an axial [1] or polar [2] spin-triplet state, respectively. However, the identification of the proper symmetry of the superconducting order parameter is still an open problem.

In recent years, it has been shown that experiments based on Josephson tunneling between conventional and potentially unconventional superconductors are powerful tools to directly probe the phase of the superconducting order parameter [3, 4].

At this time, no phase sensitive tunneling measurements have been performed to directly probe the symmetry of the superconducting order parameter of UBe_{13}. As a natural starting point, it seems necessary to investigate Josephson tunneling into this material.

In this work, we report on preliminary results of measurements of the $I - V$ characteristics of UBe_{13}–UBe_{13} and Nb–UBe_{13} tunnel contacts.

2. Samples and Experiments

The preparation of the polycrystalline sample of UBe_{13} used in these experiments is described in detail elsewhere [5]. The surface of the polycrystal has been polished and the crystal was exposed to air for less than one hour to avoid extensive oxidation of the surface. The UBe_{13}-tip was fabricated from

11

M. Ausloos and S. Kruchinin (eds.), Symmetry and Pairing in Superconductors, 11–18.
© 1999 *Kluwer Academic Publishers. Printed in the Netherlands.*

a small piece of a crushed single crystal. AC-susceptibility measurements of this UBe$_{13}$-tip revealed a relatively sharp superconducting transition at a T_c of about $0.75K$. For the fabrication of the Nb-tips with typical tip-radius of the order of $10\mu m$, Nb-wire of 1 mm diameter was electro-edged.

All the experiments have been performed in a home-built closed-cycle ^{3}He cryostat. Magnetic shields were installed to reliably cancel the earth's magnetic field to less than $510^{-3}G$.

The contacts were mounted inside an X-band resonator, which was connected by a coaxial waveguide to a Gunn-diode ($\nu = 10.52GHz$), to allow rf-irradiation of the tunnel contacts. The contacts were formed by pressing the tip onto the polished surface of the counter electrode. Adjustments of the contact could be made using a screwdriver-type arrangement which can be operated from outside the cryostat. The vertical resolution of the tip motion was typically better than $1\mu m$.

3. Results and Discussion

3.1. UBe$_{13}$–UBe$_{13}$

In Fig. 1, the measured current-voltage (IV) characteristics of a UBe$_{13}$-UBe$_{13}$ tunnel contact is shown for temperatures $0.35 < T < 0.85K$ (upper panel).

From this set of IV curves, we calculated the derivatives of the voltage-drop across the junction with respect to the bias-current, dV/dI. The result of these calculations is plotted as a function of the bias-current in the lower panel of the same figure. The IV curve at the lowest temperature of this experiment is indicated as open circles. With increasing bias-current from zero, the slope of the IV curve changes significantly at a particular current value and the differential resistance dV/dI across the junction increases by at least a factor of 4. For temperatures $T \geq 0.64K$, this anomaly disappears and the IV curve at this temperature is plotted as full squares in the figure.

In Ref. [6], the simultaneous occurrence of Josephson effects and series resistances in SIS junctions has been reported. In that work, the finite differential resistance (dV/dI) of the junction, even at zero bias-current, has been attributed to the polycrystalline nature of the polished surface of one of the involved superconductors. An oxide layer between two grains may give rise to a non-zero resistance between these two grains. In the UBe$_{13}$-UBe$_{13}$ tunnel junction reported on in this work, similar behavior is observed and might be due to the fact, that the tunnel contact was established at a breakpoint inside the tip, and possibly a non-zero resistance at the interface between the tip and the counter electrode was measured.

For a given temperature T, we can extract the current, at which the slope of the IV curve changes most from the lower panel of Fig. 1, and

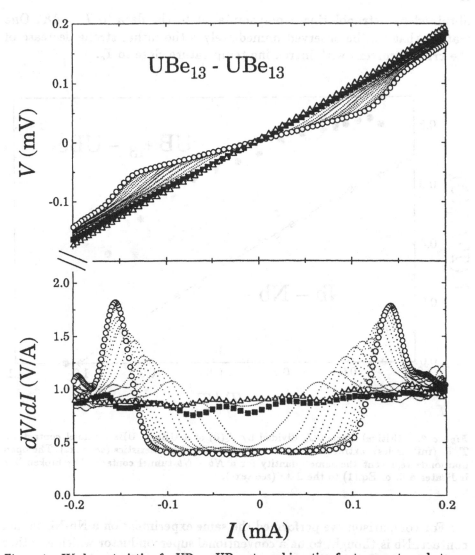

Figure 1. *IV* characteristic of a UBe_{13}- UBe_{13} tunnel junction for temperatures between 0.35 and $0.85 K$ (upper panel). The *IV* curve at the lowest temperature is shown by open circles, and at the highest temperature by open triangles. The full squares indicate the *IV* curve at $T = 0.64 K$, above which the anomaly is not observed anymore (see text). The lower panel shows (dV/dI) vs. I for the same set of temperatures.

we identify this quantity as the "critical current $J_c(T)$" of the junction arrangement. The result of this procedure is shown in Fig. 2, where the reduced critical current $J_c(T)/J_c(0)$ is plotted as a function of the reduced temperature T/T_c (full circles). The critical current at $T = 0 K$, $J_c(0)$, was obtained by extrapolating a polynomial fit to the data to $T = 0K$. One feature that can be observed immediately is the rather strong decrease of the critical current with increasing temperature close to T_c.

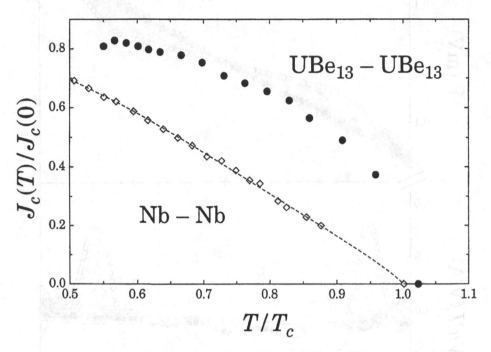

Figure 2. Critical current J_c divided by $J_c(0)$ of a UBe$_{13}$- UBe$_{13}$ tunnel contact vs. T/T_c (full circles), extracted from the measured IV characteristics (see text). The open diamonds represent the same quantity for a $Nb - Nb$ tunnel contact. The broken line indicates a fit of Eq.(1) to the data (see text).

For comparison, we performed the same experiment on a Nb-Nb tunnel contact. Nb is thought to be a conventional superconductor which is rather well described by the BCS approximation. The same procedure to extract the critical current from the IV curves as described above was used, and the result is shown by open diamonds in Fig. 2. We note, that the temperature dependencies of the critical currents J_c of the two investigated tunnel contacts are substantially different. The renormalized $J_c(T)$ values of the Nb-Nb contacts are consistently lower than those of the UBe$_{13}$–UBe$_{13}$ contacts and vary almost linearly with T for $0.5 < T/T_c < 1$. This characteristic differences of $J_c(T)$ of these two tunnel contacts may be an additional hint, that the superconductivity in UBe$_{13}$ is indeed of unconventional nature.

The critical current of a Josephson-junction of two identical conventional superconductors is given by [7]:

$$J_c(T) = \frac{\pi \Delta(T)}{2R_n} \tanh(\frac{\Delta(T)}{2k_B T_c}), \tag{1}$$

where $\Delta(T)$ denotes the energy-gap, k_B the Boltzmann constant and R_n the resistance of the junction for currents higher than the critical current. Using this expression and the measured energy-gap of Nb [8], we calculated the expected critical current for a Nb–Nb tunnel contact. The result of this calculation is shown in Fig. 2 as a broken line. We note that the J_c-data of the Nb–Nb tunnel contact can be described quite well by Eq. (1), proving the reliability of our experimental setup.

3.2. Nb-UBe$_{13}$ TUNNEL CONTACTS

In Fig. 3 we show the derivatives dV/dI of the measured IV characteristics of two characteristically different Nb–UBe$_{13}$ tunnel contacts. With our experimental setup, it was possible to vary the coupling-strength between the two superconductors by varying the applied pressure to the Nb-tip contacting the crystal. For both low and high pressure contacts, an anomaly becoming more pronounced with decreasing temperature was observed. Therefore, we may define a critical current $J_c(T)$ in a similar way as for the UBe$_{13}$–UBe$_{13}$ or the Nb-Nb tunnel contacts as discussed above. The resistance of the tunnel contact for currents exceeding J_c, R_n, increases with decreasing applied pressure.

3.2.1. *Low resistance tunnel contact*
In the upper panel of Fig. 3, we show dV/dI vs. I of a Nb–UBe$_{13}$ tunnel contact with a rather low contact resistance R_n for temperatures between 0.45 and $8.9K$. An anomaly of the dV/dI curve is observed for temperatures below $T \approx 8.9K$, i.e. slightly below T_c of Nb, but of the order of ten times higher than T_c of UBe$_{13}$. A substantial change in the behavior of the contact was observed at $T^{ast} = 5.25K$. For $T > T^{ast}$, the shape of the anomaly is that of a triangle, whereas for $T < T^{ast}$, a region, centered at $I = 0$, where dV/dI is constant, appears. For $T < T^{ast}$, the behavior of the contact changes dramatically under microwave irradiation ($\nu = 10.25GHz$) where a Shapiro-step like feature alters the IV characteristics as illustrated in Fig 4. This rather unusual feature has previously been observed and was interpreted as a "proximity-induced Josephson effect" [9, 10].

3.2.2. *High resistance tunnel contact*
The lower panel of Fig. 3 shows the differential resistance of a Nb–UBe$_{13}$ contact for a set of temperatures, $0.32 < T < 0.84K$, where only very little pressure was applied to the Nb-tip. For temperatures $T < 0.84K$, a V-shaped anomaly centered at zero bias-current was observed. We note,

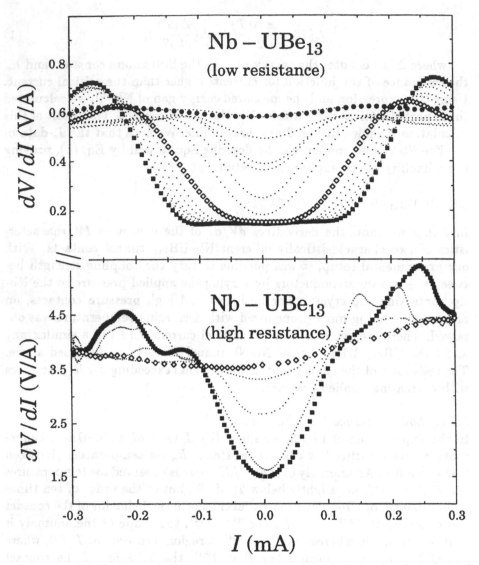

Figure 3. dV/dI vs.I of a Nb–UBe$_{13}$ contact with a moderately low R_n (upper panel) for temperatures between 0.45 (full squares) and 8.9 (full circles). The open diamonds represent the data at $T = 5.25K$, where the behavior of the tunnel contact changes significantly. The lower panel shows the same quantity for a Nb–UBe$_{13}$ contact with higher resistance R_n at temperatures between 0.32 (full squares) and 0.84K (empty diamonds).

that similar to the behavior of the low resistance contact above T^{ast}, the contact resistance at zero bias-current depends on temperature and increases towards R_n with increasing temperature towards T_c of the UBe$_{13}$ counter electrode. Other remarkable features of this contact are the ab-

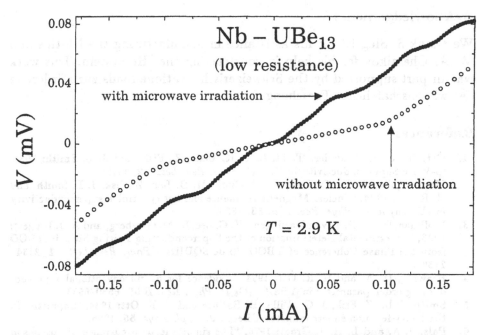

Figure 4. IV curves of a Nb–UBe₁₃ tunnel junction at $T = 2.9K$ (open circles). When applying microwaves to the junction arrangement, a Shapiro-step like feature is observed (full circles).

sence of Shapiro step like features and the existence of a certain current, I^*, where dV/dI is independent of the temperature.

4. Conclusions

The IV characteristics of UBe₁₃–UBe₁₃ and Nb–UBe₁₃ tunnel contacts have been measured. The temperature dependence of the critical current J_c of the UBe₁₃–UBe₁₃ contact has been extracted and compared with the also measured and well known behavior of a Nb–Nb contact. We found the renormalized $J_c(T)$ curves of these two tunnel contacts to be characteristically different. For Nb–UBe₁₃, two different kind of contacts have been obtained, which were distinguishable by their normal state resistance R_n. For the high resistance contacts, an anomaly at zero bias-current was observed for temperatures $T < 0.84K$, whereas for the low resistance contact, an anomaly was observed up to $T = 8.9K$. The shape of the anomaly of the low resistance contact changes significantly at $T = 5.25K$. In the low temperature regime ($T < 5.25K$), Shapiro-step like features were observed while irradiating the contact with microwaves.

18

Acknowledgements

We thank S. Siegrist for his assistance in manufacturing the Nb-tips and M. A. Chernikov for his help in constructing the ^{3}He system. This work was in part supported by the Schweizerische Nationalfonds zur Förderung der wissenschaftlichen Forschung.

References

1. Ott, H. R., H. Rudigier, T. M. Rice, K. Ueda, Z. Fisk, and J. L. Smith: 1984, 'p-Wave Superconductivity in UBe$_{13}$'. *Phys. Rev. Lett.* **52**, 1915.
2. MacLaughlin, D. E., C. Tien, W. G. Clark, M. D. Lan, Z. Fisk, J. L. Smith, and H. R. Ott: 1984, 'Nuclear Magnetic resonance and heavy fermion superconductivity in (U, Th)Be$_{13}$'. *Phys. Rev. Lett.* **53**, 1833.
3. Wollman, D. A., D. J. Van Harlingen, W. C. Lee, D. M. Ginsberg, and A. J. Leggett: 1993, 'Experimental Determination of the Superconducting Pairing State in YBCO from the Phase Coherence of YBCO-Pb dc SQUIDs'. *Phys. Rev. Lett.* **71**, 2134–2137.
4. Brawner, D. A. and H. R. Ott: 1994, 'Evidence for an unconventional superconducting order parameter in YBa$_2$Cu$_3$O$_{6.9}$'. *Phys. Rev. B* **50**, 6530–6533.
5. Smith, J. L., Z. Fisk, J. O. Willis, B. Batlogg and H. R. Ott: 1984, 'Impurities in the heavy-fermion superconductor UBe$_{13}$'. *J. Appl. Phys.* **55**, 1996.
6. Pals, J. A. and L. H. J. Graat: 1976, 'The simultaneous occurrence of Josephson effects and series resistance in Nb-U$_6$Fe point contacts'. *Phys. Lett. A* **56A**, 487–8.
7. Ambegaokar, V. and A. Baratoff: 1963, 'Tunnelling Between Superconductors'. *Phys. Rev. Lett.* **10**, 486.
8. Hendriks, J. W., H. R. Ott, J. H. M. Stoelinga and P. Wyder: 1977, 'Far-infrared absorption of superconducting films with spatially varying order parameter'. *Solid State Commun.* **21**, 555-559.
9. Han, S., K. W. Ng, E. L. Wolf, H. F. Braun, L. Tanner, Z. Fisk, J. L. Smith, and M. R. Beasley: 1985, 'Anomalous s-wave proximity-induced Josephson effects in UBe$_{13}$, CeCu$_2$Si$_2$, and LaBe$_{13}$: a new probe of heavy-fermion superconductivity'. *Phys. Rev. B* **32**, 7567–70.
10. Han, S., K. W. Ng, E. L. Wolf, A. Millis, J. L. Smith, and Z. Fisk: 1986, 'Observation of negative s-wave proximity effect in superconducting UBe$_{13}$'. *Phys. Rev. Lett.* **57**, 238–41.

INFLUENCE OF INELASTIC QUASIPARTICLE SCATTERING ON THERMODYNAMIC AND TRANSPORT PROPERTIES OF HIGH-TEMPERATURE OXIDES

A. I. VOITENKO AND A. M. GABOVICH
Crystal Physics Department, Institute of Physics, NASU,
prospekt Nauki 46, 252650 Kiev, Ukraine

1. Introduction

The temperature dependences of many ceramic metal oxide parameters both in the normal and superconducting states are quite different from those for traditional superconductors. One of the most important distinctions from the BCS behavior is the almost rectangular temperature, T, dependence of those characteristics which are measured by resistive (tunnel and point-contact) [1, 2, 3, 4, 5, 6, 7], infrared [8], or Raman [9, 10, 11] spectroscopies, and are generally identified [6] with the superconducting order parameter Δ or the energy gap Δ_g in the quasiparticle spectrum. The other unusual feature of the high-T_c superconductors is the absence [12, 13, 14, 15] of the Hebel-Slichter peak [16] in the nuclear spin-lattice relaxation rate R_s below T_c. This peak is a clear manifestation of the s-wave Cooper pairing and traditionally serves as a check of its realization for various specific superconductors.

It is natural to suggest that the unconventional dependences $\Delta_g(T)$ and $R_s(T)$ are the results of high temperatures when superconductivity exists here. As a consequence, the thermal phonons begin to play crucial role in oxides near T_c, the situation being different in low-T_c metals. This point of view is not new [17, 18, 19] and was applied to the substances considered in the framework of the Eliashberg theory [20]. However, we think that in order to incorporate other superconducting pairing mechanisms (e.g., excitons or plasmons) and to take into account the unusual normal state properties, one should use a more phenomenological approach based on the BCS scheme.

Hereafter, we consider Cooper pairing to be of the s-type. The clarification of the gap symmetry in high-T_c and other "exotic" superconductors is

19

M. Ausloos and S. Kruchinin (eds.), Symmetry and Pairing in Superconductors, 19–30.
© *1999 Kluwer Academic Publishers. Printed in the Netherlands.*

still on agenda [21, 22, 23]. Nevertheless, some of the high-T_c oxides, especially with electron-like carriers, are proven to be of the s-type. It should also be stressed that the main idea of our work will survive with necessary technical modification even for anisotropic s- or d-order parameters.

2. Theory

In this paper we assume that inelastic scattering processes result in destructive action on Cooper pairs. The validity of such a consideration was demonstrated long ago by Appel [17] for the particular case of the interaction of electrons with thermal phonons. In the framework of the BCS theory the role of these scattering processes is twofold: (i) they renormalize the pairing electron-electron attraction and (ii) suppress superconductivity in a way similar to the pair-breaking factor of Abrikosov-Gor'kov theory [24]. One can expect [25] that the result of Ref. [17] may be fully applied to other types of inelastic interactions as well. The key point of the presented theory is the description of the T-dependent pair-breaking factor $\nu = (\tau T_{c0})^{-1}$, where τ is the inverse inelastic relaxation time and T_{c0} is the critical temperature of the metal with the inelastic thermal scattering switched off (Boltzmann and Planck constants $k_B = \hbar = 1$).

In the Abrikosov-Gor'kov theory the superconductivity suppression is caused by the spin-flip of one of the Cooper pair components at a paramagnetic impurity. This process was described by the pair-breaking parameter ν, and the actual value of the order parameter Δ obeys the equation

$$2x_0 \sum_{n=0}^{\infty} \left\{ \frac{1}{\sqrt{u_n^2 + 1}} - \frac{1}{x_n + \frac{\nu}{\delta}} \right\} - \psi \left[\frac{1}{2} \left(1 + \frac{\nu}{\pi t} \right) \right] + \psi \left(\frac{1}{2} \right) = \ln t. \quad (1)$$

Here $\delta = \Delta/T_{c0}, t = T/T_{c0}, \psi(y)$ is the logarithmic derivative of the gamma function,

$$x_n = u_n \left[1 - \frac{\nu}{\delta \sqrt{1 + u_n^2}} \right], \quad (2)$$

and u_n is determined from the equation

$$u_n = \frac{1}{\delta} \left[(2n + 1)\pi t + \frac{\nu u_n}{\sqrt{1 + u_n^2}} \right]. \quad (3)$$

One should note that the quantities Δ and Δ_g do not coincide:

$$\delta_g(t) = \delta(t) \left\{ 1 - \left[\frac{\nu(t)}{\delta(t)} \right]^{2/3} \right\}^{3/2}, \quad (4)$$

where $\delta_g = \Delta_g/T_{c0}$.

The essence of the pair-destructive effect of the inelastic electron-phonon scattering consists in changing the electron energy rather than its spin polarization. But, if the electron-phonon coupling constant is treated in the spirit of the BCS theory as a parameter, then the equation for the order parameter $\Delta(T)$ formally coincides with the Abrikosov-Gor'kov one (1). Nevertheless, there is a substantial distinction. Since the scattering in the Abrikosov-Gor'kov theory is due to paramagnetic impurities, the relevant parameter ν depends only on their concentration, being temperature-independent. On the contrary, in our theory the pair-breaking factor strongly depends on T. At the same time, it is reasonable to assume that quasiparticle recombination processes below T_c make the dominant contribution into inelastic scattering [26] They are characterized by the activation energy Δ_a. It is obvious that Δ_a is connected with Δ or Δ_g, which makes the problem self-consistent. The preexponential multiplier can be determined taking into account the temperature behavior of the scattering above T_c, and the most general here is the power law, so we take the pair-breaking factor in the form

$$\nu(t) = At^\beta \exp\left[-\delta_a(t)/t\right], \tag{5}$$

where $\delta_a = \Delta_a/T_{c0}$. It seems rather reasonable that the superconducting gap Δ_g should be selected as the activation energy, but there could be other choices. For instance, an assumption $\Delta_a = \Delta$ was made in Ref. [25] where the critical current of bulk superconductors was investigated. Moreover, experimentally by measuring critical vortex velocities in epitaxial films of $La_{1.85}Sr_{0.15}CuO_{4-x}$, $YBa_2Cu_3O_{7-x}$, and amorphous Mo_3Si it was shown [27] that the observed T-dependence of the quasiparticle-energy relaxation rate could be well approximated if the values $\beta = 0$ and $\Delta_a = 2\Delta$ in Eq.(5) are selected. However, non-self-consistent fitting without allowance for the feedback influence of ν on Δ failed near T_c in Mo_3Si, supporting our self-consistent constrain

The exponent β is a parameter of the theory and can be found from theoretical considerations as well as from resistivity measurements. In the case of the thermal phonon scattering for T lower than the characteristic (Debye) temperature T_D, the β value is 3 [16, 17, 28, 29]. The same value will hold also for the scattering of other Bose quasiparticles with an acoustic dispersion law, e.g., for plasmons in two-component electron-hole systems. The exponent β equals 2 when phonons are strongly damped [28] or the electron-electron interaction prevails over the electron-phonon one [29], as well as for the electron-magnon scattering in ordered [30] or disordered [31] ferro- (antiferro-) magnets. When itinerant electrons are scattered by localized ones we have a modified relationship [32], $\tau^{-1} \propto T^2(\ln T)^{d+1}$, d being the system's dimensionality. Finally, β can be equal to 1 in the cases:

(i) for $T \geq T_D$ in the case of the electron-phonon interaction [28], (ii) in "dirty" two-dimensional systems due to electron-electron scattering [32], and (iii) for the electron-electron interaction in the presence of the nesting Fermi surface sections [29]. The measurements show that $\beta = 3$ in ordinary metals [28] (so that the Bloch-Grüneisen law for the electrical resistance holds). The values $\beta = 1$ and $\beta = 2$ are observed in many other cases, in particular, for high-T_c oxides [28, 33]. The value $\beta = 1$ for $YBa_2Cu_3O_7$ was also obtained indirectly from the light reflection spectra [34]. On the other hand, when the metals are strongly disordered the electron-phonon scattering processes with $\beta = \frac{7}{2}$ and $\Delta_a = 0$ become the main factor [35]. The choice $\Delta_a = 0$ is a particular case that does not need self-consistency. In the general case there is always a definite contribution to $\nu(t)$ from the terms of the type (5).

In Refs. [19, 36] it is stated that for high-T_c superconductors $\Delta_a = 0$ too, but for another reason. Namely: the quasiparticle spectrum in a super-conductor with the strong electron-phonon coupling is gapless due to the presence of the imaginary part in its self-energy which within a numerical factor coincides with τ^{-1}. Strictly speaking, thermal processes partially fill a gap in the density of states even for weak coupling metals and small $T \ll T_c$ (for $T = 0$ the damping is exactly zero). But this does not mean that the conclusion made in Refs. [19, 36] is true. The gap can be detected in a huge variety of different experiments for all superconductors. The important question here is: what is the magnitude of thermal smearing? We think that the calculation results in Refs. [19, 36] do not give grounds for the categorical conclusions made there, since even for coupling constant $\lambda \gg 5$ but low enough T the gap structure in the electron state density is quite pronounced (see plots in Ref. [19]). For intermediate $\lambda = 2$ the results of Ref. [19] do not differ qualitatively from the BCS picture. There are also a lot of experimental evidence, demonstrating the existence of the energy gap in high-T_c superconductors [6]. Hence, we shall assume that strong-coupling effects do not change drastically the electron spectrum character and lead only to the usual renormalization factor $1 + \lambda$ for the quantity τ^{-1} which may be incorporated into the constant A without loosing generality. Of course, all the speculations above are based on the assumption of the s-wave pairing. Meanwhile, during last years many direct and indirect measurements in high-T_c oxides of the superconducting order parameter and gap indicate that their symmetry is rather of d- than of s-type [22, 37]. This problem is far from being resolved and is not considered here.

Transport properties of superconductors with inelastic quasiparticle scattering in our approach can be determined similarly to the case of paramagnetic impurity scattering [38]. The nuclear spin-lattice relaxation rate R_s in the superconducting state is the most popular among these, because the BCS theory leads to the non-monotonic behavior of $R_s(T)$, whereas in experiments for high-T_c ceramics $R_s(T)$ is monotonic [12, 14, 15, 16]. The ratio $\rho = R_s/R_n$ of the nuclear spin-lattice relaxation rates in superconducting and normal states is

$$\rho(t) = \int_0^\infty \frac{dx}{2t} \cosh^{-2}\frac{x}{2t} \left\{ \left[\mathrm{Im}\frac{u(x)}{\sqrt{1-u^2(x)}} \right]^2 + \left[\mathrm{Im}\frac{1}{\sqrt{1-u^2(x)}} \right]^2 \right\}. \quad (6)$$

Here $u(x)$ is the analytical continuation of the discrete function u_n (see Eq.(3)) to real frequencies and is given implicitly by the equation

$$x = u\delta \left(1 - \frac{\nu}{\delta\sqrt{1-u^2}} \right). \quad (7)$$

When $\nu \to 0$ the integral in the right hand side of Eq.(6) diverges logarithmically at the lower limit. Actually, in ordinary superconductors the function $R_s(T \leq T_c)$ is finite, but there is a broad so-called Hebel-Slichter peak below T_c [16]. For the case of a weak electron-phonon coupling the singularity suppression may be due, e.g., to the electron state density smearing when the energy gap is anisotropic or there is a spread of the gap values. When the coupling becomes strong, quasiparticle energy levels broaden due to the interaction with thermal phonons [39], so that the peak height becomes finite. Below it is shown that this interaction or the analogous processes involving other Bose excitations can wipe out the peak in $R_s(T)$ altogether.

3. Results

3.1. ORDER PARAMETER AND SUPERCONDUCTING GAP

In Fig.1(a) the superconducting order parameter Δ is shown versus the reduced temperature $t = T/T_{c0}$ for various types of the pair-breaking factor. The dashed curve is the Mühlschlegel curve [16] within a factor $T_{c0}/\Delta(0) = \gamma/\pi$, where $\gamma = 1.7810\ldots$ is the Euler constant. Curve 1 corresponds to our model with a non-self-consistent temperature-dependent pair-breaking scattering. We remark at once that even in this simplest version of the theory the ratio $2\Delta(0)/T_c$ is far in excess of its BCS value $2\pi/\gamma$, the conclusion being in agreement with the majority of experimental data for high-T_c oxides [6]. Curves 2 and 3 were calculated self-consistently and only in the gap region defined by the condition $\nu(t) \leq \delta(t)$. One can see that the order parameter becomes double-valued when treated self-consistently. The dashed portions of curves 2 and 3 describe the lower branches connecting the branching points and the points where the condition $\nu < \delta$ fails. Thus, the gapless region is not considered at all.

24

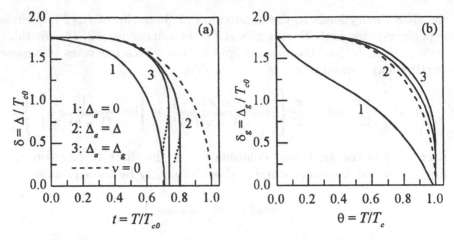

Figure 1. (a) Dependences $\delta(t)$ for various models of the dynamical pair-breaking factor $\nu = 0.5(T/T_{c0})\exp(-\Delta_a/T)$. (b) The same for $\delta_g(\theta)$ dependences.

At first glance, the double-valued character of $\Delta(T)$ resembles that for the gap edge $\Delta_0^{sc}(T)$ obtained in Refs. [40, 41] within Eliashberg theory [20]. However, the calculations of these authors led to the single-valued behavior of $\Delta(T)$ and $\Delta_g(T)$, which vanished continuously when $T \to T_c$ in line with the theory of second-order phase transitions. The basic difference of our approach from those of Refs. [40, 41] or [42] is that, according to them, the electron density of states is non-zero for any energy when $T \neq 0$, whereas our calculations for experimentally justified [28, 33] parameters A and β show that the quantities Δ and Δ_g are real functions of T and the gap Δ_g does exist for all the temperatures up to T_c.

We have also shown that the double-valued nature of the order parameter $\Delta(T)$ and hence the gap $\Delta_g(T)$ survives for arbitrary small A and $\beta > 0$, but the branching point rapidly shifts towards the close vicinity of T_c when $A \to 0$. Therefore, the predicted effect would not be observed for low-T_c superconductors since it goes there beyond the accuracy of the experiment.

The lower branches of curves 2 and 3 correspond to the energetically less favorable unstable states. Therefore, the continuous behavior of $\delta(t)$ is interrupted at the branching point and the phase transition of the first kind into the normal state occurs, so that the order parameter goes discontinuously to zero at $T = T_c$. The transition is depicted by the vertical solid parts of curves 2 and 3.

In Fig. 1(b) the corresponding values for Δ_g are shown as functions of normalized temperature $\theta = T/T_c$, which are more convenient for comparison with experiment. The unstable branches of curves 2 and 3 are omitted. The profile of the curve 1 distinctly marks the failure of the non-self-consistent model.

In reality, the hysteretic phenomena inherent to the first order phase transitions should occur. The corresponding maximal overheating, T_h, and overcooling, T_l, temperatures can be obtained from the absolute instability investigations. But the kinetic factors (heterogeneous and homogeneous nucleation) can narrow down the range $T_h - T_l$ up to its complete disappearance, so that our original picture shown in Fig.1(a) can remain unaltered. The problem of superconducting phase nucleation is yet unsolved.

The transformation of the phase transition order in a similar situation, when the instability is driven by a non-equilibrium external electromagnetic radiation, was obtained in Ref. [43]. At the same time, quasiparticle injection by tunnel currents leads to the coexistence of low and larger gaps in the sample and a hysteresis above a certain voltage [44]. The superconducting transition also becomes here of the first kind.

It should be noted that our conclusion about the first order superconducting phase transition is by no means exotic from the purely theoretical standpoint. It has been obtained for the first time in Ref. [45] and confirmed in Ref. [46] as a result of the order parameter fluctuations, although more recent calculations [47] have thrown doubt on the effect, claiming that the first kind is an artifact of the ϵ- and $1/n$-renormalization group expansions. A phenomenological model [48] has been also proposed where the free energy functional is a generalization of the BCS one and can describe the superconducting first order transition. On the other hand, the high-T_c oxide properties are favorable to detect this phenomenon.

Recently two phenomenological approaches [42, 49] were suggested which led to superficially similar curves. Here the imaginary part of the polarization operator is used as one of the constituting elements, being the same as in the marginal Fermi liquid theory [50]. But the authors of Ref. [42] think that the abrupt decrease of their $\Delta(T)$ is an artifact of non-self-consistency. Thus, little remains from their results and no agreement with the experiment is achieved. At the same time, in our model, as we have already stated above, the changeover of the phase transition order in our model is a consequence of the calculation self-consistency. That is why the self-consistent set of equations in Ref. [49] based on the marginal Fermi liquid concept leads to the dependences $\Delta(T)$ which are very similar to ours. The authors of Ref. [49] also support the idea of the first-order superconducting phase transition.

The superconducting transition of the first kind due to the inelastic current carrier scattering by thermal excitations was also obtained [51] in the framework of the marginal Fermi liquid model. But later the authors abandoned their conclusion [52], because their calculations led to the dependence of the phase transition order on the boson spectrum form. It should be emphasized that our approach results in the first order transition for any reasonable quasiparticle scattering law, i.e., for the wide class of excitations. The only condition is the self-consistent interdependence of $\nu(T)$ and $\Delta(T)$ or $\Delta_g(T)$.

We have also made calculations of dependences $\Delta(T)$ and $\Delta_g(T)$ for other $\beta > 0$. It turned out that the curves for the same A deviate more from the BCS curve when β increases, thus approaching closer to the rectangular form. Therefore, the predicted features are common for various possible inelastic scattering mechanisms. The self-consistency plays the leading role in determining the behavior of $\Delta(T)$ and $\Delta_g(T)$.

Since the superconducting order parameter Δ and the associated energy gap Δ_g are the most important parameters which control all other superconducting state properties, a large number of experiments were carried out since 1987, where the electron density of states was measured in high-T_c oxides by the methods of tunnel, point-contact, infrared, and Raman spectroscopies. However, a direct comparison of theory and experiment is often hampered. Tunnel measurements [7] of pure and partially Ni-substituted $Bi_2Sr_2CaCu_2O_{8+x}$ samples revealed "more square" $\Delta(T)$ features than according to the BCS theory. There is quite a number of papers which give conclusive evidence that the gap-like features in optical spectra depend weakly on T up to T_c and then decrease abruptly in accordance with our results. These are measurements of infrared reflection or absorption [8] as well as Raman spectra [9, 10, 11, 53] for $RBa_2Cu_3O_{7-x}$ ($R = Y$, Dy, $Sm_{0.5}Y_{0.5}$, $Sm_{0.5}Ho_{0.5}$) and $Bi_2Sr_2CaCu_2O_{8+x}$. The temperature-dependent superconducting gap Δ_g is defined from Raman spectra as the peak position below T_c. The relevant experimental points for $Bi_2Sr_2CaCu_2O_{8+x}$ [10] and $YBa_2Cu_3O_7$ [53] single crystals are depicted in Fig.2 together with our theoretical curves obtained for different values of A and $\beta = 1$. One can readily see that the agreement exists between the experiment and our self-consistent theory in contrast to BCS dependence.

3.2. NUCLEAR SPIN-LATTICE RELAXATION RATE

In Fig. 3 the dependences $\rho(\theta)$ calculated according to Eq. (6) are shown for the model $\beta = 1$ and $\Delta_a = \Delta_g$. One can see that the increase of the scattering magnitude A drastically changes the character of the curves. The coherent peak disappears and the temperature dependence of ρ becomes monotonic with jump to unity at $T = T_c$ already for $A > 1$, which corresponds to $T/T_{c0} < 0.6$. At the same time, the superconducting gap does not vanish up to T_c.

27

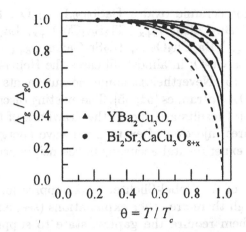

Figure 2. Dependences of δ_g on θ for $\nu = At\exp(-\delta_g/t)$ and $A = 0$ (dashed curve), 0.5, 1, 3, 5 and 10 (from bottom to top). The experimental results are shown for comparison.

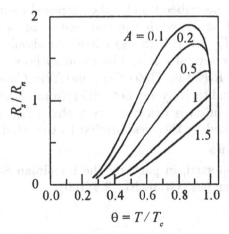

Figure 3. Dependences of ρ on θ for $\nu = At\exp(-\delta_g/t)$ and various A.

It should be emphasized that just below T_c the curves $\rho(\theta)$ corresponding to large enough A's possess an upwards curvature in agreement with the experiment [12, 13, 14, 15], whereas the non-self-consistent theory [18] also involving strong inelastic scattering retains the downwards curvature appropriate to the BCS theory [16], although destroying the Hebel-Slichter peak.

The experimental investigation of the nuclear spin-lattice relaxation was carried out for ceramic oxides $BaSb_{0.25}Pb_{0.75}O_3$, $BaBi_{0.25}Pb_{0.75}O_3$, $Ba_{0.6}K_{0.4}BiO_3$, $Tl_2Ba_2CaCu_2O_{8+x}$, $Tl_2Ba_2CuO_{6+x}$, $La_{1.85}Ca_{0.15}CuO_{4-x}$, $La_{1.85}Sr_{0.15}CuO_{4-x}$, $YBa_2Cu_3O_{7-x}$, $Bi_2Sr_2CaCu_2O_{8+x}$, Pb-Sr(Y-Ca)-Cu-O, and Bi-Pb-Sr-Ca-Cu-O; in almost all cases the Hebel-Slichter peak was absent [12, 13, 14, 15]. Nevertheless, some measurements show the peak to exist in $YBa_2Cu_3O_{7-x}$ ceramics [54, 55]. The existing discrepancies between various experimental results are due to the drawback of the NMR method itself, which requires injection of extra microwave energy. It is especially dangerous for the experimental accuracy in the neighborhood of the phase transition.

The possibility of the Hebel-Slichter peak suppression agrees well with our theory, although there are other explanations (see, e.g., Ref. [18]). But the majority of them require the gapless state to suppress the peak, as opposed to our treatment.

We would like to call attention to very interesting measurements of the nuclear spin-lattice relaxation rate in Chevrel phases $TlMo_6Se_{7.5}$ ($T_c = 12.2\,K$) and $Sn_{1.1}Mo_6Se_{7.5}$ ($T_c = 4.2\,K$) [56]. The nuclear relaxation rate in these compounds is described well by the proposed theory: $R_s(T)$ decreases exponentially for $T \ll T_c$, while the coherent peak exists only in the substance with lower T_c ($Sn_{1.1}Mo_6Se_{7.5}$) where one should expect the thermal phonon role to be essentially less. The same authors carried out the comparative studies of $R_s(T)$ in $TlMo_6Se_{7.5}$ and $YBa_2Cu_3O_7$ [57]. The results for $YBa_2Cu_3O_7$ qualitatively do not differ from those obtained in other works, but the attention is drawn to very sharp fall in $R_s(T)$ at $T \leq T_c$ which is not described by theories familiar to the cited authors.

Acknowledgements

This work was supported, in part, by the Ukrainian State Foundation for Fundamental Researches (Grant No. 2.4/100).

References

1. J. Geerk, X. X. Xi and G. Linker, Zeit. Phys. B **73**, 329 (1988).
2. T. Ekino and J. Akimitsu, J. Phys. Soc. Jpn. **58**, 2135 (1989).
3. H. Sato, H. Takagi and S. Uchida, Physica C **169**, 391 (1990).
4. Y. Enomoto, K. Moriwaki and K. Tanabe, J. Appl. Phys. **68**, 5735 (1990).
5. I. K. Yanson, Fiz. Nizk. Temp. **17**, 275 (1991) [Sov. J. Low Temp. Phys. **17**, 143 (1991)].
6. T. Hasegawa, H. Ikuta and K. Kitazawa, in *Physical Properties of High Temperature Superconductors III*, edited by D. M. Ginsberg (World Scientific Publishing, Singapore, 1992) p.525.
7. H. Hancotte, R. Deltour, D. N. Davydov, A. G. M. Jansen and P. Wyder, Phys. Rev. B **55**, 3410 (1997).
8. D. van der Marel, M. Bauer, E. H. Brandt, H.-U. Habermeier, D. Heitmann, W. Konig and A. Wittlin, Phys.Rev.B **43**, 8606 (1991).
9. M. Boekholt, M. Hoffmann and G. Güntherodt, Physica C **175**, 127 (1991).

29

10. A. Yamanaka, H. Takato, F. Minami, K. Inoue and S. Takekawa, Phys.Rev.B **46**, 516 (1992); J. Phys. Chem. Sol. **53**, 1627 (1992).
11. A. Hoffmann, P. Lemmens, L. Winkeler, G. Güntherodt, J. Low Temp. Phys. **99**, 201 (1995).
12. A. V. Bondar', S. M. Ryabchenko, Yu. V. Fedotov and A. A. Motuz, Pis'ma Zh. Eksp. Teor. Fiz. **50**, 133 (1989).
13. S. E. Barrett, J. A. Martindale, D. J. Durand, C. H. Pennigton, C. P. Slichter, T. A. Friedmann, J. P. Rice and D. M. Ginsberg, Phys. Rev. Lett. **66**, 108 (1991).
14. L. Reven, J. Shore, S. Yang, T. Duncan, D. Schwartz, J. Chung and E. Oldfield, Phys.Rev.B **43**, 10466 (1991).
15. K. Asayama, G.-Q. Zheng, Y. Kitaoka, K. Ishida and K. Fujiwara, Physica C **178**, 281 (1991).
16. M. Tinkham, *Introduction to Superconductvity* (McGraw Hill, New York, 1975).
17. J. Appel, Phys. Rev. Lett. **21**, 1164 (1968).
18. P. B. Allen and D. Rainer, Nature **349**, 396 (1991).
19. A. E. Karakozov, E. G. Maksimov and A.A.Mikhailovskii, Zh. Eksp. Teor. Fiz. **102**, 132 (1992) [Sov. Phys. JETP **75**, 70 (1992)].
20. J. P. Carbotte, Rev. Mod. Phys. **62**, 1027 (1990).
21. B. Brandow, Phys. Repts. **296**, 1 (1998).
22. J. F. Annett, N. D. Goldenfeld and A. J. Leggett, in *Physical Properties of High Temperature Superconductors*, edited by D. M. Ginsberg (World Scientific, Singapore, 1996) Vol. 5, p. 375.
23. R. A. Klemm, Phys. Rev. B **55**, 3249 (1997).
24. A. A. Abrikosov and L. P. Gor'kov, Zh. Eksp. Teor. Fiz. **39**, 1781 (1960) [Sov. Phys. JETP **12**, 1243 (1961)].
25. T. R. Lemberger and L. Coffey, Phys.Rev. B **38**, 7058 (1988).
26. S. B. Kaplan, C. C. Chi, D. N. Langenberg, J. J. Chang, S. Jafarey and D. J. Scalapino, Phys.Rev.B **14**, 4854 (1976).
27. S. G. Doettinger, S. Kittelberger, R. P. Huebener and C. C. Tsuei, Phys. Rev. B **56**, 14157 (1997).
28. P. B. Allen, Comments Cond.Mat.Phys. **15**, 327 (1992).
29. P. A. Lee and N. Read, Phys.Rev.Lett. **58**, 2691 (1987).
30. T. Moria, *Spin Fluctuations in Itinerant Electron Magnetism* (Springer Verlag, Berlin, 1985).
31. V. S. Lutovinov, in *High Temperature Superconductivity and Localization Phenomena*, edited by A. A. Aronov, A. I. Larkin and V. S. Lutovinov (World Scientific, Singapore, 1992) p.41.
32. Y. Imry, Phys. Rev. B **42**, 927 (1990).
33. K. Levin, Ju. H. Kim, J. P. Lu and Q. Si, Physica C **175**, 449 (1991).
34. Z. Schlesinger, R. T. Collins, F. Holtzberg, C. Feild, S. H. Blanton, U. Welp, G. W. Crabtree, Y. Fang and J. Z. Lee, Phys. Rev. Lett. **65**, 801 (1990).
35. T. P. Deveraux and D. Belitz, Phys.Rev.B **43**, 3736 (1991).
36. A. A. Mikhailovsky, S. V. Shulga, A. E. Karakozov, O. V. Dolgov and E. G. Maksimov, Solid State Commun. **80**, 511 (1991).
37. J. Lesueur, B. Leridon, M. Aprili and X. Grison, in *The Gap Symmetry and Fluctuations in High Temperature Superconductors*, edited by J. Bok and G. Deutscher (Plenum Press, New York, 1998).
38. K. Maki, in *Superconductivity*, edited by R. D. Parks (Dekker, New York, 1969) Vol.2, p.1035.
39. M. Fibich, Phys. Rev. Lett. **14**, 561 (1965).
40. C. R. Leavens, Phys. Rev. B **29**, 5178 (1984).
41. F. Marsiglio and J. P. Carbotte, Phys. Rev. B **43**, 5355 (1991).
42. C. Bandte, P. Hertel and J. Appel, Phys.Rev. B **45**, 8026 (1992).
43. C. S. Owen and D. J. Scalapino, Phys. Rev. Lett. **28**, 1559 (1972).
44. G. Shön and A.-M. Tremblay, Phys. Rev. Lett. **42**, 1086 (1979).

30

45. B. I. Halperin, T. C. Lubensky and Sh.-K. Ma, Phys.Rev.Lett. **32**, 292 (1974).
46. A. S. Zeltser, A. E. Filippov, Zh. Eksp. Teor. Fiz. **106**, 1117 (1994); A. E. Filippov, A. V. Radievsky, A. S. Zeltser, Phys. Lett. A **192**, 131 (1994).
47. L. Radzihovsky, Europhys. Lett. **29**, 227 (1995).
48. A. P. Ivashin, V. V. Krasil'nikov and S. V. Peletminskii, Fiz. Nizk. Temp. **19**. 1295 (1993).
49. M. L. Horbach, F. L. J. Vos and W. van Saarlos, Phys. Rev. B **48**, 4061 (1993).
50. C. M. Varma, P. B. Littlewood, S. Schmitt-Rink, E. Abrahams and A. E. Ruckenstein, Phys.Rev.Lett. **63**, 1996 (1989).
51. P. B. Littlewood and C. M. Varma, J. Appl. Phys. **69**, 4979 (1991).
52. P. B. Littlewood and C. M. Varma, Phys. Rev. B **46**, 405 (1992).
53. R. Hackl, P. Muller, D. Einzel and W. Glaser, Physica C **162-164**, 1241 (1989).
54. E. Lippmaa, E. Joon, I. Heinmaa, A. Miller, V. Midel, R. Stern and S. Vija, Physica C **162-164**, 263 (1989).
55. Y. Kitaoka, K. Ishida, K. Fujiwara, K. Asayama, H. Katayama-Yoshida, Y. Okabe and Y. Takahashi, IBM J. Res. Develop. **33**, 277 (1989).
56. S. Ohsugi, Y. Kitaoka, M. Kyogaku, K. Ishida, K. Asayama and T. Ohtani, J. Phys. Soc. Jpn. **61**, 3054 (1992).
57. Y. Kitaoka, S. Ohsugi, K. Asayama and T. Ohtani, Physica C **192**, 272 (1992).

ANISOTROPIC 3D FLUCTUATION CONDUCTIVITY IN $Y_1Ba_2Cu_3O_{7-y}$ THIN FILMS

B. SORKIN, P. KONSIN
Institute of Physics, University of Tartu, Riia 142, EE-2400 Tartu, Estonia

S. K. PATAPIS, S. AVGEROS
Solid State Section, Dept. of Physics, University of Athens, Panepistimiopolis, Zografos, Greece

AND

L. MÉCHIN
Dept. of Materials Science and Metallurgy, University of Cambridge, Pembroke Str., Cambridge, CB2 3QZ, UK

Abstract. In zero applied magnetic field the paraconductivity is analyzed for different high-quality c-axis oriented perpendicular to the substrate epitaxial thin films of $Y_1Ba_2Cu_3O_{7-y}$ grown on various substrates. The obtained for the two-band model formulas [1] for the temperature dependence of the anisotropic three-dimensional (3D) paraconductivity in the vicinity of a superconducting phase transition are used. One of the two correlation lengths is critical and determines the fluctuation conductivity $\Delta\sigma(T)$. The theory is compared with the experiment and we obtained reasonable values and the slope of $\Delta\sigma(T)$, the superconducting transition temperature T_c and the coherence length $\xi(0)$.

1. Introduction

The crystal structure of the new high-T_c Y-compounds consists of CuO_2 planes and CuO chains between them. This structure favours superconductivity. It is expected that the thermodynamic fluctuations are affected by the structure in high-T_c superconductors. Ginzburg showed that the superconducting fluctuations increase the heat capacity already above T_c [2].

31

M. Ausloos and S. Kruchinin (eds.), Symmetry and Pairing in Superconductors, 31–39.

The relative width of fluctuation region where the fluctuation correction to heat capacity of a bulk conventional superconductor is important was estimated as being equal [1]

$$\Delta T_f/T_c \sim \left(\frac{k_B T_c}{E_F}\right)^4 \sim \left(\frac{a}{\xi}\right)^4 \sim 10^{-14} \div 10^{-16} , \tag{1}$$

where a is the interatomic constant, E_F is the Fermi energy, ξ is the coherence length.

The Ginzburg-Levanyuk parameter (number) Gi in conventional superconductors was estimated as [3]

$$Gi \sim \left(\frac{a}{\xi}\right)^4 \sim 10^{-8} \div 10^{-16} , \tag{2}$$

where $a \sim 1 \overset{\circ}{A}, \xi \sim 10^2 \div 10^4 \overset{\circ}{A}$.

The order parameter can be presented as the sum of equilibrium (Δ_0 in the one band model) and fluctuation part (Δ_1) : $\Delta(\vec{r}) = \Delta_0 + \Delta_1(\vec{r})$. Above T_c the equilibrium Δ_0 is equal to zero and $\left(\frac{\Delta_1(\vec{r})}{E_F}\right)^2$ is also very small quantity.

From diamagnetic susceptibility of aluminium powder it follows $\Delta T_f/T_c \sim 0.05$ [4]. In high-T_c superconductors the Ginzburg region is of order $1.01 T_c - 1.1 T_c$ [5].

Interesting results were found for dirty and low dimensional superconducting systems. The power of the ratio (a/ξ), which enters in (1)-(2), drastically decreases as the effective dimensionality of the electron spectrum diminishes (for $\xi > a$). Another possibility to increase the strength of the fluctuation effects is to decrease the coherence length. That really happens in dirty superconductors because of the diffuse character of the electron scattering. It means that the fluctuation phenomena can be more easily observable in amorphous materials with reduced dimensionality like films.

Hight-T_c superconductors present a special interest in this sense, because their electron spectrum is extremely anisotropic and their coherence length is very small. As a result the temperature range in which the fluctuations are important may be measured by tens of degrees [6]. The magnitude of the superconducting fluctuations is essentially governed by a characteristic length, the Ginzburg-Landau coherence length ξ [3, 7] and provide a method for it determination [1, 8].

There have been many papers on the broadening of the superconducting transition or appearance of the paraconductivity [3, 7], the enhancement of the electrical conductivity near T_c in zero applied magnetic field, in bulk samples, single crystals and thin films. Two different processes are evoked by thermodynamic fluctuations of the superconducting order parameter(s). The direct process due to the acceleration of superconducting pairs, created in the thermal nonequilibrium, with finite lifetime proposed by Aslamazov and Larkin (AL) [9]. Indirect process results from the scattering of normal-state quasiparticles with the superconducting pairs discovered by Maki [10] and Thompson (MT) [11]. This scattering leads to a decay of the fluctuation pairs into quasiparticles which continue to be much accelerated. The analysis in $Y_1Ba_2Cu_3O_{7-y}$ was based on the AL, MT or on the formula of Lawrence and Doniach (LD) for two-dimensional, layered superconductors [9], which interpolate between the limiting cases of two- and three-dimensional fluctuations in the AL theory. The direct AL contribution can also be derived in the framework of the time-dependent Ginzburg-Landau theory [1, 12] and therefore should represent a rather universal feature of any type of superconducting mechanism. After the discovery of high-T_c superconductors there have been various two-band approaches to describe these systems [13-16]. Using a model of two overlapping hole bands a simple description of high-T_c values and other properties vs. doping for some representative systems have been already obtained [13-16]. Extremely short in-plane coherence length implies that superconductivity will be much more sensitive to structural or chemical imperfections in high-T_c materials.

In this work we investigate the fluctuation conductivity of $Y_1Ba_2Cu_3O_{7-y}$ thin films on various substrates experimentally and theoretically.

2. Model

We use the formulas for the temperature dependence of the paraconductivity in the vicinity of the superconducting phase transition have been derived by Konsin et al. [1] using the linearized Ginzburg-Landau equation for a two-band model [17] taking into account interband scattering of pairs within the Kubo formalism.

In general, two correlation lengths exist in the two-band model. Only one of them is critical and determines the fluctuation conductivity $\Delta\sigma(T)$.

For the anisotropic three-dimensional ($3D$) fluctuation conductivity we have [1]

$$\Delta\sigma = \frac{e\alpha^{-1/2}}{12}\frac{T}{T - T_c}\frac{1}{B(T)}\left(1 - \frac{\alpha}{B(T)^2}\right), \tag{3}$$

where

$$\alpha = \frac{\hbar^2}{8}\left(\frac{1}{m_1} + \frac{\gamma}{m_2}\right) \tag{4}$$

$$\gamma = \frac{m_1\rho_2\eta_1(0, T_c)}{m_2\rho_1\eta_2(0, T_c)}\left(\frac{p_{F2}}{p_{F1}}\right)^2, \tag{5}$$

$$B^2 = \beta + c\frac{m_1}{m_2}\xi^2(0)\epsilon^{-1} , \tag{6}$$

$$\beta = c\frac{3}{2}(\xi_{10}^2 + \xi_{20}^2) , \quad c = \beta_1^{-1}\eta_1(0, T_c) , \tag{7}$$

$$\xi^2(0) = (\xi_{10}^2 + \xi_{20}^2)\frac{\eta_1(0, T_c)\eta_2(0, T_c)}{A_1(T_c)\eta_2(0, T_c) + A_2(T_c)\eta_1(0, T_c)} , \tag{8}$$

$$\xi_{\sigma 0}^2 = \frac{\hbar^2}{4m_\sigma}\frac{\beta_\sigma}{\eta_\sigma(0, T_c)} , \tag{9}$$

$$\beta_\sigma = \frac{7\zeta(3)p_{F\sigma}^2}{12\pi^2 m_\sigma}\frac{\rho_\sigma}{(k_B T_c)^2} , \tag{10}$$

$$\eta_\sigma(0, T_c) = \sum_{\vec{k}}\frac{tanh(\tilde{\epsilon}_\sigma(\vec{k})/2k_B T_c)}{\tilde{\epsilon}_\sigma(\vec{k})} , \tag{11}$$

$$A_\sigma(T_c) = \frac{1}{2k_B T_c}\sum_{\vec{k}} cosh^{-2}\left[\frac{\tilde{\epsilon}_\sigma(\vec{k})}{2k_B T_c}\right] . \tag{12}$$

Here $\tilde{\epsilon}_\sigma(\vec{k}) = \epsilon_\sigma(\vec{k}) - \mu$, where $\epsilon_\sigma(\vec{k})$ are the energies of electrons in the bands $\sigma = 1, 2$ and μ is the chemical potential; $m_{1,2}$ are the effective masses of the bands [13-15], $\rho_{1,2}$ are the densities of states, $p_{F1,2} = m_{1,2}v_{1,2}$ the Fermi momenta. The superconducting transition temperature T_c is determined by the equation

$$1 - W^2\eta_1(0, T_c)\eta_2(0, T_c) = 0 , \tag{13}$$

where W describes the interband pair-transfer scattering.

In Aslamazov-Larkin 3D case $\Delta\sigma \sim \epsilon^\nu$, where $\nu = -1/2$ and $\epsilon = (T - T_c)/T_c$. From the analysis of the formula (3) for the anisotropic 3D fluctuation conductivity in the two-band model it follows that the effective ν corresponding $\Delta\sigma \sim \epsilon^\nu$ in (3) is in the interval -0.5 to -1.0. Therefore Eq.(3) for the anisotropic 3D fluctuation conductivity in the two-band model allows us to describe both 3D ($\nu = -0.5$) and 2D ($\nu = -1$) AL cases as well.

3. The material, experimental and data-analysis procedure

In this paper, measurements of conductivity for three different high-quality thin films of $Y_1Ba_2Cu_3O_{7-\delta}$ are presented. The films, grown on four different substrates (LaAlO$_3$, KTaO$_3$, YSZ and SrTiO$_3$), were all c oriented. The experimental method for making these measurements is the standard four-point DC technique. Currents of 1 μA, corresponding to a current density in the range of 5 A/cm^2 were systematically reversed to eliminate thermal emf's. These films which are 1000-2000 Å in thickness with the c-axis perpendicular to the substrate surface were photolithographically patterned (LaAlO$_3$, KTaO$_3$ and YSZ) in a proper bridge arrangement [18] (3 mm×10 μm). In SrTiO$_3$ the quadratic arrangement of contact areas was used.

This YBaCuO epitaxial thin film, grown on the (100) surface of a single-crystal substrate of LaAlO$_3$ has been fabricated by using the BaF$_2$ method [19]. The sample with the lowest flux pinning was chosen for the fluctuations conductivity measurements. The resistivity at 100 K is 160 $\mu\Omega$cm and the transition width, defined at the half maximum of the temperature derivative, is 0.9 K. For T$_c$ we took the midpoint temperature of half-width of derivative $d\rho/dT$. Its value is 90.25 K which coincides within the experimental error with the midpoint value.

Thin films of YBaCuO composition have been also produced epitaxially on KTaO$_3$. These films were grown in situ on the (100) surface of KTaO$_3$ substrates [20]. The lattice mismatch is now 3.5%. The characteristics of these film include: sharp superconducting transition, transition temperature width 0.3 K, and a T$_c$ defined equal to 92.15 K. The value of the normal resistivity, $\rho(100\ K) = 250\mu\Omega cm$, comes from slight cracks in the film near the substrate caused from the lattice mismatch leading to errors in geometrical cross-sections and hence to a larger normal-state resistivity.

The polycrystalline YBaCuO thin film with c-axis perpendicular to the film surface has been grown on polycrystalline yttria-stabilized zirconia (YSZ) by pulsed laser ablation [18]. This film is weak-linked. The transition is more than 2 K wide and the critical temperature is equal to 89.2 K.

The epitaxial YBaCuO thin film with c-axis perpendicular orientation to substrate has been grown on SrTiO$_3$. The transition temperature equals 90.3 K and $\Delta T_c \sim 0.7$ K.

Above T$_c$, where the resistivity deviates from the linear temperature dependence, the paraconductivity should be defined as $\Delta\sigma = 1/\rho_m(T) - 1/\rho_n(T)$, where ρ_m is the measured resistivity and ρ_n is the normal linear resistivity extrapolated from high-temperature data (we used 130–250 K region).

4. Calculations

Next we calculate the T$_c$ and fluctuation conductivity on the base of Eqs.(3) to (12) for Y$_1$Ba$_2$Cu$_3$O$_{7-y}$ thin films on various substrates. For Y$_1$Ba$_2$Cu$_3$O$_{7-y}$ thin films (on LaAlO$_3$, KTaO$_3$, YSZ and SrTiO$_3$ substrates) we use the parameters $\rho_1 = 0.9$ (eV)$^{-1}$, $\rho_2 = 2.2$ (eV)$^{-1}$, $W = 0.23$ eV, $m_1 = 10^{-2}m_0$ and $v_2 = 3.5 \cdot 10^6$ cm·s^{-1}. For Y$_1$Ba$_2$Cu$_3$O$_{7-y}$ thin film on LaAlO$_3$ single-crystal substrate we use $m_2 = 3 \cdot 10^{-1}m_0$, $v_1 = 1.7$ cm·s^{-1}.

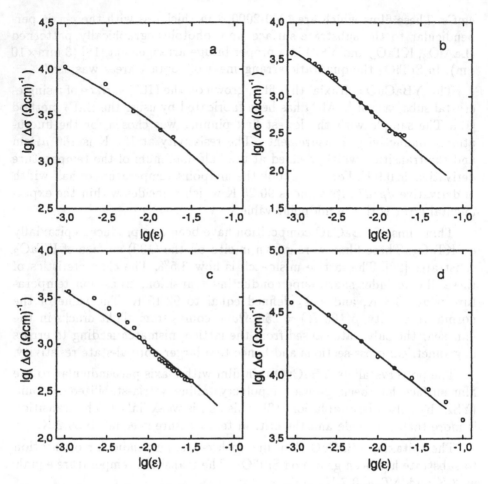

Fig.1a. The dependence of the fluctuation conductivity $log(\Delta\sigma)$ versus $log(\epsilon)$ in $Y_1Ba_2Cu_3O_{7-y}$ thin film on $LaAlO_3$ substrate. The full line is from theory, circles are from experiment.

Fig.1b. The same as Fig.1a but for YBaCuO film on $KTaO_3$.

Fig.1c. The same as Fig.1a but for YBaCuO film on YSZ.

Fig.1d. The same as Fig.1a but for YBaCuO film on $SrTiO_3$.

Then for the carrier concentration $n = 0.1374$ per CuO_2 plane the calculated transition temperature (Eq.(13)) $T_c = 90.25$ K in agreement with the experimental value of $T_c = 90.25$. For the coherence length we obtain the value $\xi(0) = 3.2\mathring{A}$. Fig.1a shows the calculated dependence of $\Delta\sigma(log(\epsilon))$ in the interval $-3 < log(\epsilon) < -1$ in comparison with the experimental data (open circles). The agreement of the theory with the experimental data is good concerning the slope of $\Delta\sigma(log(\epsilon))$ and with respect to the true resistivity values.

For $Y_1Ba_2Cu_3O_{7-y}$ thin film with slight cracks in the film on $KTaO_3$ substrate we use $m_2 = 5m_0$, $v_1 = 6$ cm·s^{-1}. Then for the carrier concentration $n = 0.137$ per CuO_2 plane the calculated transition temperature (Eq.(13)) $T_c = 90.15$ K in agreement with the experimental value of $T_c = 90.15$. For the coherence length we obtain the value $\xi(0) = 3.2\text{Å}$. Fig.1b shows the calculated dependence of $\Delta\sigma(log(\epsilon))$ in the interval $-3 < log(\epsilon) < -1$ in comparison with the experimental data (open circles) in accord with the experiment.

For $Y_1Ba_2Cu_3O_{7-y}$ thin film on YSZ substrate we use $m_2 = 3\cdot10^{-1}m_0$, $v_1 = 1.0$ cm·s^{-1}. Then for the carrier concentration $n = 0.134$ per CuO_2 plane the calculated transition temperature (Eq.(13)) $T_c = 89.2$ K in agreement with the experimental value of $T_c = 89.2$. The coherence length equals $\xi(0) = 3.3\text{Å}$. Fig.1c shows the calculated dependence of $\Delta\sigma(log(\epsilon))$ in the interval $-3 < log(\epsilon) < -1$ in comparison with the experimental data (open circles) in accord with the experimental data.

For $Y_1Ba_2Cu_3O_{7-y}$ thin film on $SrTiO_3$ substrate we use $m_2 = 3\cdot10^{-2}m_0$, $v_1 = 6.3$ cm·s^{-1}. Then for the carrier concentration $n = 0.1376$ per CuO_2 plane the calculated transition temperature (Eq.(13)) $T_c = 90.31$ K in agreement with the experimental value of $T_c = 90.31$. For the coherence length we obtain the value $\xi(0) = 3.3\text{Å}$. Fig.1d shows the calculated dependence of $\Delta\sigma(log(\epsilon))$ in the interval $-3 < log(\epsilon) < -1$ in comparison with the experimental data (open circles). The agreement of the theory with the experimental data is good.

As it follows from our estimates the films grown on various substrates have different effective masses m_2 and Fermi velocities v_1. The lower band is narrow and its effective mass m_2 is sensitive to the growing conditions of the films. The small values of v_1 are also dependent on these conditions. The obtained parameters correlate with the ones used in [1].

5. Conclusions

Calculations of amplitudes using a two-band model allows us to obtain real values and the slope of the measured in this work paraconductivity $\Delta\sigma(log(\epsilon))$, the superconducting transition temperature T_c and the coherence length of $Y_1Ba_2Cu_3O_{7-y}$ thin films on various substrates. A simple analysis shows that this model is sensitive to structural changes caused by cracks in films. The two-component scenario of superconductivity [13-16,21] for a system with a narrow band interpenetrating the wide one is valid for all described cases.

6. Acknowledgements

This work was supported by Greek Ministry of National Economy, the program of NATO Research Fellowship and partially supported by the Estonian Science Foundation grant No. 1929.

References

1. Konsin, P., Sorkin, B., and Ausloos, M. (1998) 3D fluctuation conductivity in the two-band model: theory and experiment, *Supercond. Sci. Technol.* **11**, 1–3.
2. Ginzburg, V.L. (1960) Some remarks on second order phase transitions and the microscopic theory of ferroelectricity, *Fiz. Tverdogo Tela* **2**, 2031–2043.
3. Varlamov, A.A. and Ausloos, M. (1997) Fluctuation phenomena in superconductors in: Fluctuation Phenomena in High Temperature Superconductors, *Kluwer Academic Publishers*, Dordrecht, The Netherlands, 3–41.
4. Tinkham, M. (1975) Introduction to superconductivity, McGRAW-HILL Book Company, *International Series in Pure and Applied Physics*, p. 310.
5. Carrington, A., Walker, D.J.C., Mackenzie, A.P., and Cooper, J.R. (1993) Hall effect and resistivity of oxygen-deficient $YBa_2Cu_3O_{7-\delta}$ thin films, *Phys. Rev. B* **48**, 13051–13059.
6. Lobb, C.J. (1987) Critical fluctuations in high-T_c superconductors, *Phys. Rev. B* **36**, 3930–3932.
7. Ausloos, M., Patapis, S.K., and Clippe, P., in Physics and Materials Science of High Temperature Superconductors II, Proc. NATO-ASI on High Temperature Superconductors, Porto Curras, 1991, Ed. by Kossowsky, R., Raveau, B., Wohlleben, D., and Patapis, S.K. (Kluwer, Dordrecht, 1992), p.755–785 and refs. therein.
8. Sekirnjak, C., Lang, W., Proger, S., and Schwab, P. (1995) Novel approach for the analysis of the fluctuation magnetoconductivity in $YBa_2Cu_3O_7$ thin films, *Physica C* **243**, 60–68.
9. Aslamazov, L.G. and Larkin, A.I. (1968) Influence of fluctuations on the properties of superconductor above critical temperature, *Sov. Phys. Solid State* **10**, 1104–1111.
10. Maki, K. (1968) The critical fluctuation of the order parameter in type-II superconductors, *Progr. of Theor. Physics* **39**, 897–906.
11. Thompson, R.S. (1970) Microwave, flux flow and fluctuation resistance of dirty type-II superconductors, *Phys. Rev. B* **1**, 327.
12. Plakida, N.M. (1995) High-temperature superconductivity (Berlin, Springer).
13. Konsin, P., Kristoffel, N., and Örd, T. (1988) The interband interaction as a possible cause of high-temperature superconductivity, *Phys. Lett. A* **129**, 339–342.
14. Konsin, P., Kristoffel, N., and Örd, T. (1989) The role of the Fermi level position in high-T_c superconductivity, *Phys. Lett. A* **137**, 420–422.
15. Konsin, P., Kristoffel, N., and Örd, T. (1990) Application of the interband model to the dependence of T_c on oxygen in $Y_1Ba_2Cu_3O_{7-y}$, *Phys. Lett. A* **143**, 83–84.
16. Kristoffel, N., Konsin, P., and Örd, T. (1994) Two-band model for high-temperature superconductivity, *Rivista Nuovo Cimento* **17**, 1–41.
17. Konsin, P. (1995) Ginzburg-Landau equations and upper critical magnetic field in graphite intercalation compounds, *phys. stat. sol. b* **189**, 185–191.
18. Jones, E.C., Christen, D.K., Thompson, J.R., Feenstra, R., Zhu, S., Lowndes, D.H., Phillips, J.M., Siegel, M.P. and Budai, J.D. (1993) Correlation between the Hall coefficient and the superconducting transport properties of oxygen-deficient $YBa_2Cu_3O_{7-y}$ epitaxial thin films, *Phys. Rev. B* **47**, 8986–8995.
19. Siegal, M.P., Phillips, J.M., Hebard, A.F., van Dover, R.B., Farrow, R.C., Tiefel, T.H., and Marshal, J.H. (1991) Correlation of structural quality with supercon-

ducting behavior in epitaxial thin films of $YBa_2Cu_3O_{7-\delta}$ on $LaAlO_3$ (100), *J. Appl. Phys.* **70**, 4982–4988.

20. Patapis, S.K., Jones, E.C., Phillips, J.M., Norton, D.P., Lowndes, D.H. (1995) Fluctuation conductivity of YBaCuO epitaxial thin films grown on various substrates, *Physica C* **244**, 198-206.

21. Mihailovic, D., Mertelj, T., Voss, K.F., Heeger, A.J., and Herron, N. (1992) Origin of different critical temperatures in oxide superconductors: A comparison of $(Tl,Cd)_2(Ba,La)_2CuO_6$ with $(La,Ba)_2CuO_3$ by Raman and infrared absorption spectroscopy, *Phys. Rev. B* **45**, 8016–8020.

QUANTUM MONTE CARLO SIMULATIONS OF THE TWO-DIMENSIONAL ATTRACTIVE HUBBARD MODEL: PHASE DIAGRAM AND SPECTRAL PROPERTIES

J. M. SINGER, T. SCHNEIDER AND P. F. MEIER
Physikinstitut, Universität Zürich,
CH-8057 Zürich, Switzerland

1. Introduction

The problem of a crossover from a weak coupling BCS (Bardeen-Cooper-Schrieffer, [1]) picture of Cooper pair formation and condensation at a critical temperature to a Bose-Einstein condensation (BEC, [2]) of preformed (local) pairs has recently attracted great attention. The motivation to study

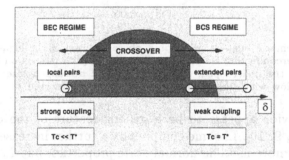

Figure 1. "Crossover" from a BCS-like endpoint to a BEC endpoint along the superconducting phase transition line. δ symbolices the control parameter driving the crossover, which is in this diagram associated with the (inverse) coupling strength in case of the attractive Hubbard model and the electronic carrier doping in case of the cuprates.

this problem comes from experimental observations regarding unusual properties of the high-T_c cuprate superconductors. Particularly interesting in this respect are recent experiments showing a pseudo-gap structure in the normal-state density of states of underdoped cuprates that persists almost up to room temperature [3, 4, 5, 6, 7]. A further unusual property different from conventional BCS-type superconductors is the extreme short

M. Ausloos and S. Kruchinin (eds.), Symmetry and Pairing in Superconductors, 41–70.

coherence length (of the order of some lattice constants) of the pairs in the superconducting state, much smaller than in usual superconductors (where it is of the order of several thousand Å). Consequently, the pairing in the cuprates is much less mean-field like. Many physical properties of cuprate superconductors depend on hole doping, thus, leading to a rather universal doping dependence of the transition temperature T_c. At a certain doping level x_u, the so called underdoped limit, the materials undergo at $T > 0$ an insulator to anomalous metal transition, and at $T = 0$ an insulator to superconductor transition. As x is increased T_c rises rapidly and attains a maximum at x_m (optimum doping).

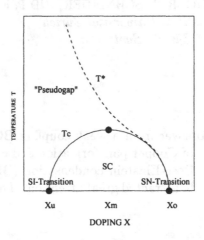

Figure 2. Schematic temperature-doping phase diagram of the high-Tc cuprates; SC denotes the superconducting state, enclosed by the T_c phase transition line, with x_u, x_m and x_o marking the underdoped, optimal doped and overdoped case. T^* indicates the temperature, below which the opening of a pseudogap is observed.

This behavior appears to be a generic feature of the cuprate superconductors [8, 9, 10]. In some compounds a further increase of the doping level leads to more metallic normal state properties, but T_c now decreases and falls to zero in the overdoped limit x_o. Here, these materials undergo at $T = 0$ a superconductor to normal metal transition (Fig. 2. Although the schematic phase diagram looks quite symmetric, there are fundamental differences between the underdoped and the overdoped regime.

To gain insight into the electronic properties of the normal and superconducting states we explore the effect of electron correlations, and in particular pair formation and appearance of superconducting correlations, on the density of states, the spectral densities and the band structure in a rather simplified lattice model system, the two-dimensional Hubbard model with local attractive interaction (2AHM). Although this model is unlikely to provide a microscopic description of high-temperature superconductiv-

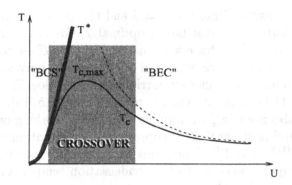

Figure 3. 2AHM: 'Sketch' of the crossover regime along the coupling strength axis U for fixed density ρ. Shown are qualitative curves for the critical temperature T_c, as obtained e.g. by QMC (solid line), as well as the BCS and BEC limits (dashed lines). Additionally, a T^* curve is drawn (dark gray curve), indicating the pair formation scale in contrast to the condensation temperature T_c.

ity, it serves rather well to reveal the effect of correlations on measurable properties. It is evident that the local attractive onsite density-density interaction term favors double occupancy of sites and hence the formation of (s-symmetric) pairs. If these pairs of charge 2e are mobile, superconductivity will occur below a certain temperature. This model exhibits a pronouced crossover (see illustration in Fig. 3) from a weak coupling regime, where the essential features of the superconducting state are well described by a BCS theory, to a strong coupling regime, properly described by a BE condensation of local, preformed (real-space) pairs. Because of its conceptual simplicity, the attractive Hubbard model allows an easy investigation of the crossover from weak to strong coupling, from extended to local pairs, from BCS to BEC, just by tuning the interaction parameter U (Fig. 3), which acts as a control parameter for the phase transition line in analogy to the carrier doping in e.g. the high temperature superconducting cuprates. At sufficiently low T an instability of the Fermi sea towards superconductivity occurs, the transition is essentially mean-field in character, see e.g. [11, 12, 13] and references therein. This link of the weak coupling regime with BCS superconductivity has been provided by Schmitt-Rink and Noziéres [14]. The evolution from Cooper-pair superconductivity for small U to local pair superconductivity for large U is smooth.

In the present work the electronic carrier density ρ ('electron doping', charge carriers per site and spin) is assumed to be finite and fixed, away from half-filling. In Fig. 3 we sketch a qualitative T-U-phase diagram, as discussed e.g. in [13], and indicate the $T_c(U)$ curve as obtained by QMC together with the BCS and BEC limits. Moreover, we plot in this figure also a T^*-curve, giving the pair formation energy scale. Since at both endpoints of this phase transition line, the $U = 0$ and the $U \to \infty$ limit, T_c vanishes, it is evident that there must be an optimal T_c for an intermediate value of U [13]. For $U = 0$ we have a normal metal. For $U \to \infty$ the electrons form bound pairs which are immobile since they can only move via virtual ionization with an infinite energy barrier. On the lattice, T_c vanishes in this limit, while in the continuum limit it remains finite [15, 16]. This difference is due to the absence of a pair hopping term when working on a lattice. For increasing coupling strength T_c is suppressed by fluctuations down from the BCS value T_c^{BCS}, and T_c^{BCS} becomes rather a mere pair formation scale. T_c^{BCS} is no longer connected to the condensation temperature, where long range order is established.

A large number of studies has been published discussing several aspects of this crossover. Here we will provide an extension of the earlier work, we put special emphasis on the discussion of the spectral properties and their temperature evolution in the intermediate coupling regime, which is particularly difficult to access by any kind of approximative method. In the intermediate coupling regime the physics will be dominated by the interplay between quasiparticles and bound pairs, leading to non-trivial behavior. We therefore concentrate on the investigation of the attractive Hubbard model with the quantum Monte Carlo (QMC) approach, which has the potential to treat this type of strongly correlated system on a lattice numerically exactly, allowing us to go far beyond certain approximative methods.

2. The Attractive Hubbard Model

We consider the 2D attractive Hubbard model (2AHM, 'negative-U model') on a square lattice:

$$\mathcal{H} = -t \sum_{\langle ij \rangle \sigma} (c_{i\sigma}^{\dagger} c_{j\sigma} + h.c.) - U \sum_{i} n_{i\uparrow} n_{i\downarrow} - \mu \sum_{i\sigma} n_{i\sigma} , \qquad (1)$$

where $c_{i\sigma}^{\dagger}$ ($c_{i\sigma}$) denote fermionic creation (annihilation) operators at site i with spin σ, and t is the kinetic term between two neighboring sites, which serves as an energy unit throughout the paper. The limit $\langle ij \rangle$ restricts the sum to next-neighbors, U denotes the interaction ('coupling'), which is chosen to be attractive, and μ is the chemical potential. We consider the intermediate coupling regime $W/2 \le U \le W$, W being the bandwidth ($W = 8t$ in $D = 2$). Away from half-filling the 2AHM is believed to undergo a Kosterlitz-Thouless (KT) [17] transition.

In the free case ($U/t = 0$) we have the well-known dispersion relation for the D-dimensional system,

$$\epsilon(k) = -2t \sum_{\alpha=1}^{D} \cos(k_\alpha) - \mu \quad . \tag{2}$$

With the exception of the critical endpoints and $\rho = 0.5$ (ρ is the density of electrons per site and per spin, i.e. $\rho = 1$ corresponds to the fully occupied lattice with 2 electrons of opposite spin at each site), the phase transition line is supposed to be a line of D-dimensional XY critical points, while the special point $\rho = 0.5$ corresponds to a D-dimensional XYZ critical point. For this reason, the transition temperature vanishes in 2D at $\rho = 0.5$ (half-filling). There one finds a coexistence of superconducting and long-range charge-density correlations, which, in 2D, drive the effective KT transition temperature to zero ($T_{KT}^{2D}(\rho = 0.5) = 0$), accounting for another critical point $\rho_{c,2D} = 0.5$. At zero temperature, $T = 0$, there are two critical endpoints, namely $\rho_c = 0$ and $\rho_c = 1$, where the model undergoes an insulator to superconductor transition.

In a strictly two-dimensional superconductor, thermal fluctuations will destroy 'true' long-range order for all $T > 0$, but a KT transition may still separate two phases. The low temperature phase is characterized by a finite superfluid density and correlations that decay algebraically with distance. In a finite, periodic lattice, the correlations might well level off similar to a 'conventional' phase transition, thus enabling us to study effects of a 'quasi-conventional' transition even in 2D, with a size-dependent critical temperature. A proper scaling leads in fact to the 'real' KT temperature [18]. We will make use of this (in a strict sense) 'size' effect when we study the temperature evolution of the spectral properties of a 2D system.

In addition to the density-driven transitions at fixed U there are also interaction driven transitions at fixed ρ. These have been discussed earlier on from a phase transition point of view [13]: As a function of coupling strength U and fixed ρ there is a phase transition line with critical endpoints $U = 0$ and $U \to \infty$. At $T = 0$ a normal metal to superconductor transition occurs at $U = 0$, while for $U \to \infty$ there is a superconductor-localization (insulator) transition. In the strong coupling regime this model can be mapped onto hard core bosons on a lattice, in which Cooper pairs are treated as conserved particles obeying Bose statistics.

The nature of the phase transition is quite different in the two limits: In the weak coupling, BCS-like limit a formation of Cooper pairs and their condensation takes place simultaneously at T_c. A first deviation from this scenario is usually described in terms of superconducting fluctuations. In the preformed pair (BEC) regime, however, the pair formation at T^* and their condensation at T_c are independent processes. T^* and T_c are widely separated, with T^* being only a characteristic energy scale, not a phase transition temperature. For $T > T^*$ the pairs are thermally dissociated. In the weak coupling limit, below T_c, we have a BCS-type condensate of strongly overlapping Cooper pairs. Thermodynamics and T_c are determined by single particle excitations (broken Cooper pairs, quasi-particles). In the opposite, strong coupling regime one has a Bose condensate of tightly bound local pairs, and thermodynamics and T_c are governed by the collective modes. For a review see e. g. [19], the crossover has been studied by a series of authors [2, 11, 12, 13, 14, 19, 20, 21, 22, 23, 24, 25, 26, 27, 28].

3. Technique

We present numerical studies of the described model using a particular type of quantum Monte Carlo (QMC) method, the temperature dependent QMC formulation in the grand canonical ensemble (after Hirsch et al. [29, 30, 31, 32]). We emphasize that the QMC approach has the potential to treat these types of strongly correlated systems, allowing us to go far beyond certain approximative methods. It provides an approximation-free, numerically exact ansatz (besides controllable statistical errors), unlike most standard analytical techniques, and yields information about systems much larger than those accessible by exact diagonalization algorithms. The QMC method uses the Suzuki-Trotter and the Hubbard-Stratonovich transformation to 'break up' the quantum-mechanical many-particle system. Applying these methods to the attractive Hubbard model frees us from the central drawback of fermion QMC calculations, the so-called 'sign problem'. This allows us to perform reliable and stable calculations over a vast parameter range

To extract spectral properties of the Hubbard model we retrieve time-dependent correlation (Green's) functions (in imaginary time), which may be written

$$G_\sigma(i-j, \tau) = -\langle T_\tau c_{i\sigma}(\tau) c_{j\sigma}^\dagger(0) \rangle . \tag{3}$$

As we are merely interested in dynamical properties at finite temperatures, we restrict ourselves to the evaluation of this quantity in the grand canonical algorithm; a scheme for the application in the $T = 0$-formalism projector QMC was introduced by von der Linden [33] and applied in [30]. The analytic continuation seeks to extract real frequency, dynamical information from these imaginary-time correlation functions computed in QMC simulations. The imaginary-time Green's functions $G(k, \tau)$ in k-space are intimately connected to the spectral function $A(k, \omega)$ via

$$G(k, \tau) = \begin{cases} -\displaystyle\int_{-\infty}^{\infty} d\omega \frac{\exp(-\omega\tau)}{\exp(-\beta\omega) + 1} A(k, \omega) & \text{if } \tau > 0 \\[2ex] +\displaystyle\int_{-\infty}^{\infty} d\omega \frac{\exp(-\omega\tau)}{\exp(+\beta\omega) + 1} A(k, \omega) & \text{if } \tau < 0 \end{cases} \tag{4}$$

from which we can easily obtain the density of states as a summation over all k-states:

$$N(\omega) = \frac{1}{N} \sum_k A(k, \omega) . \tag{5}$$

We seek these spectral densities because linear-response theory relates these functions to experimentally measurable quantities.

But, these simple equations pose a serious problem: The correlation functions are easy to obtain, but it is rather difficult to extract the spectral properties from the computed QMC data because an analytic continuation, or rather an inverse Laplace transformation from imaginary to real time, is required. This inversion of QMC data (and usually of all statistically computed data) is extremely numerically illposed due to two obvious reasons: Data are available only for a limited set of imaginary times and the data are usually more or less noisy. As a consequence, the solution might not be unique in general.

The method of choice is the maximum entropy (MaxEnt) ansatz, as proposed by Gubernatis, Jarrell and co-workers [34, 35] for a similar type of data. The method we used to obtain the presented information is a slight variation of this ansatz after von der Linden's work [33]. The MaxEnt technique approaches statistical data analysis within the concepts of conditional probabilities (Bayesian logic), where the spectral density is regarded as a probability function, and what is extracted from the data is the most probable spectral density [34]. Unique about the MaxEnt approach is the specification of the prior probability function of the solution in terms of the information-theory definition of entropy.

4. Results

In the following section we will discuss our results for the density of states $N(\omega)$, the k-resolved spectral density $A(k, \omega)$ and the band structure $\Omega(k, \omega)$. We concentrate on three different parameter sets in the intermediate coupling (crossover) regime:

1. $U/t = 4$, $\rho = 0.25$ ('quarter-filling') at $N = 16 \times 16$,
2. $U/t = 6$, $\rho = 0.4$ (close to half-filling) at $N = 16 \times 16$
3. $U/t = 8$, $\rho = 0.1$ (low-density limit) at $N = 10 \times 10$.

Additionally, to allow a comparison along the U / fixed ρ-axis, we provide some data for two further parameter sets at only one fixed temperature value $T/t = 0.5$: $U/t = 4.0$, $\rho = 0.4$ and $U/t = 8.0$, $\rho = 0.4$.

These systems have been chosen, since they allow us to present a wide overview on the behavior in the intermediate regime. Simultaneously, we can use our computational resources in the most economical way. The sys-

48

tem size 16×16 [1] represents the uppermost system size giving reasonable computer times for this type of simulation (we use single nodes of an IBM SP2 system), i.e. giving us still the chance to produce excellent statistics (i.e. long enough runs to reduce statistical noise). For the low-density ($U/t = 8$, $\rho = 0.1$) run we even lowered the size in favor of an enlarged number of independent MC bins and prolonged Markov chains, and thus even smaller statistical errors. The reason is that for $\rho \to 0$ the MaxEnt procedure becomes unusually difficult, whereas for larger U the statistical fluctuations along the MC time axis increase and thus we need longer runs to produce statistical error bars comparable to the lower U simualtions. Of particular interest is certainly the regime in parameter space close to half-filling, since there almost all techniques besides QMC/MaxEnt fail to produce reasonable results. The critical temperatures of the three systems – as simply determined from the inflection point in the temperature dependence of the long-range plateau of the equal-time s-wave pair correlation function – are approximately in the same range[2], for example, $T_c/t(U/t = 4.0, \rho = 0.25) \simeq 0.11$. These values are determined for one particular lattice size, they are not scaled, and therefore present only an upper bound. A phase diagram using these data is shown in [13]. For a more appropriate determination of the KT transition temperature we have to do either a scaling analysis similar to the one proposed by Moreo et al. [18], or we follow a slightly different approach recently presented in a paper by Engelbrecht et al. [36]. We we will briefly sketch the latter: the starting point is the equal-time s-wave pair correlation function $\chi^{os}(r)$ calculated in the 16×16 lattices,

$$\chi^{os}(r) = \langle \Delta^{\dagger}(0)\Delta(r) + \Delta(0)\Delta^{\dagger}(r) \rangle, \tag{6}$$

where

$$\Delta^{\dagger}(l) = c_l^{\dagger} c_l^{\dagger}. \tag{7}$$

The correlation length ξ at each temperature is defined through the scaling relation [17]

$$\chi^{os}(r) = \text{const.} \cdot \frac{1}{r^{\eta}} \cdot \exp(-r/\xi) \quad , \tag{8}$$

at large distances r. For the XY universality class the exponent has the theoretical value $\eta = 0.25$. To extact T_{KT} we fit our QMC results to the XY scaling form

$$\xi(T) = a \exp(A/\sqrt{\beta_{KT} - \beta}) \quad , \tag{9}$$

with $\beta = 1/T$. The circles in Figs. 4 and 5 are QMC data points, the solid lines are fits to eqn. (9). To obtain the fit curve, we used only data points

[1] periodic boundary conditions are applied
[2] Technically, these temperatures can be reached by our stabilized QMC algorithm

with $\xi < L/2 = 8$ (due to the imposed periodic boundary conditions; they are marked with filled circles in the figures). Results for two parameter sets are presented:

- $U/t = 4.0$, $\rho = 0.25$ (electrons per site and spin, i.e. quater-filling, corresponding to $n = 0.5$) , yielding $T_{KT}/t = 0.049$, Fig. 4.
- $U/t = 6.0$, $\rho = 0.4$, yielding $T_{KT}/t = 0.053$, Fig. 5.

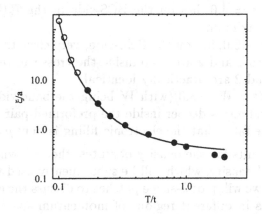

Figure 4. Pair correlation length ξ vs. temperature T, $U/t = 4.0$, $\rho = 0.25$. QMC data (circles) and best fit to eqn. (9) (solid line, using only the data points marked by filled circles), yielding $T_{KT}/t = 0.049$.

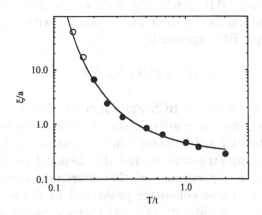

Figure 5. Pair correlation length ξ vs. temperature T, $U/t = 6.0$, $\rho = 0.4$. QMC data (circles) and best fit to eqn. (9) (solid line, using only the data points marked by filled circles), yielding $T_{KT}/t = 0.053$.

It provides a technical advantage that this approach to determine T_{KT} is rather sensitive, in contrast to other methods.

Moreover, we present for $U/t = 8.0$ data in the low density regime $\rho = 0.1$, since in this case we find at least some results from other methods [16, 21, 37]. It allows us to compare and interpret our findings. Using the phase diagram for this model (a 'sketch' is shown in Fig. 3, quantitative data are presented in an earlier publication [13]), we would classify these parameters into three regimes:

- Regime 1, $U/t = 4.0$, lies on the BCS-side of the $T_c(U)$ curve, very close to the maximum,
- regime 2, $U/t = 6.0$, lies on the BEC-side, very close to the maximum, and, thus, both 1 and 2 are deep inside the crossover regime. The T_c's of regime 1 and 2 are practically identical.
- Regime 3, $U/t = W = 8.0$ (with W being the bandwidth of the non-interacting system) is deeper inside the preformed-pair (BEC) regime, even given the fact, that the electronic filling is only $\rho = 0.1$.

We will discuss results for the density of states, then we will switch to the k-resolved spectral density, which will be subsequently used to draw a band structure. Finally we will plot k-space patches to discuss the evolution of the spectral properties in different regions of momentum space. Comparisons with results from other techniques, in particular BCS, will be given at the appropriate sections.

4.1. DENSITIES OF STATES

To start this section on QMC results for the density of states (DOS) we will present initially a non-QMC result: Fig. 6 shows the density of states $N(\omega)$ for a 'free', non-interacting system and a system with a finite coupling, treated in the simple BCS approach,

$$E^2(k) = \epsilon^2(k) + \Delta^2(k) , \qquad (10)$$

with Δ being an s-symmetric BCS gap function. The free system shows the usual 2D lattice system density of states, with a van Hove singularity due to the characteristic saddle point. In contrast to the free system, the BCS result gives a gap structure around the chemical potential μ (i.e., at $\omega - \mu = 0$), with the appearance of characteristic 'coherence' peaks at the edge of the gap. These coherence peaks will be discussed in detail in a subsequent section. Despite of the usual thermal broadening for $T > 0$, these two curves characterize the DOS within the BCS mean-field approach. The appearance of a gap is intimately connected to the superconducting regime.

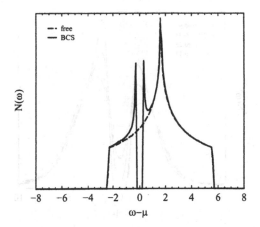

Figure 6. Density of states $N(\omega)$ using $T = 0$ BCS-theory, electronic filling $\rho = 0.25$, free system, $U = 0$ (dashed line), and superconductor $U \neq 0$ (solid line). The calculation has been done on a lattice.

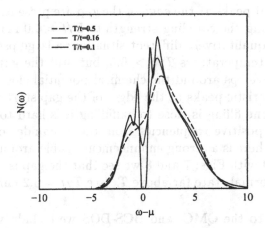

Figure 7. Density of states $N(\omega)$, QMC, coupling strength $U/t = 4.0$, electronic filling $\rho = 0.25$, lattice size $N = 16 \times 16$, periodic boundary conditions.

Fig. 7 presents similar data obtained by QMC for three different temperatures (the band filling is chosen again to be $\rho = 0.25$). Here we find already a rather different behavior: For quite high temperatures (approx. $5T_c$) an initially small 'deformation' of the DOS around the chemical potential begins to form, which, on lowering the temperature, develops into a 'pseudo-gap' like dip. At $T/t = 0.1$ the system has reached the superconducting regime ($T_c/t = 0.11$), the gap is fully developed down to zero,

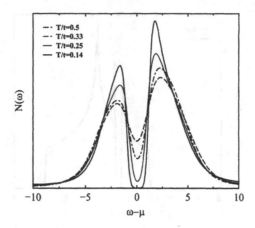

Figure 8. Density of states $N(\omega)$, QMC, $U/t = 6.0$, $\rho = 0.4$, $N = 16 \times 16$.

and – characteristic for the superconducting state and similar to the BCS DOS – the typical peaks at the edge of the gap are present. Going to Fig. 8 and thus increasing the coupling strength to $U/t = 6.0$ marks the onset of a crossover to a qualitatively different situation: a large pseudogap evolves already at high temperatures $T/t > 0.5$, but still the gap only 'touches' zero and fully develops around the chemical potential close to $T_c/t \approx 0.14$, with the characteristic peaks at the edges of the gap starting to form. Due to the fact that the filling is close to halffilling it is hard to distinguish the peak located at positive frequencies from the remainder of the van Hove singularity, but there is a strong enhancement visible around T_c. Moreover, comparing Fig. 9 with Figs. 7 and 8 we see that the gap is already fully developed in the normal state far above T_c (see $T/t = 0.2$-curve, $T_c/t \approx 0.1$).

Additionally to the QMC- and BCS-DOS we include with figure 10 a DOS for the 2AHM following the approach introduced by Schmid [38] and later on reproduced and extended by Tchernyshyov [23] ([39] follows a similar argumentation). Their approach is of particular interest, since these authors studied the 2AHM assuming that superconducting long-range order is suppressed by strong 2D fluctuations. This is in fact true in a strict sense in the thermodynamic limit, as discussed earlier. In the regime below a BCS critical temperature, which now serves only as a pair formation scale, they find still a pseudo-gap (i.e. a suppression of the DOS at the chemical potential). But now the gap has a rather 'unconventional' V-shaped form, normally attributed only to d-wave superconductors. Remind that the system under investigation is nevertheless still an s-wave superconductor. This

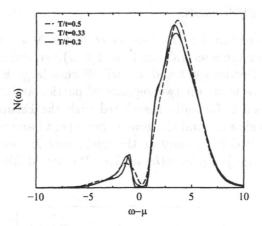

Figure 9. Density of states $N(\omega)$, QMC, $U/t = 8.0$, $\rho = 0.1$, $N = 10 \times 10$.

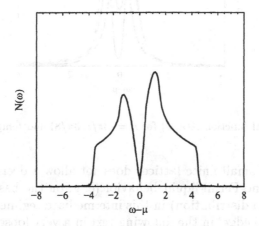

Figure 10. Density of states $N(\omega)$, Schmid's ansatz for the 2AHM, $\rho = 0.4$. The calculation uses a noninteracting density of states obtained for a discrete lattice.

pseudogap is only due to fluctuating pairs (Tchernyshyov [23] calls them 'slow' fluctuations), and resembles the above-T_c pseudogap in figure 7. Nevertheless, with increasing coupling strength U this picture becomes more and more inappropriate ($U/t \to \infty$ would cause an infinitely large fully developed gap at any temperature, with tightly bound pairs).

4.2. SPECTRAL FUNCTIONS

After describing results for the density of states $N(\omega)$, which is equal a k-summation over the spectral function $A(k,\omega)$, we will now focus on the k-resolved data. Sections 4.3 and 4.4 will discuss large k-space regions in detail. Initially we focus on two regimes of particular interest. These are the k-region which is formally associated with the Fermi surface k_F and secondly the k-region around the zone corners (π, π) and equivalent points. In fact, k_F is well defined only in the noninteracting and weak coupling regime, e.g. via the jump in $n(k)$ at k_F. We would like to remark that

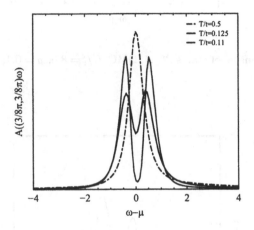

Figure 11. Spectral function $A(k,\omega)$ for $k = (3\pi/8, 3\pi/8)$ and temperatures above T_c, $U/t = 4.0$, $\rho = 0.25$.

the discussion of small finite lattices does not allow a decisive comment on Fermi-liquid vs. no-Fermi-liquid behavior (e. g. on the basis of an analysis of the momentum distribution) in this intermediate regime. We will use the expression 'Fermi edge' in the following text in a very loosely defined sense, which usually only indicates the 'filling edge', i. e. $E - \mu = 0$.

Fig. 11 presents data at k-points $(3\pi/8, 3\pi/8)$ and equivalent, which are as close as possible to the Fermi edge position of the non-interacting system for $\rho = 0.25$. There, for intermediate coupling $U/t = 4.0$ the system behaves qualitatively rather close to what we expect from the BCS theory, with some modifications around T_c. At high temperatures we find only one single peak, which is finally splitted by a gap around $\omega - \mu = 0$ below T_c. Here in our QMC results for the 2AHM this splitting initially starts in a rather small temperature window above T_c, which is probably defined to a large extent by thermal fluctuations. While at k_F there is only one single peak at e.g. $T/t = 0.5$, we would like to point to the fact that we have already two or more peaks in other k-regions (we will comment on that in sections 4.3 and 4.4). The splitting of the single peak at k_F into two maxima separated by a (superconducting) gap causes a shift of spectral weight away from $\omega - \mu = 0$ and, thus, a pile-up of two rather sharp 'resonances' at the edge of the gap. This was already seen in the discussion of the density of states $N(\omega)$. The BCS theory would give an ungapped, single peak at the Fermi egde down to T_c, with a gap $\Delta(T)$ emerging only below T_c. An

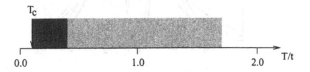

Figure 12. 'Temperature axis' for $U/t = 4.0$, $\rho = 0.25$: Whereas $T_c \approx 0.11t$, the light shaded area marks the temperature window above T_c, where at least a second band is detectable ('Pseudogap-regime'), and the narrow, dark shaded region gives the temperature, where the spectral function peak at the filling edge ('Fermi edge') starts to split ('Fluctuation regime').

enhancement of the coupling strength, Fig. 13 (due to the different value of ρ we have chosen another k-vector close to the assumed 'Fermi' edge) shows, that there the splitting starts to form at much higher temperatures (in fact, almost an order of magnitude higher than for $U/t = 4.0$). The 'fully' developed gap can still be found only at $T \approx T_c$. Again a pile-up of spectral weight at the gap edges can be detected. When we go to even larger U-values this picture changes qualitatively: The peak structure will be completely gapped at higher and higher temperatures far above T_c (see, e.g. the $U/t = 8.0$ band structure in Figs. 20 and 23), with the extreme limit of an infinitely wide gap at all temperatures for $U/t \to \infty$.

Furthermore, the region around the zone corners is of peculiar interest, in particular in the temperature window $T^* < T < T_c$. Fig. 14 shows $A(k, \omega)$-data for the zone corner $k = (\pi, \pi)$ itself as well as for its closest neighbors along the (discrete) lattice axes and the diagonal. The data are obtained at $T/t = 0.5t$. In the left column of Fig. 14 we keep the electron density fixed at $\rho = 0.4$ and increase the coupling strength from $U/t = 4$ to $U/t = 8$, going from top to bottom. The right column varies the density for the upper and lower row, thus enabling us to investigate the ρ-dependency of $A(k, \omega)$. The most striking feature is the appearance of not only two peaks, but three of them are visible, corresponding to at last three branches in the excitation spectrum. 'At least' refers to the fact, that we are not able to reproduce the band structure itself, but only the positions of spectral peaks with a weight significantly different from zero. We would like to note that we cannot split 'very close' lying peaks, in particular not if these peaks

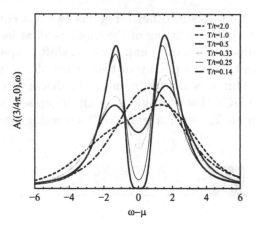

Figure 13. Spectral function $A(k, \omega)$ for $k = (3\pi/4, 0)$ and temperatures above T_c, $U/t = 6.0$, $\rho = 0.4$.

have a large difference in weight (i.e. a broad peak with large spectral weight will easily 'cover' a small one in its tails).

The three peaks differ as follows: There is a large, broad peak (its position is marked Ω_1) at frequencies $\omega - \mu > 0$ (unoccupied range), a second broad one with low weight at the opposite side of the frequency axis (occupied range), marked Ω_2, and finally a third one, located close to the position of the chemical potential. Its position will be referred to as Ω_3, it is e.g. for $U/t = 6.0$, $\rho = 0.4$ located almost exactly at the chemical potential (i.e. $\Omega_3(\pi, \pi) = 0$). Ω_1, Ω_2 and Ω_3 shift as a function of doping as well as coupling (the coupling strength dependency is plotted in Fig. 16), but for all systems shown in Fig. 14 there are clearly three peaks present. The spectral weight (relative to the Ω_1-peak) of the Ω_3-peak is lowest in the $\rho = 0.1$, $U/t = 8.0$ system, and largest in the $\rho = 0.4$, $U/t = 8.0$ system (there it accounts at its maximum for roughly 10% of the total spectral weight). The position of the Ω_3-peak shifts (a) to lower frequencies if the coupling strength U is increased and $\rho = $ const., and (b) to lower frequencies if the filling ρ is increased and $U = $ const.. The spectral weight of Ω_3 increases with increasing coupling strength, and since both, the Ω_2- and Ω_3-line become broader and simultaneously seem to approach each other with increasing coupling strength (at least for $\rho = 0.4$), it will be increasingly difficult to split them from each other. The appearance of a third peak (band) around the zone corners in the normal state of the 2AHM is not totally unexpected. Several authors, e.g. [16, 40, 41] among others, predict the appearance of at least one further feature besides the established single

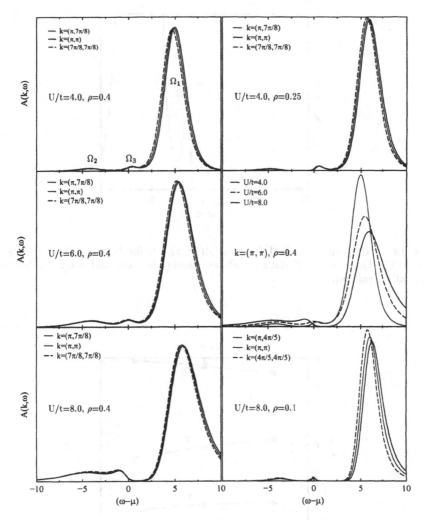

Figure 14. Spectral function $A(k, \omega)$, k-vectors around (π, π), for $T/t = 0.5$ and different U-, ρ-values.

particle band and pair band (present below T^*, see [11]). Nevertheless, to exclude artifacts of the finite lattice calculations, we studied the evolution of the Ω_3-peak with system size. Fig. 17 shows the size dependence of $\Omega_3(\pi, \pi)$ for lattice sizes N between 64 and 256 sites, for a $U/t = 6.0$, $\rho = 0.4$ system at temperature $T/t = 0.3$ (here we observed the strongest size dependency among all systems under investigation). Clearly, the position of the peak converges to a fixed value for $N \to \infty$. The peak itself becomes even more pronounced (i.e. its relative spectral weight allows an easy distinction from the rest of the structure), if we increase the system size. Figs. 14 and 16 give the impression, that the pseudogap opens in this temperature regime

58

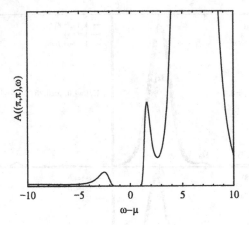

Figure 15. 'Blow-up': Spectral function $A((\pi, \pi), \omega)$, for $U/t = 4.0$, $\rho = 0.25$ and $T/t = 0.25$. This 'enlarged' data set shows clearly that the additional Ω_3 branch is numerically significant.

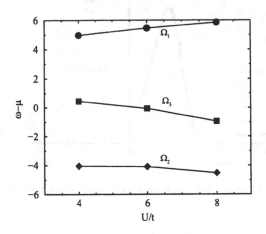

Figure 16. Peak positions Ω_1, Ω_2 and Ω_3 as a function of coupling U, $k = (\pi, \pi)$, $\rho = 0.4$, $T/t = 0.5$.

and at $k \approx (\pi, \pi)$ between the Ω_2 and Ω_3 peaks, not exactly at $\omega - \mu = 0$. This observation is at least qualitatively partially consistent with the work presented in [37]

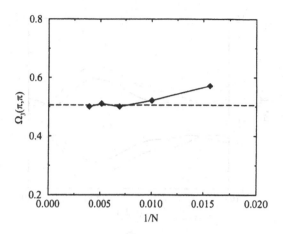

Figure 17. Position of the central peak Ω_3 at $k = (\pi, \pi)$ in the band structure for $U/t = 6.0$, $\rho = 0.4$, $T = 0.3$ as a function of inverse system size $1/N = 1/L^2$ ($\diamond$); the dashed line indicates the supposed level for $L \rightarrow \infty$.

4.3. BAND STRUCTURE

After having identified quite unconventional structures in $A(k, \omega)$ in section 4.2 we want to discuss rather extended regions in k-space. For this purpose we plot $A(k, \omega)$ along the triangle $(0, 0) \rightarrow (\pi, 0) \rightarrow (\pi, \pi) \rightarrow (0, 0)$ in momentum space, which allows us an examination along the main lattice directions between zone center and corner. In principle, we would be interested in a 'real band structure calculation' for the 2AHM, which would give us all bands and there dispersion. Using a special type of plot, we can extract some aspects of the bandstructure from the calculation of $A(k, \omega)$. We use a gray shade coding to plot the spectral function in the k-ω-plane, along the described k-path. The gray shades code the value of $A(k, \omega)$, a quartic folding form is used to get a non-linear coding (i.e. small values of $A(k, \omega)$ are strongly enhanced to visualize structures resulting from small peaks). Nevertheless, this type of plotting gives *not* the full band structure, since only branches with non-vanishing weight are visible. The reader should keep this in mind when comparing our data with 'real' band structure calculations. As a starting point, Fig. 18 depicts a band structure for the 2AHM, resulting from a simple BCS dispersion form $E^2(k) = \epsilon^2(k) + \Delta^2$, with a BCS gap Δ. This BCS band structure is shown for two different values of the chemical potential, one is close to the lower edge of the noninteracting band and the other one corresponds roughly to quarter-filling.

In contrast, Figs. 19-23 present QMC data. Fig. 19 shows the band structure (the name is used in the following discussion as a synomym

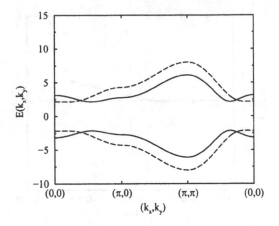

Figure 18. Band structure resulting from a BCS calculation, $E^2(k) = \epsilon^2(k) + \Delta^2$. The dashed curve corresponds to a system where the chemical potential is close to the lower band edge of the non-interacting system (i.e. $\rho \to 0$), and the solid curve represents data, where μ is deep inside the noninteracting band.

for the described gray coded $A(k, \omega)$ plot) for $U/t = 4.0$, quarter-filling $\rho = 0.25$ and a series of temperatures down to $T \approx T_c$. The $T/t = 2.0$ and $T/t = 0.11$ panels are the easiest ones to describe: $T/t = 2.0 \gg T^*$ and $T/t = 0.11 \approx T_c$. At a temperature $T \gg T^*$ we have only one band (which we denote in accordance to section 4.2 Ω_1), corresponding to the noninteracting system, pairs are thermaly dissociated. Lowering the temperature ($T/t = 1.0$) causes the appearance of a second band, pairs start to form and occupy a pair band Ω_2 ('two-particle bound state') [11]. This second band is particularly pronounced at the zone center and zone corners. If we decrease the temperature further down to $T/t = 0.5$, we see a splitting of the Ω_2-band into two distinct branches in a region around the zone corner (π, π). This splitting starts for $T/t = 0.5$ at $k \approx (\pi, \pi/4)$, and the region expands (visible as some kind of 'bubble' extending from (π, π)) if we go to lower temperatures $T/t = 0.3$, $T/t = 0.25$. For the last temperature we find three branches extending over the whole k-regime, besides the immediate neighborhood of the Fermi surface vector k_F. During this process, Ω_3 seems to shift towards higher frequencies. In the regime close to the transition region around T_c we get another splitting, a fourth band evolves out of Ω_2, again initially around the zone corner. At T_c and below, the band structure becomes 'conventional' again, the dispersion of the spectral weights is consistant with a band structure predicted for a usual (BCS-type) superconductor below T_c (compare Fig. 18).

Similar data are provided for $\rho = 0.4$, $U/t = 6.0$ (Fig. 20) and $\rho = 0.1$,

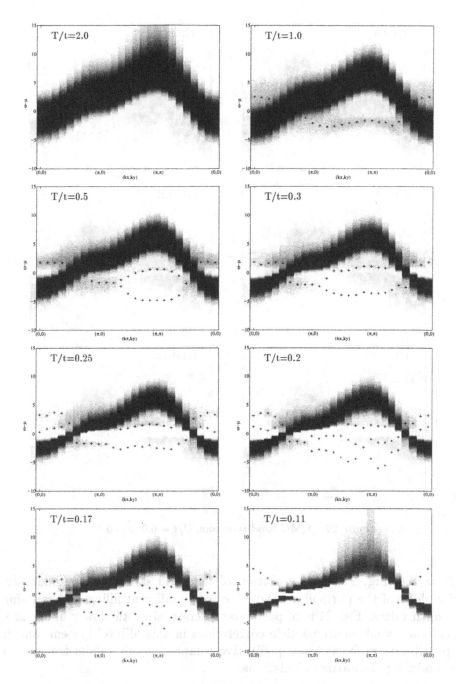

Figure 19. QMC band structure, $U/t = 4.0$, $\rho = 0.25$. The crosses mark low peaks which otherwise are hardly visible in this plot.

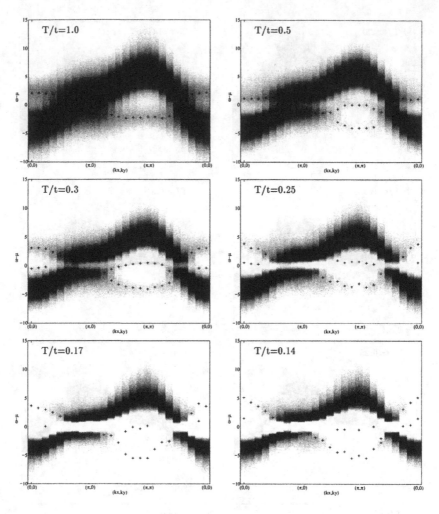

Figure 20. QMC band structure, $U/t = 6.0$, $\rho = 0.4$.

$U/t = 8.0$ (Fig. 21). The evolution of the band structure is similar, despite shifts of the particular bands due to the different filling and coupling strength values. Fig. 21 is of particular interest, since the lower filling produces only weak many particle correlations in this 'dilute' system, and it is possible to make at least qualitative comparsions to results from other methods, e.g. T-matrix calculations.

An interesting result obtained by Kagan et al. [16] is reproduced in Fig. 22. There, we show their T-matrix band structure for a system with $\rho = 0.5$, $U/t = 10$ and $T/t = 0.1$ in the *normal* state (their [16] approach does not produce superconducting correlations, due to a explicit implementation of

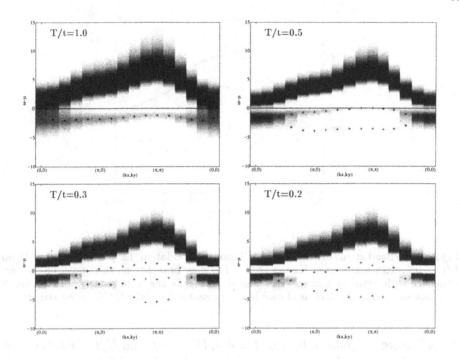

Figure 21. QMC band structure, $U/t = 8.0$, $\rho = 0.1$.

Mermin and Wagner's theorem[42]). The different branches are marked gray, if the particular spectral weight is close to zero, and thus would be invisible to our method. It is marked black, if the spectral weight is non-vanishing, so that we should be able to see the band. Unfortunately, [16] presents only data along the diagonal. Most important and interesting for a comparison to our data is the fact that these authors find a third branch resulting from the η resonance initially proposed by Yang [41], which - using our presentation technique - should be particularly pronounced around $k = (\pi, \pi)$. This mode has only a weak U-dependence. This branch comes down in energy with increasing U, similar to our results (see, e.g., Fig. 16. It is necessary to compare the results of [16] with our QMC data taken at quite high temperatures, e.g. $T/t = 0.5$, since at lower temperatures, e.g. $T/t = 0.1$, the QMC system is dominated by superconducting correlations, as discussed before, causing additional changes in the band structure.

Finally, in Fig. 23 we present a direct comparison between two bands for $U/t = 4$ and $U/t = 8$, both measured at $\rho = 0.4$ and $T/t = 0.5$. This figure shows the pronounced differences arising from the BCS to BEC crossover. In addition to the two coupling values presented in Fig. 23 we ask the reader to consider also the uppermost right panel of Fig. 20, $U/t = 6.0$, $\rho = 0.4$, $T/t = 0.5$, which can be regarded as an in-between panel for Fig.

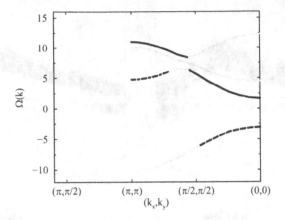

Figure 22. Band structure $\Omega(k)$ from a T-matrix calculation, for $\rho = 0.05$, $U/t = 10$ and $T/t = 0.1$, data taken from Kagan et al.. The grey branches mark regions of the band structure with virtually zero spectral weight, thus, only the black parts would be 'visible' in a plot using $A(k, \omega)$-data as shown in the presentation of the QMC 'band structures'.

23. All three coupling values $U/t = 4.0$, $U/t = 6.0$ and $U/t = 8.0$ show two distinct bands in the normal state, but only $U/t = 8.0$ has a large, fully developed (pseudo-) gap around $\omega - \mu = 0$ for *all* k-vectors. In contrast, for $U/t = 4.0$ we find a single, ungapped peak at the k-vectors corresponding to the 'Fermi'-surface (defined in the general sense as previously discussed). $U/t = 6.0$ marks somehow the crossover. Included in the right column of Fig. 23 are also the corresponding k-space maps (there coding is described in detail in section 4.4). Here, the effect of the crossover is even more easily seen: in contrast to the $U/t = 8$ case we find a clear 'signature' of the 'Fermi surface' (or rather its residues, signalled by the black pixels corresponding to ungapped k-points). Common to both cases is the 'bubble' around (π, π), where the Ω_3-branch shows up, indicated by the white patches in the k-maps.

4.4. K-SPACE MAPS

After the discussion of the spectral functions and the band structure we want to study systematically how certain features evolve in momentum space. For that purpose we use an unconventional approach: In figure 24 we plot a particular type of map of the Brillouin zone $((-\pi, \pi), (-\pi, \pi))$. We use four different levels of gray shading to code the 'type' of spectral density / band structure found at a particular k-point: A black square ('k-pixel') represents a k-point with a fully *ungapped* $A(k, \omega)$ (one peak), a dark gray k-

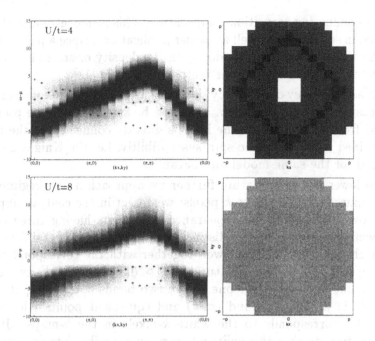

Figure 23. QMC band structure (left column), $\rho = 0.4$, $T/t = 0.5$, $U/t = 4.0$ and $U/t = 8.0$, and affiliated k-space maps (right column).

pixel a *partially* developed gap (i.e. $A(k, 0) \neq 0$, but two distinct maxima), a light gray k-pixel a *fully* gapped spectral function, and a white k-pixel marks a k-point, which has a spectral function having *more than two* distinct peaks. Using this coding scheme we start with a totally 'black' Brillouin zone for very high temperatures $(T/t = 2.0)$, shown in the uppermost left corner of figure 24. This patch is equivalent to the corresponding subpicture in Fig. 19, and represents the fact that we have only one single band at this temperature, possible pairs are thermally dissociated.

On lowering the temperature, dark gray areas appear at the corners of the zone around (π, π) and equivalent points, signalling the opening of a 'pseudo-gap' (i.e. a splitting of the single peak in $A(k, \omega)$ into two peaks). Further lowering of the temperature causes
(a) a growth of the gapped regions around (π, π),
(b) an additional 'nucleus' of gapped states at the zone center around $(0, 0)$, and
(c) finally a fully opened gap down to zero in this regions (uppermost row in figure 24). The temperature, where the initial opening of gapped k-regions is observed, will be called T^* in the further context of this paper. We would like to emphasize, that even in the case of a moderate coupling strength $U/t = 4$, this temperature T^* is about 15 times higher than T_c,

the superconducting transition temperature. This is in contrast to the value presented in section 4.1 as well as earlier publications, since a pure definition of T^* via the opening of a pseudogap in the density of states is rather less sensitive and gives a much lower temperature scale [13].

Such an 'accuracy dependency' is a typical feature of a smooth crossover in contrast to a phase transtion. It should be stressed that the pseudogap obtained from an analysis of the DOS is directly connected to the pseudo-gap obtained from the static spin susceptibiltiy, i.e. the Knight shift [43], calculated for the same model and parameters.

If we lower the temperature further we approach a new regime in the k-maps, marked by the white k-pixels: we detect in the central map of the second row $(T/t = 0.5)$ the appearance of k-points having three (and for even lower temperatures four) distinct maxima in the spectral function A. This patch again should be viewed together with the corresponding band structure, Fig. 19. For temperatures $T/t > 0.5$ we had only two distinct branches, but now the lower one of the two bands starts to split up into two bands in a region around (π, π) and equivalent points (the 'bubble' splitting Ω_2 corresponds to the white k-pixels in the k-maps). It seems important to note that this splitting starts in a similar k-region where the initial formation of a second band Ω_2 (i.e. the opening of the pseudo-gap) was observed.

Going to temperatures $T/t \leq 0.2$ we find that

(a) after an intermediate extension the white regimes finally disappear. The spectral weights on several of the peaks vanish, but not necessarily the bands itself.

(b) only a narrow 'ring structure' of not fully gapped k-points remains, reflecting the exact position of the 'Fermi surface' (defined rather in terms of a drop in the momentum distribution $n(k)$ than in a strict sense, as discussed earlier). This 'ring structure' vanishes only close to and below $T_c = 0.11t$. Below T_c, a homogenous patch is obtained. It is important to note that for this U-value the final opening of the gap *along* the Fermi surface happens only at T_c (+ a possible fluctuation window), resembling remnants of a BCS-type of transition (we again refer also to Fig. 12). In a conventional superconductor the so-called 'Fermi-surface instability' at T_c is a characteristic feature of a BCS-type of theory. The fact that the gap along the Fermi surface seems to open at different temperatures for different directions is due to the finite lattice, which gives rise to only a finite number of k-points. Obviously, for the chosen filling $\rho = 0.25$ the Fermi surface lies in between of two k-points along the main axes, whereas it 'hits' exactly along the diagonals.

Here we are really in a crossover regime, since we have a formation of a pseudogap, a second branch in the band structure and thus pairs at

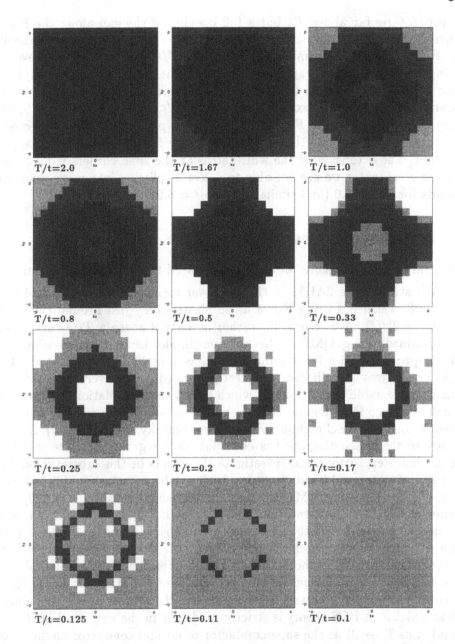

Figure 24. *k*-space maps, $U/t = 4.0$, $\rho = 0.25$. Brillouin zone ($-\pi \leq k_{x,y} \leq \pi$) with each discrete *k*-point gray coded according to the 'gap state' of it's $A(k,\omega)$. Coding scheme: BLACK (ungapped), DARK GRAY (two peaks, gap not fully developed down to zero), LIGHT GRAY (two peaks, fully developed gap) and WHITE (more than two peaks).

temperatures far above T_c, but a full opening of the gap along the Fermi surface takes place only at approximately T_c. This is particularly eminent in Fig. 23, where we compare $U/t = 4$ and $U/t = 8$, both on the level of band structure and k-map plots (left and right columns, respectively). In the normal state system for $U/t = 4.0$ the 'Fermi surface' is again clearly marked by the black k-pixels (similar to the $T/t = 0.5$-panel in Fig. 19, but larger due to $\rho = 0.4$ in Fig. 23 in contrast to $\rho = 0.25$, just as expected). In contrast, we have at the same temperature for $U/t = 8$ a fully gapped system. The extension of the white areas at the zone corners is identical, at the zone center we are not able to detect a splitting into more than two bands for $U/t = 8.0$ (and similarly for $U/t = 6.0$).

5. Discussion and Conclusions

To summarize our results we have investigated the following problem:

We studied the 2AHM in the crossover regime between $U/t = 4$ and $U/t = 8$, which is - using Fig. 3 as an illustration - just around the maximum of $T_c(U)$. A T_c-curve (or rather in a strict sense a T_{KT}-curve) can be obtained by e.g. QMC studies or even simpler using a combination of a BCS approach and a KT argumentation, as done by Denteneer et al. [44]. The latter gives a qualitatively correct picture (quantitatively it provides at least a reasonable upper bound), which allows an interpolation between the weak coupling BCS regime (with a monotonically increasing T_c) and the strong coupling BEC regime (with a monotonically decreasing T_c). Such a study of the phase diagram has been published e.g. in [13]. Here we want to go one step further and investigate the effects of this crossover on the density of states and band structure. First results have been published earlier, e.g. in [11] using QMC simulations, but also for example in [20] and other studies. Our presentation widely extends these previous data, and focuses in particular on the effects of the crossover on the density of states, the spectral densities and the band structure. We are able to show how remnants of the BCS concept (like a kind of 'Fermi surface') 'survive' into the crossover regime, whereas also precursors of the BEC (or better 'preformed pair') regime appear quite early (like the second band). We notice that – although BCS theory is strictly valid only in the very limit of $U \to 0$ (and thus $T_c \to 0$) at the superconductor to normal conductor endpoint of the $T_c(U)$ phase line – remnants of a BCS / mean field type of transition are visible up to quite large values of U ($U/t \simeq 4$). Similarly interesting is the fact that the other endpoint of the phase transition line $T_c(U)$, the superconductor - insulator point at $U \to \infty$, has an even bigger bassin of attraction. It starts at $U/t \to \infty$ and goes down well below the maximum of the $T_c(U)$ curve into the BCS regime (taking the appearance of a second branch in the excitation spectrum as a marker). This confirms our earlier findings [10] analyzing an 'Uemura'-type of plot for the 2AHM. The presented concepts have been studied over a wide parameter range, using simulations on rather large systems with a state-of-the-art QMC/MaxEnt simulation tool. It is difficult to decide on the origin of some of the observed effects (preformed, tightly bound pairs vs. fluctuations or a combination) from the spectral properties only, but we will try to extend and supplement the presented data with measurements of other thermodynamic quantities as soon as possible. A really striking discovery is the appearance of more than two bands in a certain temperature regime $T^* > T > T_c$.

6. Acknowledgement

We would like to acknowledge interesting discussions with M. H. Pedersen, H. Beck, R. Frésard, M. Capezzali, R. Micnas, J. Engelbrecht and V. Loktev on the 2AHM and the crossover topic, P. Schwaller and J. Osterwalder on spectroscopy in 'real' cuprates and H.-G. Matuttis on MaxEnt and QMC. This work was supported by the Swiss National Science Foundation.

References

1. See e.g., J. R. Schrieffer, *Theory of Superconductivity*, Addison-Wesley, Reading, 1988;
2. See e.g., J. Blatt, *Theory of Superconductivity*, Academic Press, New York, 1964; R. Micnas, J. Ranninger, S. Robaszkiewicz, Rev. Mod. Phys. **62**, 113 (1990).
3. H. Ding, T. Yokoya, J. C. Campuzano, T. Takahashi, M. Randeria, M. R. Norman, T. Mochiku, K. Kadowaki, J. Giapintzakis, Nature **382**, 51 (1996); H. Ding, J. C. Campuzano, M. R. Norman, M. Randeria, T. Yokoya, T. Takahashi, T. Takeuchi, T. Mochiku, K. Kadowaki, P. Guptasarma, D. G. Hinks, cond-mat/9712100.
4. A. G. Loeser, Z.-X. Shen, D. S. Dessau, D. S. Marshall, C. H. Park, P. Fournier, A. Kapitulnik, Science **273**, 325 (1996).
5. P. Schwaller, T. Greber, J. M. Singer, J. Osterwalder, P. Aebi, H. Berger, L. Forró, unpublished.
6. P. Coleman, Nature **392**, 134 (1998).
7. Ch. Renner, B. Revaz, K. Kadowaki, L. Maggio-Aprile, Ø. Fischer, unpublished; Ch. Renner, B. Revaz, J.-Y. Genoud, K. Kadowaki, Ø. Fischer, Phys. Rev. Lett. **80**, 149 (1998).
8. T. Schneider, J. M. Singer, Proc. Adriatico Res. Conf. "Fluctuation Phenomena in High Temperature Superconductors", Ed. A. Ausloos et al., Kluwer Academic, Dordrecht, 1996.
9. T. Schneider, Proc. Europhys. Conf. "Physics of Magnetism 96", Ed. R. Micnas, Acta Physica Polonica A, 1996.
10. T. Schneider, J. M. Singer, Europhys. Lett. **40**, 79 (1997).
11. J. M. Singer, M. H. Pedersen, T. Schneider, H. Beck, H.-G. Matuttis, Phys. Rev. B **54**, 1286 (1996).
12. J. M. Singer, M. H. Pedersen, T. Schneider, Physica B **230-232**, 955 (1997).
13. J. M. Singer, T. Schneider, M. H. Pedersen, Eur. Phys. J. B **2**, 17 (1998).
14. P. Noziéres and S. Schmitt-Rink, J. Low Temp. Phys. **59**, 95 (1985).
15. M. Drechsler and W. Zwerger, Ann. Phys. (Leipzig) **1**, 15 (1992).

70

16. M. Y. Kagan, R. Frésard, M. Capezzali, H. Beck, Phys. Rev. B **57**, 5995 (1998).
17. J. M. Kosterlitz and D. J. Thouless, J. Phys. C **6**, 1181 (1973).
18. A. Moreo and D. J. Scalapino, Phys. Rev. Lett. **66**, 946 (1991).
19. R. Micnas and T. Kostyrko, *Recent Developments in High Temperature Superconductivity*, J. Klamut et al. (Eds.), Springer, Berlin 1996.
20. R. Micnas, M. H. Pedersen, S. Schafroth, T. Schneider, J. J. Rodríguez-Núñez, H. Beck, Phys. Rev. B **52**, 16223 (1995).
21. M. H. Pedersen, J. J. Rodríguez-Núñez, H. Beck, T. Schneider, S. Schafroth, cond-mat/9702173.
22. T. Schneider, H. Beck, D. Bormann, T. Meintrup, S. Schafroth, A. Schmidt, Physica C **216**, 432 (1993).
23. O. Tchernyshyov, Phys. Rev. B **56**, 3372 (1997).
24. A. J. Leggett, J. Phys. (Paris) Colloq **41**, C7-19 (1980).
25. R. Haussmann, Z. Phys. B. **91**, 291 (1993) and Phys. Rev. B **49**, 12975 (1994).
26. V. Loktev, S. G. Sharapov, Cond. Mat. Phys. **11**, 131 (1997); V. P. Gusynin, V. M. Loktev, S. G. Sharapov, cond-mat/9709034; V. M. Loktev, V. M. Turkowski, cond-mat/9707191.
27. M. Randeria, *Bose-Einstein Condensation*, A. Griffin et al. (Eds.), Cambridge University Press, Cambridge, England, 1994; M. Randeria, Proc. Adriatico Res. Conf. *Fluctuation Phenomena in High Temperature Superconductors*, Ed. M. Ausloos et al., Kluwer Academic, The Netherlands, 1996; M. Randeria et al., Phys. Rev. Lett. **41**, 327 (1990).
28. N. Trivedi and M. Randeria, Phys. Rev. Lett. **75**, 312 (1995); M. Randeria, N. Trivedi, A. Moreo, R. Scalettar, Phys. Rev. Lett., **69**, 2001 (1992).
29. J. E. Hirsch, Phys. Rev. B **31**, 4403 (1985);
30. H.-G. Matuttis, PhD. Thesis, University of Regensburg, 1995.
31. E. Y. Loh Jr., J. E. Gubernatis, R. T. Scalettar, R. L. Sugar, S. R. White, Proc. *Workshop on Interacting Electrons in Reduced Dimensions* D. Baeriswyl and D. K. Campbell (Eds.), Plenum Press, New York, 1989.
32. M. Ulmke, H. Müller-Krumbhaar, Z. Phys. B – Cond. Matter **86**, 383 (1992).
33. W. von der Linden, Physics Reports **220**, 53 (1992).
34. M. Jarrell and J. E. Gubernatis, Phys. Reports **269**, 133 (1996).
35. R. N. Silver, D. S. Sivia, J. E. Gubernatis, M. Jarrell, Phys. Rev. Lett. **65**, 496 (1990); J. Gubernatis, M. Jarrell, R. N. Silver, D. S. Sivia, Phys. Rev. B **44**, 6011 (1991); R. N. Silver, D. S. Sivia, J. E. Gubernatis, Phys. Rev. B **41**, 2380 (1990); R. N. Silver, D. S. Sivia, J. E. Gubernatis, *Quantum Simulations of Condensed Matter Phenomena*, J. E. Gubernatis and J. D. Doll (Eds.), World Scientific, Singapore, 1990.
36. J. R. Engelbrecht, A. Nazarenko, cond-mat/9806223.
37. M. Letz and R. J. Gooding, cond-mat/9802107.
38. A. Schmid, Z. Phys. **231**, 324 (1970).
39. E. Abrahams, M. Redi, C. Woo, Phys. Rev. B **1**, 280 (1970).
40. A. I. Solomon, K. A. Penson, cond-mat/9712228.
41. C. N. Yang, Phys. Rev. Lett. **63**, 2144 (1989).
42. N. D. Mermin and H. Wagner, Phys. Rev. Lett. **17**, 1133 (1966); P. C. Hohenberg, Phys. Rev. **158**, 383 (1967).
43. J. M. Singer, P. F. Meier, Physica C **302**, 183 (1998).
44. P. J. H. Denteneer, Guozhong An, J. M. J. van Leeuwen, Phys. Rev. B **47** (1993) 6256. P. J. H. Denteneer, Guozhong An, J. M. J. van Leeuwen, Europhys. Lett. **16** (1991) 5.

RHOMBIC VORTEX LATTICE
IN d-WAVE SUPERCONDUCTIVITY

JUN'ICHI SHIRAISHI AND MAHITO KOHMOTO
Institute for Solid State Physics,
University of Tokyo, Roppongi, Minato-ku, Tokyo 106, Japan

AND

KAZUMI MAKI
Department of Physics and Astronomy,
University of Southern Calfornia Los Angeles, Cal. 90089-0484,
USA

Abstract. The rhombic vortex lattice (*i.e.* the square lattice tilted by 45° from the a axis) is stable in the wide field region in d-wave superconductors in a magnetic field parallel to the c axis. The triangular lattice in a weak magnetic field transforms into the rhombic vortex lattice at $H_{cr}(T) \sim \kappa^{-1} H_{c2}(T)$ with the second order transition where κ is the Ginzburg-Landau parameter. This transition is the standard second order transition as described by Landau and we expect the jump in the specific heat $\Delta C \sim T(\partial H_{cr}/\partial T)^2$.

1. Introduction

Since the discovery of the high T_c cuprate superconductors by Bednorz and Müller [1] in 1986, the most important event was the realization that the superconductivity in hole doped high T_c cuprates is of d-wave type [2, 3]. One of the characteristics of d-wave superconductivity is seen in the vortex state in a magnetic field parallel to the c axis. In this configuration the stable vortex lattice appears to be rhombic in the most part of the $B - T$ phase diagram as shown in Fig.1. However before considering the vortex lattice in a d-wave superconductor, we will summarize briefly what is known about the vortex state in a classic s-wave superconductor. The whole field

71

M. Ausloos and S. Kruchinin (eds.), Symmetry and Pairing in Superconductors, 71–82.

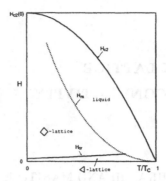

Figure 1. The phase diagram with the rhombic vortex lattice is shown schematically. In reality the rhombic lattice becomes stable around $B = 10^{-2} H_{c2}(t)$. $H_m(T)$ specifies the the vortex lattice melting transition.

of type II superconductivity and vortex state is created by Abrikosov[4] in 1957. Although his theory is based on the Ginzburg-Landau theory of superconductivity [5], it is now understood that the notion of quantized flux and regular vortex lattice apply in a much wider context. First of all, after the appearance of the theory of superconductivity by Bardeen, Cooper and Schrieffer [6], Gor'kov was able to derive the Ginzburg-Landau theory from the microscopic theory in the vicinity of $T = T_c$ the transition temperature [7]. Further Abrikosov's analysis is extended in the whole temperature region [8]. Also clearly Abrikosov was inspired by the beautiful magnetization data from Pb-Tl and Pb-In systems obtained by Shubnikov *et al.* [9]. In the original paper Abrikosov thought the square vortex lattice would be most stable. Later it was discovered that the triangular vortex lattice is more stable [10]. The triangular vortex lattice is first observed by small angle neutron scattering (SANS) by Cribier *et al.* [11] in pure Nb crystal. Later Essmann and Träuble [12] were able to take the photograph of the vortex lattice by decoration technique.

After development of the scanning tunneling microscope (STM), the vortex lattice in NbSe$_2$ has been seen by STM by Hess *et al.* [13]. Hess *et al.* have not only succeeded in imaging the vortex lattice but also studied the local quasi-particle density of states around a vortex. They discovered a large density of states for $|E| < \Delta$ around a vortex core as first predicted by Caroli, de Gennes and Matricon [14], where Δ is the energy gap. Later Gygi and Schlüter [15] have shown by solving the Bogoliubov-de Gennes equation numerically that there are a large localized density of states around the vortex core. In the 2D model in particular the bound state spectrum around the core is given by $E_m = \frac{2}{\pi}(|m| + \frac{1}{2})\Delta(p_F\xi)^{-1}$ with $m = 0, \pm1, \pm2, \cdots$ and p_F is the Fermi momentum and ξ is the superconducting coherence length [15]. In a classical *s*-wave superconductor, we have $p_F\xi \sim 10^3 \sim 10^4$, which implies that there are thousands of the bound states in the vortex core.

One of the big surprises about the vortex state in high T_c cuprate superconductors is that there is only a few bound states as in the optimally doped YBCO [16] or no bound state at all as in the optimally doped Bi2212 [17]. This suggests strongly that the superconductors in high T_c cuprates are in the quantum limit (*i.e.* $p_F\xi \sim 1$) as shown explicitly in [18, 19]. In particular for $p_F\xi = 1.33$ we find $E_0/\Delta = 0.55$ [19] where E_0 is the energy of the lowest bound state energy. This suggests that there will be no bound state for $p_F\xi \leq 1$. Note since Δ in Bi2212 is almost twice as large as the one in YBCO, we expect $p_F\xi \simeq 1$ for Bi2212. In addition we discovered the low energy ($E/\Delta \leq 0.3$) extended states with four legs of the local density of states stretched into four diagonal directions [19]. These extended states are unique to d-wave superconductors and perhaps they stabilize the rhombic vortex lattice which we are considering.

Earlier we have shown that the rhombic vortex lattice becomes stable in the vicinity of $B = H_{c2}(T)$ for $T \leq 0.81T_c$ [20, 21]. More recently we analyzed the vortex lattice in the lower field region making use of the extended Ginzburg-Landau equation which includes the fourth order terms in ∂_x and ∂_y, where ∂_x and ∂_y are the gauge invariant differential operators [22, 23]. First of all we note that the higher order differential operator generates a fourfold distortion around the vortex core in $\Delta(\mathbf{r})$ the superconducting order parameter. A similar result has been previously obtained by Enomoto et al. [24]. However we find this core distortion gives rise to the long range directional force proportional to $\cos(4\theta)r^{-4}$ between two vortices where r is the distance between the two vortices and θ is the angle $\mathbf{r}$ makes from the a axis [23]. Therefore the stable vortex lattice configuration is found from the competition between two forces; the usual magnetic interaction which favors the triangular lattice and the core interaction which favors that two neighboring vortices align parallel to two diagonal directions $(1,1,0)$ and $(1,-1,0)$. The real surprise is that the critical field $H_{cr}(T)$, which gives the line of the second order transition to the rhombic lattice is rather small $H_{cr}(T) \simeq 0.41\kappa^{-1}(1-t)^{-1/2}H_{c2}(t)$. This implies that the rhombic lattice is stable in the most part of the B-T phase diagram. For example for optimally doped YBCO, we have $H_{cr}(0) \simeq 1$ Tesla. So this is fully consistent with the observation of the rhombic lattice though elongated somewhat in the a direction by SANS [25] and STM [16] at low temperatures and in a small magnetic field of 3 Tesla. We believe that the distortion of the rhombic lattice is due to the $d+s$ superconductivity [26] specific to YBCO, though there is an alternative interpretation [27].

Earlier we thought that the vortex lattice is elongated in the b direction since $\xi_b/\xi_a \sim 1.5$ where ξ_a and ξ_b are the coherence lengths in the a and b directions. On the other hand a recent study of the quasi-particle spectrum around a single vortex line shows that there are extended state with $|E| \leq$

0.3Δ with their four legs stretched along the nodal directions of $\Delta(\mathbf{k})$ [18, 19]. If this is the case we expect the strong attractive force between two vortices lying along the nodal directions and this suggests that the vortex lattice is elongated in the a direction.

Now coming back to the second order transition, the jump in the specific heat is obtained as $\Delta C = -T\frac{\partial M}{\partial T}\frac{\partial H_{cr}}{\partial T}$ where the T-derivative is taken along the transition line. Earlier a similar vortex lattice transition is studied by Franz et al. [28] and Kogan et al. [29] in terms of the generalized London equation. Although qualitatively they predicted a very similar vortex lattice transition (i.e. the second order transition), we predict a much sharper rise of the apex angle (see Fig. 4). Further they could not predict $H_{cr}(T)$ since their models were purely phenomenological. On the other hand our theory is based on the microscopic theory of d-wave superconductivity, though we calculate the relevant coefficients in terms of the weak-coupling theory [30]. One of our surprises is that the weak-coupling theory works very well. Perhaps the results may not be reliable quantitatively. But they are reliable at semi-quantitative level [31].

It is now well established that the vortex lattice melts due to the thermal fluctuation [32]. Recently the jump in the entropy associated with this melting transition $H = H_m(T)$ has been observed in both optimally doped Bi2212 [33] and in YBCO [34]. The observed jump in the entropy is a fraction of k_B per pancake vortex, though there is no theoretical estimate for this jump in the entropy. So naturally our transition line $H_{cr}(T)$ terminates at $H_m(T)$ as shown in Fig. 1.

2. Distorted vortex core in d-wave superconductor

We consider a weak-coupling model for d-wave superconductor[30]. Extending the procedure used by Ren et al. [35], we obtain

$$
\left(-\ln t + \frac{7\zeta(3)}{2(4\pi T)^2}v^2(\partial_x^2 + \partial_y^2) + \frac{31\zeta(5)}{16(4\pi T)^4}v^4\left[5(\partial_x^2 + \partial_y^2)^2\right.\right.
$$
$$
\left.\left. +2(\partial_x^2 - \partial_y^2)^2\right]\right)\Delta(\mathbf{r}) = \frac{21\zeta(3)}{(4\pi T)^2}|\Delta(\mathbf{r})|^2\Delta(\mathbf{r}), \tag{1}
$$

which is converted into the dimensionless form

$$
\left(1 + (\partial_x^2 + \partial_y^2) + \epsilon\left[5(\partial_x^2 + \partial_y^2)^2 + 2(\partial_x^2 - \partial_y^2)^2\right]\right)f(\mathbf{r})
$$
$$
= |f(\mathbf{r})|^2 f(\mathbf{r}). \tag{2}
$$

where we introduced

$$
\xi(T)^2 = \frac{7\zeta(3)v^2}{2(4\pi T)^2(-\ln t)}, \quad \Delta(T)^2 = \frac{(4\pi T)^2(-\ln t)}{21\zeta(3)}, \tag{3}
$$

$t = T/T_c$ and rescaled $\mathbf{r} \to \mathbf{r}/\xi(T)$, $\Delta(\mathbf{r}) = \Delta(T)f(\mathbf{r})$. Here ∂_x and ∂_y are gauge invariant differential operators and we define the small symmetry breaking parameter

$$\epsilon \equiv \frac{31\zeta(5)(-\ln t)}{196\zeta(3)^2} \sim 0.11(1-t). \tag{4}$$

Eq.(1) is written down basically in [24], though we ignore a few terms of the order of $(-\ln t)^2$, which are of secondary importance.

For a single vortex ar $\mathbf{r} = 0$, the solution is written as

$$f(\mathbf{r}) \;=\; g(r)e^{i\phi} + \epsilon\left(e^{4i\phi}\alpha(r) + e^{-4i\phi}\beta(r) + \gamma(r)\right)e^{i\phi}. \tag{5}$$

In particular for $r \gg 1$, the solution is found

$$g(r) \;=\; 1 - \frac{1}{2}r^{-2} - \frac{9}{8}r^{-4} - \frac{161}{16}r^{-6}\cdots, \tag{6}$$

$$\begin{aligned}
\alpha(r) \;=\;& \frac{5}{2}r^{-2} + \left(c - \frac{55}{4}\log r\right)r^{-4} \\
&+ \left(\frac{-2873 - 456\,c}{80} + \frac{627}{8}\log r\right)r^{-6}\cdots,
\end{aligned} \tag{7}$$

$$\begin{aligned}
\beta(r) \;=\;& -\frac{5}{2}r^{-2} + \left(\frac{5 - 2\,c}{2} + \frac{55}{4}\log r\right)r^{-4} \\
&+ \left(\frac{-6627 - 184\,c}{80} + \frac{253}{8}\log r\right)r^{-6}\cdots,
\end{aligned} \tag{8}$$

and

$$\gamma(r) \;=\; -9r^{-4} - \frac{297}{2}r^{-6} - \frac{5313}{8}r^{-8}\cdots. \tag{9}$$

Where c is a free parameter, which should be fixed by studying the asymptotic behaviors of $\alpha(r)$ and $\beta(r)$ for $r \to 0$. For later purpose it is convenient to introduce the interpolation expressions, which gives the correct asymptotics for $r \to 0$;

$$g(r) \;=\; \tanh\frac{r}{c_0} - \frac{1}{2r^2}\left(1 - c_1\mathrm{sech}\frac{r}{c_0}\right)\tanh^5\frac{r}{c_0} - \cdots, \tag{10}$$

$$\alpha(r) \;=\; \frac{5}{2}r^{-2}\tanh^7\frac{r}{c_2} + \left(\frac{5}{4} - \frac{55}{4}\log r\right)r^{-4}\tanh^{11}\frac{r}{c_2}\cdots, \tag{11}$$

$$\beta(r) \;=\; -\frac{5}{2}r^{-2}\tanh^5\frac{r}{c_2} + \left(\frac{5}{4} + \frac{55}{4}\log r\right)r^{-4}\tanh^9\frac{r}{c_2}\cdots, \tag{12}$$

where $c_0 = 1.71$, $c_1 = 0.80$, $c_2 = 2.5$.

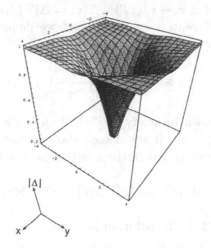

$|\Delta|$

Figure 2. $|\Delta(\mathbf{r})|$ in the x-y plane. we used $\epsilon = 1$ for the clarity of the figure, while $\epsilon \simeq 0.1$ in the real situation.

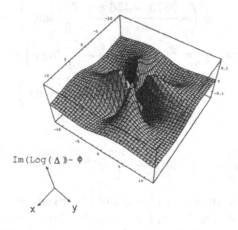

Im(Log(Δ))$-\phi$

Figure 3. $\Phi(\mathbf{r}) =$ phase of $\Delta(\mathbf{r}) - \phi$ is shown in the x-y plane for $\epsilon = 1$.

In particular $\alpha(r)$ and $\beta(r)$ constructed this way agree quite well with the corresponding functions numerically obtained by Enomoto et al. [24].

We show in Fig. 2 and Fig. 3, $|\Delta(\mathbf{r})|$ and phase of $\Delta(\mathbf{r}) - \phi$. For clarity we enhanced ϵ by a factor of 5 to the usual value. The fourfold distortion of $\Delta(\mathbf{r})$ are seen very clearly. In particular $|\Delta(\mathbf{r})|$ along the diagonal directions is somewhat larger than the one parallel to the a or b directions.

3. Interaction energy between two vortices

Before studying the vortex lattice, let us first consider two vortices, which are placed at $(0,0)$ and $d(\cos\theta, \sin\theta)$. Also we limit ourselves when $\kappa \gg d \gg 1$ (or $\lambda \gg r \gg \xi$ in the natural unit). First the free energy in the dimensionless unit is given

$$F_s - F_n = \int d^2r \left(-\frac{1}{2}|\Delta|^4 + \frac{1}{8\pi}b^2 \right), \tag{13}$$

where $b = b(\mathbf{r})$ is the local field. Now making use of the usual approximation

$$\Delta(\mathbf{r}) = \Delta \prod_i f(\mathbf{r} - \mathbf{r}_i) \tag{14}$$

where $f(\mathbf{r})$ is the solution for a single vortex, we obtain

$$\Omega_{\text{two-vortex}} \simeq -\frac{1}{2}\left(A - 2a_1 - 2a_1\epsilon(\alpha(d) + \beta(d))\cos 4\theta \right), \tag{15}$$

where A is the area and

$$a_1 = \frac{8\pi}{3}\left(\ln 2 + \frac{1}{8} \right) \simeq 6.854. \tag{16}$$

Here we have neglected the magnetic interaction. The interaction energy between the two vortices is given by

$$\Omega_{ij} = \frac{4\pi\xi^2}{\kappa^2}\left(K_0\left(\frac{r}{\lambda}\right) + \epsilon a_2\xi^2\kappa^2\frac{\cos 4\theta_{ij}}{r_{ij}^4} \right), \tag{17}$$

where $a_2 = 20(\ln 2 + 1/8)/3 = 5.451$ and θ_{ij} is the angle vector $\mathbf{r}_{ij}$ makes from the a axis. The first term is the magnetic interaction between two vortices and $K_0(z)$ is the Hankel function. As is readily seen the second term give rise to the interaction energy which favors two vortices aligned along the nodal directions $(1,1)$ and $(1,-1)$.

4. Vortex lattice transition

We have carried out the lattice sum over vortex where vortices are set on $\mathbf{r}_{l,m} = r_{l,m}(\cos\theta_{l,m}, \sin\theta_{l,m}) = ld(\cos\theta, \sin\theta) + md(\cos\theta, -\sin\theta)$, where $d = \sqrt{\phi_0/\sin(2\theta)B}$, ϕ_0 is the flux quantum, and l and m run over all integers. Here θ is the Apex angle of the vortex lattice as shown in Fig. 4. The lattice sum is carried using the Poisson transform as in [36]. Ω for the vortex lattice is given by

$$\Omega = \Omega_0 + \frac{\pi\xi^2 B}{\kappa^2\phi_0}\psi\left(\frac{B}{H_{cr}}\right), \tag{18}$$

78

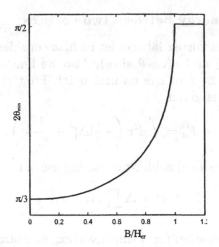

Figure 4. Appex angle $2\theta_{min}$ as a function of B/H_{cr} where $2\theta_{min} = \pi/2$ and $\pi/3$ correspond to the square lattice and the triangular lattice with hexagonal symmetry, respectively.

where

$$\Omega_0 = -\frac{A}{2} + \frac{a_1\xi^2}{2\phi_0}B$$

$$+\frac{2\pi\xi^2}{\kappa^2\phi_0}B\left[\frac{2\pi\lambda^2}{\phi_0}B + \frac{1}{2}\log\frac{\phi_0}{2\pi\lambda^2 B} - \frac{1}{2}(1-\gamma)\right], \quad (19)$$

$$\psi\left(\frac{B}{H_{cr}}\right) = f\left(\theta_{min}\left(\frac{B}{H_{cr}}\right)\right), \quad (20)$$

$$f(\theta) = \left(\frac{B}{H^*(t)}\right)^2 \sum_{l,m}' \frac{\sin^2 2\theta \cos 4\theta_{l,m}}{((l+m)^2 \sin^2\theta + (l-m)^2 \cos^2\theta)^2}$$

$$+\sum_{l,m}' \left(E_1\left(\pi(l^2\tan\theta + m^2\cot\theta)\right)\right.$$

$$\left.+\frac{(-1)^{l+m} + \exp\left(-\pi(l^2\cot\theta + m^2\tan\theta)\right)}{\pi(l^2\cot\theta + m^2\tan\theta)}\right), \quad (21)$$

and

$$H^*(t) = \left(\frac{98\zeta(3)^2(2\pi)^3}{155a_1\zeta(5)(-\ln t)}\right)^{1/2}\frac{H_{c2}(t)}{\kappa}$$

$$= 5.64667(-\ln t)^{-1/2}\frac{H_{c2}(t)}{\kappa}. \quad (22)$$

By minimizing $f(\theta)$ in θ, we find the apex angle θ_{min} as function of B, which is shown in Fig.4. θ_{min} increases gradually from 30° (corresponding

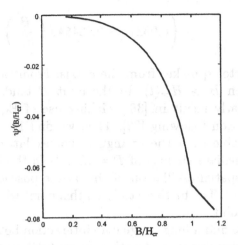

Figure 5. Singular part of the magnetization $-M_{singl} = \psi'(B/H_{cr})$ as a function of B/H_{cr}.

to the triangular vortex lattice) as B increases and rises very sharply to 45° (rhombic vortex lattice) at at $B = H_{cr}$ where

$$H_{cr} = 0.5796(1 - t)^{-1/2}\kappa^{-1}H_{c2}(t). \tag{23}$$

Earlier a similar θ-B curve is obtained within generalized London equation [28, 29]. But the rise of θ to 45° is much slower than the present prediction.

Further we obtain the magnetization from $M = -\partial\Omega/\partial B$. The singular part of the magnetization

$$-M_{singl} = \psi'(B/H_{cr}) = H_{cr}\frac{d\psi(B/H_{cr})}{dB}, \tag{24}$$

is shown in Fig. 5. M has a kink at $B = H_{cr}(t)$ indicating clearly the second order transition. From the kink in the magnetization we can calculate the jump in the specific heat ΔC as mentioned in the introduction.

5. Concluding Remarks

More recently we have studied the vibration spectrum of the rhombic vortex state within the hydrodynamic equation à la Landau [37] and Onsager [38] and found that the rhombic vortex lattice is stable for $B > H_{cr}(t)$. The more realistic model for high T_c cuprate superconductors, the time dependent Ginzburg-Landau equation [39] gives the same stability condition.

Now the vibration frequency is given by

$$\omega^2 \simeq \left(\frac{eB}{mc}\right)^2 q^4\lambda^2 d^2\left[\left(0.01989 + 0.007257\left(\frac{B}{H_{cr}}\right)^2\right)\right.$$

$$-\left(0.03176 - 0.004546\left(\frac{B}{H_{cr}}\right)^2\right)\cos 4\chi\right], \qquad (25)$$

where χ is the vector $\mathbf{q}$ makes from the a axis. From Eq. (25) it is readily seen $\omega^2 \geq 0$ when $B > H_{cr}(t)$. In the limit B tends to zero Eq. (25) agrees with the early result in [36]. Making use of Eq. (25), we analyze the melting transition following [32]. Then we find 1) the rhombic vortex lattice is more stable than the triangular vortex lattice against melting except in the immediate vicinity of $B = H_{cr}$. 2) In the immediate vicinity of $B = H_{cr}$, the spatial oscillation of the lattice position $\langle|\delta\mathbf{u}|^2\rangle$ diverges like $-\ln(B/H_{cr}-1)$. The further study on this point will clarify the nature of the tricritical point.

In summary we find a long range core interaction between two vortices, which favors the rhombic vortex lattice. In contrast to the vortex state in a classical s-wave superconductor, the rhombic vortex lattice is stable in the major part of the B-T phase diagram. Therefore for the quantitative study of the vortex lattice and related fluctuation effect it is crucial to consider the rhombic lattice. Recently a number of experiments found the rhombic vortex lattice in quarterly intermetallic compounds borocarbides like $ErNi_2B_2C$, YNi_2B_2C and $LuNi_2B_2C$ [40, 41, 42], which may suggest the superconductivity in borocarbides is of d-wave. Indeed a recent specific heat measurement of YNi_2B_2C does not exhibit the exponential term associated with the s-wave superconductor [43]. Further there is a $\sqrt{B}$-term in the specific heat in the presence of a magnetic field indicating the node in $\Delta(\mathbf{k})$ [44]. Therefore the possibility of d-wave superconductivity in borocarbides should be seriously explored.

We believe the study of the vortex lattice in novel superconductors open a new window to look into the symmetry of the underlying superconductivity.

Acknowledgements

One of us (KM) is thankful for the hospitality of the CNRS-MPI at Grenoble, ISSP, University of Tokyo, that of Peter Fulde at Max-Planck Institut für Physik der Komplexer Systeme at Dresden, and the support of CREST during the course of the present work. This work is in part supported by NSF under grant number DMR9531720.

References

1. J.G. Bednorz and K.A.Müller, *Z. Phys.* B **64** 189 (1986).
2. C.C. Tsuei, J.R. Kirtley, *Physica* C **282-287** 4 (1997).
3. D. J. von Harlingen, *Physica* C **282-287** 128 (1997).
4. A.A. Abrikosov, *Soviet Phys. JETP* **5** 1174 (1957).
5. V.L. Ginzburg and L.D. Landau, Zh. Eksperim. i Teor. Fiz. **20** 1064 (1950).

6. J. Bardeen, L.N. Cooper and J.R. Schrieffer, *Phys. Rev.* **108** 1175 (1957).
7. L.P. Gor'kov, *Zh. Eksperim. i Teor. Fiz* **34** 735 (1958); **36** 1918 (1959); *Soviet Phys. JETP* **7** 505 (1958); **9** 1364 (1959).
8. K. Maki, *Physics* **1** 21 (1964); G. Eilenberger, *Phys. Rev.* **153** 584 (1967).
9. V.L. Shubnikov, V.I. Khotkevich, Yu.D. Shepelev and Yu.N. Riabinin, *J. Exper. Theor. Phys. (USSR)* **7** 221 (1937).
10. W.H. Kleiner, L.M. Roth and S.H. Autler, *Phys. Rev.* **133** A1226 (1964); G. Eilenberger, *Z. Phys.* **180** 32 (1064); D. Saint-James and G. Sarma, *Conference on the Physics of Type II Superconductivity, Cleveland*, unpublished (1964).
11. D. Cribier, B. Jacrot and L.M. Rao and B. Farnoux, *Phys. Letters* **9** 106 (1964).
12. U. Essmann and H. Träuble, *Phys. Letters* **24A** 526 (1967); H. Träuble and U. Essmann, *Phys. Stat. Sol.* **95** 373 and 395 (1968).
13. H. Hess, R.B. Robinson and J.V. Waszczak, *Phys.Rev.Lett.* **64** 2711 (1990); *Physica B* **169** 422 (1991).
14. C. Caroli, P.G. de Gennes and J. Matricon, *Phys. Lett.* **9** 307 (1964); C. Caroli and J. Matricon, *Phys. Kondens. Mater.* **3** 380 (1965).
15. F. Gygi and M. Schlüter, *Phys. Rev. B* **43** 7609 (1991).
16. I. Maggio-Aprile, Ch. Renner, A. Erb, E. Walker and Ø. Fischer, *Phys. Rev. Lett.* **75** 2754 (1995).
17. Ch. Renner, B. Revaz, K. Kadowaki, I. Maggio-Aprile and Ø. Fischer, *Phys. Rev. Lett.* **80** 3606 (1998).
18. Y. Morita, M. Kohmoto and K. Maki, *Phys. Rev. Lett.* **78** 4841 (1997); *Phys. Rev. Lett.* **79** 4514(c) (1997).
19. Y. Morita, M. Kohmoto and K. Maki, *Europhys. Lett.* **70** 207 (1997).
20. H. Won and K. Maki, *Europhys. Lett.* **30** 421 (1995); *Phys. Rev. B* **53** 5927 (1996)
21. K. Maki and H. Won, *Physica B* **244** 22 (1998).
22. J. Shiraishi, M. Kohmoto and K. Maki, submitted to Phys. Rev. B, submitted to Phys.Rev. Lett.
23. J. Shiraishi, M. Kohmoto and K. Maki, submitted to Europhys. Lett.
24. N. Enomoto, M. Ichioka and K. Machida, *J. Phys. Soc. Jpn.* **66** 204 (1997)
25. B. Keimer, W.Y. Shih, R.W. Lynn, F. Dogan and I.A. Aksay, *Phys. Rev. Lett.* **73** 3459 (1994).
26. K. Maki and M.T. Béal-Monod, *Phys. Lett. A* **218** 365 (1995); *Phys. Rev. B* **55** 11730 (1997).
27. A.J. Berlinsky, A.L. Fetter, M. Franz, C. Kallin and P.I. Soininen, *Phys. Rev. Lett.* **75** 2200 (1995); M. Franz, C. Kallin, P.I. P.I. Soininen, A.J. Berlinsky and A.L. Fetter, *Phys. Rev. B* **53** 5795 (1996).
28. M. Franz, I. Affleck and H. Sharifzaheh Amin, *Phys. Rev. Lett.* **79** 1555 (1997); I. Affleck, M. Franz and M. H. Sharifzaheh Amin, *Phys. Rev. B* **55** R704 (1997).
29. V. G. Kogan, P. Miranović, Lj. Dobrosavljević-Grujić, W.E. Pickett and D.K. Christen, *Phys. Rev. Lett.* **79** 741 (1997); V. G. Kogan, M. Bullock, B. Harmon, P. Miranović, Lj. Dobrosavljević-Grujić, *Phys. Rev. B* **55** R8693 (1997).
30. H. Won and K. Maki, *Phys. Rev. B* **49** 1397 (1994)
31. K. Maki Y. Sun and H. Won, *Czech J. Phys.* **46** s6 3151 (1996)
32. A. Houghton, R. A. Pelcovits and A. Sudbø, *Phys. Rev.* **B40** 6763 (1989).
33. R.A. Doyle, B. Khaykovich, M. Konczykowski, E. Zeldov, N. Morozov, D. Majer, P. H. Kes and V. Vinokur, *Physica C* **282-287** 323 (1997).
34. A. Schilling, R. A. Fisher, N. E. Phillips, U. Welp, W. K. Kwok and G. W. Crabtree, *Physica C* **282-287** 327 (1997).
35. Y. Ren, J.-H. Xu and C. S. Ting, *Phys.Rev. Lett.* **74** 3680 (1995).
36. A. L. Fetter, P. C. Hohenberg and P. Pincus, *Phys.Rev.* **147** 140 (1966); A. L. Fetter, *Phys. Rev.* **147** 153 (1966).
37. L. D. Landau, *J. Phys. (USSR)* **5** 71 (1941).
38. L. Onsager, *Nuvo Cimento* **6** suppl II 249 (1949).
39. A. Schmid, *Phys. Kond. Materie.* **5** 302 (1996); C. Caroli and K. Maki, *Phys. Rev.*

159 306 (1967).

40. U. Yaron, P. L. Gammel, A. P. Ramirez, D. A. Huse, D. J. Bishop, A. I. Goldman, C. Stassis, P. C. Canfield, K. Mortensen and M. R. Eskildsen, *Nature (London)* **382** 236 (1966); M. R. Eskildsen, P. L. Gammel, B. P. Barber, A. P. Ramirez, D. J. Bishop, N. H. Andersen, K. Mortensen, C. A. Bolle, C. M. Lieber and P. C. Canfield, *Phys. Rev. Lett.* **78** 1968 (1997); **79** 487 (1997).

41. M. Yethiraj, D. McK. Paul, C. V. Tomy and E. M. Forgan, *Phys. Rev. Lett.* **78** 4849 (1997).

42. Y. De Wilde, M. Iavarone, U. Welp, V. Metlushko, A. E. Koshelev, I. Aranson and G. W. Crabtree, *Phys. Rev. Lett.* **78** 4273 (1997).

43. M. Nohara, M. Issiki, H. Takagi and R. J. Cava, *J. Phys. Soc. Jpn.* **66** 1888 (1997).

44. G. E. Volovik, *JETP lett.* **58** 469 (1993).

PSEUDOGAPS AND MAGNETIC PROPERTIES
OF THE t-J MODEL

A. SHERMAN
Institute of Physics, University of Tartu, Riia 142,
EE-2400 Tartu, Estonia

AND

M. SCHREIBER
Institut für Physik, Technische Universität,
D-09107 Chemnitz, Federal Republic of Germany

Abstract. We apply the spin-wave theory with zero staggered magnetization to investigate normal-state properties of the 2D t-J model in the paramagnetic state. In the ranges of hole concentrations $0.02 \lesssim x \lesssim 0.17$ and temperatures $T \lesssim 100$ K where this approach is applicable the hole spectrum is nonmetallic and contains a pseudogap with properties analogous to those observed in photoemission of cuprate perovskites. Temperature dependencies of the correlation length, spin-lattice relaxation time, and static susceptibility are typical for the quantum disordered regime with a pseudogap in the spectrum of magnetic excitations.

Photoemission and magnetic measurements in cuprate perovskites revealed the existence of two pseudogaps, spectral and magnetic, in their energy spectrum. In this paper we show that the occurrence of these pseudogaps and the related properties of cuprates follow from the 2D t-J model widely used for the description of CuO_2 planes of these crystals. As known [1], for $x \gtrsim 0.02$ the long-range antiferromagnetic order is destroyed in these planes. Therefore to describe low-lying magnetic excitations and their interaction with holes we use the version of the spin-wave theory formulated in Refs. [2, 3] for the Heisenberg antiferromagnet with zero staggered magnetization. As shown in Refs. [2, 3], in the absence of holes this approach reproduces results obtained in Refs. [4, 5] with the mean-field Schwinger boson and renormalization group theories and is remarkably accurate, as

83

M. Ausloos and S. Kruchinin (eds.), Symmetry and Pairing in Superconductors, 83–90.
© 1999 *Kluwer Academic Publishers. Printed in the Netherlands.*

follows from the comparison with exact diagonalization and Monte Carlo results. Here we reformulate this approach to simplify the inclusion of holes in the Hamiltonian.

We use the Holstein-Primakoff transformation for the spin operators s_l^i to introduce boson operators of spin waves b_l,

$$s_l^z = e^{i\mathbf{\Pi}l}\left(\frac{1}{2} - b_l^\dagger b_l\right), \quad s_l^\sigma = P_l^\sigma \phi_l b_l + P_l^{-\sigma} b_l^\dagger \phi_l, \tag{1}$$

where l labels sites of a plane square lattice, $\mathbf{\Pi} = (\pi, \pi)$ (the lattice spacing is taken as the unit of length), $P_l^\sigma = [1 + \sigma \exp(i\mathbf{\Pi}l)]/2$, $\sigma = \pm 1$, and $\phi_l = (1 - b_l^\dagger b_l)^{1/2}$. We substitute Eqs. (1) into the Heisenberg Hamiltonian $\mathcal{H} = (J/2)\sum_{la} s_{l+a} s_l$ (a labels nearest neighbors, J is the exchange constant), expand ϕ_l and keep terms up to the quartic order, using the mean-field approximation in these latter terms. To account for the constraint of zero staggered magnetization,

$$\sum_l e^{i\mathbf{\Pi}l} s_l^z = 0 \quad \text{or} \quad \sum_l b_l^\dagger b_l = \frac{N}{2}, \tag{2}$$

we add the term $2J\lambda\langle b_0 b_{\bar{a}}\rangle \sum_l b_l^\dagger b_l$ with the Lagrange multiplier λ to the obtained magnon Hamiltonian:

$$\mathcal{H}_m = -2J\langle b_0 b_{\bar{a}}\rangle\left[(2 - \lambda)\sum_l b_l^\dagger b_l + \frac{1}{4}\sum_{la}\left(b_{l+a}^\dagger b_l^\dagger + b_{l+a} b_l\right)\right] \tag{3}$$

Here N is the number of sites, angular brackets denote averaging over the grand canonical ensemble, and $\bar{a}$ is one of the vectors of nearest neighbors (the four correlations $\langle b_0 b_{\bar{a}}\rangle$ are supposed to be equal). Hamiltonian (3) is diagonalized by the unitary transformation

$$U = \exp\left[\frac{1}{2}\sideset{}{'}\sum_{k\sigma} \alpha_k \left(b_{k\sigma} b_{-k,-\sigma} - b_{k\sigma}^\dagger b_{-k,-\sigma}^\dagger\right)\right] \tag{4}$$

with $\alpha_k = \ln[(1 + \eta\gamma_k)/(1 - \eta\gamma_k)]/4$, $\eta = 2/(2 - \lambda)$, $\gamma_k = \sum_a \exp(ika)/4$, $b_{k\sigma} = \sqrt{2/N}\sum_l \exp(-ikl) b_l P_l^\sigma$ (due to P_l^σ the summation is performed over one sublattice), and the primed sum sign indicates that the summation is restricted to the magnetic Brillouin zone. As a result, we obtain

$$H_m = U^\dagger \mathcal{H}_m U = \sideset{}{'}\sum_{k\sigma} \omega_k^0 b_{k\sigma}^\dagger b_{k\sigma}, \quad \omega_k^0 = -\frac{4J}{\eta}\langle b_0 b_{\bar{a}}\rangle\sqrt{1 - \eta^2\gamma_k^2}, \tag{5}$$

$$\langle b_0 b_{\bar{a}}\rangle = \frac{2}{N}\sideset{}{'}\sum_{k\sigma} \frac{\gamma_k}{\sqrt{1 - \eta^2\gamma_k^2}}\left[\langle b_{-k,-\sigma} b_{k\sigma}\rangle_U - \eta\gamma_k\left(\langle b_{k\sigma}^\dagger b_{k\sigma}\rangle_U + \frac{1}{2}\right)\right],$$

where for the following discussion in $\langle b_0 b_{\overline{a}} \rangle$ we keep the anomalous correlation $\langle b_{-k,-\sigma} b_{k\sigma} \rangle_U$ which is nonzero at $x \neq 0$. The subscript U means that the averaging is performed with the Hamiltonian transformed with operator (4). In the case $x = 0$ we have $\langle b_{k\sigma}^{\dagger} b_{k\sigma} \rangle_U = [\exp(\omega_k^0/T) - 1]^{-1}$ and $\langle b_{-k,-\sigma} b_{k\sigma} \rangle_U = 0$, while for $x > 0$ the correlations are calculated from the magnon Green's function. In Eq. (5), η is determined from the condition

$$\frac{2}{N} \sum_{k}' \frac{1}{\sqrt{1 - \eta^2 \gamma_k^2}} \left(\langle b_{k\sigma}^{\dagger} b_{k\sigma} \rangle_U + \frac{1}{2} - \eta \gamma_k \langle b_{-k,-\sigma} b_{k\sigma} \rangle_U \right) = 1 \qquad (6)$$

which follows from Eq. (2) and ensures zero site magnetization, $\langle s_l^z \rangle = 0$. Analogous equations for the Heisenberg antiferromagnet without holes were obtained in a somewhat different manner in Refs. [2, 3]. As shown in these works, in states without long-range antiferromagnetic ordering and for finite lattices one obtains $\eta < 1$ which introduces a gap in magnon spectrum (5).

When $x > 0$ the hopping term of the t-J Hamiltonian becomes nonzero. On the antiferromagnetic background its action — the transfer of a hole to neighboring sites — leads to the appearance of flipped spins and may be described by the following operator:

$$\mathcal{H}_t = t \sum_{la} h_l h_{l+a}^{\dagger} (b_{l+a}^{\dagger} + b_l), \qquad (7)$$

where $h_l^{\dagger}$ creates a hole in one of the Néel states and t is the hopping constant. After unitary transformation (4) the total Hamiltonian reads

$$H = U^{\dagger} \mathcal{H}_t U + H_m = \sum_{kk'\sigma}' (g_{kk'} h_{k\sigma}^{\dagger} h_{k-k',-\sigma} b_{k'\sigma} + \text{H.c.}) + \sum_{k\sigma}' \omega_k^0 b_{k\sigma}^{\dagger} b_{k\sigma}, \qquad (8)$$

where $g_{kk'} = -4t\sqrt{2/N}(\gamma_{k-k'} u_{k'} + \gamma_k v_{k'})$, $u_k = \cosh(\alpha_k)$, and $v_k = -\sinh(\alpha_k)$. At $\eta = 1$ Eq. (8) reduces to the Hamiltonian obtained in Ref. [6] for the case of long-range antiferromagnetic order. The number of magnons is not conserved by Hamiltonian (8) and at $x > 0$ the anomalous correlation $\langle b_{-k,-\sigma} b_{k\sigma} \rangle_U$ in Eqs. (5), (6) is nonzero.

The energy spectrum is found by solving the self-energy equations for magnon and hole Green's functions. The magnon and hole self-energies are described by bubble and sunrise diagrams, respectively [7]. The first correction to the bare vertices is exactly zero [8]. This suggests the use of the Born approximation in which the full vertices are substituted with the bare ones. In this approximation the self-energy equations become self-consistent and can be solved iteratively. In these calculations we used the parameters $J = 0.1$ eV and $t = 0.5$ eV which approximately correspond to p-type cuprate perovskites. The calculation procedure was the following:

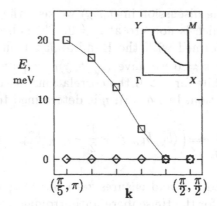

Figure 1. The position of the spin-polaron peak along the Fermi surface, shown in the inset, for $x = 0.121$ (□) and $x = 0.172$ (◇) at $T = 116$ K. Points Γ, M, X, and Y in the inset correspond to $\mathbf{k} = (0,0)$, (π,π), $(0,\pi)$, and $(\pi,0)$, respectively.

for given values of the hole chemical potential μ and T some starting value $\eta < 1$ was selected; after convergence of the iterations condition (6) was checked and η was appropriately changed for the next iteration cycle, until the condition was fulfilled with an accuracy of 10^{-3}. In the calculations a 20×20 lattice was used. We found that condition (6) can be satisfied only in a limited region of the T-x plane below the line which approximately connects the points $T = 250$ K, $x = 0.02$ and $T = 100$ K, $x = 0.17$ (the used approach is certainly applicable for $x < 0.02$ also, however in this work we did not consider this region).

For $x < 0.17$ the obtained hole spectrum contains a pseudogap. The spectrum consists of a persistent part of a narrow ($\lesssim J$) spin-polaron band, which is the distinctive feature of the low-concentration spectrum, and of a broad ($\sim t$) band which arises from $x \gtrsim 0.02$ [7, 9]. In the hole spectral function narrow intensive peaks correspond to the former band and broad less intensive maxima to the latter. The intensive peaks of the spin-polaron band determine the leading edge of the photoemission spectrum. For two hole concentrations the calculated position of these peaks and of the leading edge is shown in Fig. 1 along the line depicted in the inset. This line is the Fermi surface of the broad band and the locus of the least distance between the spin-polaron band and the Fermi level. In the hole spectral function which determines the photoemission spectrum, just at the Fermi level crossing the maximum of the broad band is completely lost within the foot of a more intensive spin-polaron peak. As seen in Fig. 1, the peak touches the Fermi level near $\mathbf{k} = (\pi/2, \pi/2)$ and for $x < 0.17$ bends upward (in the hole picture) along the line in the inset. Due to the mentioned hidden Fermi level crossing the situation looks like a part of the Fermi surface disappears and a gap opens between a band and the Fermi level near $(0, \pi)$. The retained Fermi level crossing of the broad band manifests itself in a finite intensity at the Fermi level, thus transforming the gap into the pseudogap. As follows from Fig. 1, the size and symmetry of the calculated pseudogap are in agreement with experimental observations in underdoped Bi2212 [10, 11]. Moreover, as in experiment the pseudogap is closed in optimally doped crystals with $x \approx 0.17$. The symmetries of the pseudogap and of the superconducting gap are similar, since both are determined by the same hole-magnon interaction. It should be emphasized that the discussed pseudogap has no relevance to superconducting fluctuations and is a consequence of strong electron correlations.

The discussed energy structure differs essentially from the spectrum of a conventional metal. This manifests itself also in the violation of Luttinger's theorem which concerns metallic bands of states with quasiparticle weights equalling 1. In spite of a large bandwidth of several t, the obtained broad band is not metallic as the quasiparticle weights of its states are less than 1 and change with x. It is this change rather than the change in the volume encircled by the Fermi surface in Fig. 1 which determines the hole concentration. Its observed variation for a practically unchanged Fermi surface [10] may be attributed to this behavior.

For finite x the decay of the spin correlations $\langle s_l^z s_m^z \rangle$ with the distance $|l - m|$ is nonexponential which may be partly connected with finite-size effects. To estimate the correlation length we used the formula $\xi^2 = \sum_l |l|^2 \exp(i\Pi l)\langle s_l^z s_0^z \rangle / [2 \sum_l \exp(i\Pi l)\langle s_l^z s_0^z \rangle]$ (the used spin-wave approximation is not rotationally invariant and the correlations $\langle s_l^+ s_m^- \rangle$ are zero [2, 3]). For $T = 0.02t = 116$ K the product $\xi\sqrt{x}$ is nearly constant and approximately equal to the lattice spacing in the range $0.017 \lesssim x \lesssim 0.09$, in agreement with experiment in La$_{2-x}$Sr$_x$CuO$_4$ [1]. For larger x at this temperature ξ becomes of the order of the lattice spacing. When $x \lesssim 0.06$ ξ is nearly independent of temperature in the considered range of T which agrees with experiment [1]. For larger x ξ is also weakly temperature dependent at low T and decreases more rapidly to the value of the order of the lattice spacing as the boundary of the considered T-x region is approached. The obtained saturation of ξ with decreasing T is the distinctive property of the quantum disordered regime [4, 12].

In the considered T-x region the hole contribution to the susceptibility is negligibly small in comparison with the spin contribution

$$\mathrm{Im}\chi^z(\mathbf{q}\omega) = \frac{4\mu_B^2}{N} \sum_\mathbf{k} \int_{-\infty}^{\infty} \frac{d\omega'}{\pi} [n_B(\omega' - \omega) - n_B(\omega')]$$

$$\times [S_1(\mathbf{k}, -\omega')S_1(\mathbf{k} + \boldsymbol{\kappa}, \omega - \omega') + S_2(\mathbf{k}, -\omega')S_2(\mathbf{k} + \boldsymbol{\kappa}, \omega - \omega')],$$

$$(9)$$

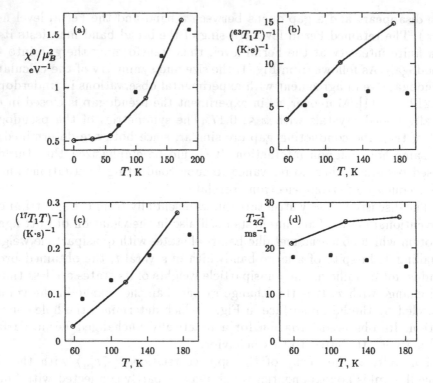

Figure 2. The static spin susceptibility (a), $1/(T_1 T)$ at the Cu (b) and O (c) sites for $\mathbf{H}\|\mathbf{c}$, and $1/T_{2G}$ (d) calculated for $x \approx 0.05$ (open circles). The experimental data for $YBa_2Cu_3O_{6.63}$ are shown by filled circles. χ^0 was inferred from the Knight shift data [12], $1/(T_1 T)$ and $1/T_{2G}$ are from Refs. [13] and [14], respectively.

$$S_1(\mathbf{k}\omega) = u_\mathbf{k}^2 \mathrm{Im} D_{11}(\mathbf{k}\omega) + 2u_\mathbf{k} v_\mathbf{k} \mathrm{Im} D_{12}(\mathbf{k}\omega) - v_\mathbf{k}^2 \mathrm{Im} D_{11}(\mathbf{k}, -\omega),$$

$$S_2(\mathbf{k}\omega) = u_\mathbf{k} v_\mathbf{k}[\mathrm{Im} D_{11}(\mathbf{k}\omega) - D_{11}(\mathbf{k}, -\omega)] + (u_\mathbf{k}^2 + v_\mathbf{k}^2)\mathrm{Im} D_{12}(\mathbf{k}\omega),$$

where μ_B is the Bohr magneton, $n_B = [\exp(\omega/T) - 1]^{-1}$, $\kappa = \mathbf{q} - \Pi$, $D_{ij}(\mathbf{k}\omega)$ are the Fourier transforms of the magnon Green's functions $D_{11}(\mathbf{k}t) = -i\theta(t)\langle[b_{\mathbf{k}\sigma}(t), b_{\mathbf{k}\sigma}^\dagger]\rangle_U$, $D_{12}(\mathbf{k}t) = -i\theta(t)\langle[b_{-\mathbf{k},-\sigma}(t), b_{\mathbf{k}\sigma}]\rangle_U$, and the summation is performed over the full Brillouin zone. The susceptibility is strongly peaked around $\mathbf{q} = \Pi$.

We applied the obtained magnetic susceptibility for calculating the spin-lattice relaxation times at the Cu and O sites $^{67}T_1$ and $^{17}T_1$ and the Cu spin-echo decay time T_{2G} by using form factors from Ref. [12]. Results for $x \approx 0.05$ are shown in Fig. 2 together with the calculated static spin susceptibility $\chi^0 = \chi(0,0)$ and the respective experimental results obtained in $YBa_2Cu_3O_{6.63}$ [12, 13, 14]. As seen from the figure, the t-J model is able to describe correctly the temperature dependences of the depicted quantities and to give their proper orders of magnitude in the quantum disordered regime. Besides, the calculated concentration dependences of χ^0 and T_{2G}^{-1} agree with experimental observations [12, 14, 15].

As seen in Fig. 2, $(T_1 T)^{-1}$ and χ^0 decrease with decreasing T. Such behaviour observed at $T > T_c$ in underdoped cuprates [15, 16] is connected with a pseudogap in the spectrum of magnetic excitations [12, 15, 16, 17]. This pseudogap in the conducting state appears as a successor of the gap in the insulator with the short-range antiferromagnetic order. As follows from Eq. (5), the gap is located near the $(0,0)$ and (π, π) points. We found that at $x \gtrsim 0.02$ the hole-magnon interaction leads to the appearance of overdamped magnons in these regions of k-space [18]. These magnons indicate the destruction of the long-range antiferromagnetic order, which is supposed to exist in the 2D case at $T = 0$, by holes and the appearance of short-range order. The overdamped magnons transform the gap into the pseudogap. Notice that even at $x \approx 0.17$ and $T \approx 100$ K usual magnons with essentially softened frequencies and increased damping continue to exist at the periphery of the magnetic Brillouin zone along with the overdamped magnons in its central part.

As indicated, the considered phase below the line connecting the points $T = 250$ K, $x = 0.02$ and $T = 100$ K, $x = 0.17$ has a nonmetallic hole spectrum. Outside of this region condition (6) cannot be fulfilled and the used spin-wave approximation is inapplicable. Nevertheless, we note that for $x \gtrsim 0.17$ we found a hole energy spectrum which is in many respects similar to the spectrum of the usual metal. This spectrum is close in shape to the band $E_{\mathbf{k}} = 2t[\cos(k_x) + \cos(k_y)]$, produced by the hopping term of the t-J Hamiltonian at complete absence of correlations, with quasiparticle weights of states nearly equal to 1. In the entire Brillouin zone magnons are overdamped. Hence the considered nonmetallic phase is intermediate between the low-concentration ($x \lesssim 0.02$) insulator phase and the metallic phase for $x \gtrsim 0.17$, suggesting the complex nature of the metal-insulator transition in strongly correlated systems.

In summary, we used the modified spin-wave approximation with the additional constraint of zero staggered magnetization to consider the energy spectrum and magnetic properties of the 2D t-J model at finite hole concentrations x and temperatures T. The approximation is applicable in the region of the T-x plane below the line which approximately connects the points $T = 250$ K, $x = 0.02$ and $T = 100$ K, $x = 0.17$. In this region the hole spectrum is nonmetallic, with quasiparticle weights of states varying with x, and contains a pseudogap with properties analogous to those observed in photoemission of cuprate perovskites. The obtained temperature dependencies of the correlation length, spin-lattice relaxation times, and static spin susceptibility are typical for the quantum disordered regime with a pseudogap in the spectrum of magnetic excitations. At $x \approx 0.17$ the considered nonmetallic phase borders the phase of the conventional metal.

References

1. Keimer, B., Belk, N., Birgeneau, R. G., Cassanho, A., Chen, C. Y., Greven, M., Kastner, M. A., Aharony, A., Endoh, Y., Erwin, R. W. and Shirane, G. (1992) Magnetic excitations in pure, lightly doped, and weakly metallic La_2CuO_4, *Phys. Rev. B* **46**, 14034–14053.
2. Takahashi, M. (1989) Modified spin-wave theory of a square-lattice antiferromagnet, *Phys. Rev. B* **40**, 2494–2501.
3. Tang, S., Lazzouni, M. E. and Hirsch, J. E. (1989) Sublattice-symmetric spin-wave theory for the Heisenberg antiferromagnet, *Phys. Rev. B* **40**, 5000–5006.
4. Arovas, D. A. and Auerbach, A. (1988) Functional integral theories of low-dimensional quantum Heisenberg models, *Phys. Rev. B* **38**, 316–332.
5. Chakravarty, S., Halperin, B. I. and Nelson, D. R. (1989) Two-dimensional quantum Heisenberg antiferromagnet at low temperatures, *Phys. Rev. B* **39**, 2344–2371.
6. Schmitt-Rink, S., Varma, C. M. and Ruckenstein, A. E. (1988) Spectral function of holes in a quantum antiferromagnet, *Phys. Rev. Lett.* **60**, 2793–2796.
7. Sherman, A. (1997) Magnetic transitions and superconductivity in the t-J model, *Phys. Rev. B* **55**, 582–590.
8. Krier, G. and Meissner, G. (1993) Spin-wave renormalization by mobile holes in a two-dimensional quantum antiferromagnet, *Ann. Phys.* **2**, 738–749.
9. Sherman, A. and Schreiber, M. (1997) The normal-state pseudogap in the spectrum of strongly correlated fermions, *Phys. Rev. B* **55**, R712–R715.
10. Marshall, D. S., Dessau, D. S., Loeser, A. G., Park, C.-H., Matsuura, A. Y., Eckstein, J. N., Bozovic, I., Fournier, P., Kapitulnik, A., Spicer, W. E. and Shen, Z.-X. (1996) Unconventional electronic structure evolution with hole doping in $Bi_2Sr_2CaCu_2O_{8+\delta}$: Angle-resolved photoemission results, *Phys. Rev. Lett.* **76**, 4841–4844.
11. Ding, H., Yokoya, T., Campuzano, J. C., Takahashi, T., Randeria, M., Norman, M. R., Mochiku, T., Kadowaki, K. and Giapintzakis, J. (1996) Spectroscopic evidence for a pseudogap in the normal state of underdoped high-T_c superconductors, *Nature* **382**, 51–54.
12. Barzykin, V. and Pines, D. (1995) Magnetic scaling in cuprate superconductors, *Phys. Rev. B* **52**, 13585–13600.
13. Takigawa, M., Reyes, A. P., Hammel, P. C., Thompson, J. D., Heffner, R. H., Fisk, Z. and Ott, K. C. (1991) Cu and O NMR studies of the magnetic properties of $YBa_2Cu_3O_{6.63}$ ($T_c = 62$ K), *Phys. Rev. B* **43**, 247–257.
14. Takigawa, M. (1994) ^{63}Cu nuclear transverse relaxation rate and spin correlations in $YBa_2Cu_3O_{6.63}$, *Phys. Rev. B* **49**, 4158–4162.
15. Imai, T., Yasuoka, H., Shimizu, T., Ueda, Y., Yoshimura K. and Kosuge, K. (1989) Cu spin dynamics in high T_c and related oxides investigated by nuclear spin-lattice relaxation, *Physica C* **162-164**, 169–170.
16. Rossat-Mignod, J., Regnault, L. P., Bourges, P., Burlet, P., Vettier, C. and Henry, J. Y. (1993) Inelastic neutron scattering study of the spin dynamics in the $YBa_2Cu_3O_{6+x}$ system, *Physica B* **192**, 109–121.
17. Sokol, A. and Pines, D. (1993) Toward a unified magnetic phase diagram of the cuprate superconductors, *Phys. Rev. Lett.* **71** 2813–2816.
18. Sherman, A. and Schreiber, M. (1993) Magnetic excitations of a doped two-dimensional antiferromagnet, *Phys. Rev. B* **48**, 7492–7498.

PROBING $s \pm id$ PAIRING STATE VIA THERMOELECTRIC RESPONSE OF SND JUNCTION

S.A. SERGEENKOV

Bogoliubov Laboratory of Theoretical Physics, Joint Institute for Nuclear Research, 141980 Dubna, Moscow Region, Russia

Abstract. The influence of charge imbalance between the quasiparticles and Cooper pairs (described by the corresponding chemical potentials, μ_q and μ_p) on the thermoelectric response of SND configuration (near T_c) is considered within the generalized Ginzburg-Landau theory for a homogeneous admixture of s-wave and d-wave superconductors. The resulting thermopower $\Delta S(T, \theta) = S_p(\theta) - B(\theta)(T_c - T)$ is found to strongly depend on the relative phase $\theta = \phi_s - \phi_d$ between the two superconductors, with a pronounced maximum near $\theta = \pm \pi/2$ (the so-called $s \pm id$ mixed pairing state). The experimental conditions, under which the predicted behavior of the induced differential thermopower can be measured are discussed.

1. Introduction

During the last few years the order parameter symmetry has been one of the intensively debated issues in the field of high-T_c superconductivity (HTS). A number of experiments points to its $d_{x^2-y^2}$-wave character [1]. Such an unconventional symmetry of the order parameter has also important implications for the Josephson physics because for a d-wave superconductor the Josephson coupling is subject to an additional phase dependence caused by the internal phase structure of the wave function. The phase properties of the Josephson effect have been discussed within the framework of the generalized Ginzburg-Landau (GL) [2] as well as the tunneling Hamiltonian approach [3]. It was found [4] that the current-phase relationship depends on the mutual orientation of the two coupled superconductors and their interface. This property is the basis of all the phase sensitive experiments probing the order parameter symmetry. In particular, it is possible to create multiply connected d-wave superconductors which generate half-integer

91

M. Ausloos and S. Kruchinin (eds.), Symmetry and Pairing in Superconductors, 91–99.

flux quanta as observed in experiments [5]. Various interesting phenomena occur in interfaces of d-wave superconductors. For example, for an interface to a normal metal a bound state appears at zero energy giving rise to a zero-bias anomaly in the I-V-characteristics of quasiparticle tunneling [6, 7] while in such an interface to an s-wave superconductor the energy minimum corresponds to a Josephson phase different from 0 or π. By symmetry, a small s-wave component always coexists with a predominantly d-wave order parameter in an orthorhombic superconductor such as $YBCO$, and changes its sign across a twin boundary [8]. Besides, the s-wave and d-wave order parameters can form a complex combination, the so-called $s \pm id$-state which is characterized by a local breakdown of time reversal symmetry $\mathcal{T}$ either near surfaces [9, 10, 11, 12] or near the twin boundaries represented by tetragonal regions with a reduced chemical potential [13]. Both scenarios lead to a phase difference of $\pm \pi/2$, which corresponds to two degenerate states [14, 15]. Moreover, the relative phase oscillations between two condensates with different order parameter symmetries could manifest themselves through the specific collective excitations ("phasons") [16].

At the same time, a rather sensitive differential technique to probe sample inhomogeneity for temperatures just below T_c, where phase slippage events play an important role in transport characteristics has been proposed [17] and successfully applied [18] for detecting small changes in thermopower of a specimen due to the deliberate insertion of a macroscopic SNS junction made of a normal-metal layer N insert, used to force pair breaking of the superconducting component when it flows down the temperature gradient. In particular, a carrier-type-dependent thermoelectric response of such a SNS configuration in a C-shaped $Bi_x Pb_{1-x} Sr_2 CaCu_2 O_y$ sample has been registered and its Λ-shaped temperature behavior around T_c has been explained within the framework of GL theory [18, 19].

In the present paper, we consider theoretically the case of SND junction and discuss its possible implications for the above-mentioned type of experiments. The paper is organized as follows. In Section 2 we briefly review the experimental results for SNS configuration (with both holelike and electronlike carriers of the normal-metal N insert) and present a theoretical interpretation of these results, based on GL free energy functional. The crucial role of the difference between the quasiparticle μ_q and pair μ_p chemical potentials in understanding the observed phenomena is emphasized. In Section 3, extending the early suggested [10, 13] GL theory of an admixture of s-wave and d-wave superconductors by taking into account pair-breaking effects with $\mu_q \neq \mu_p$, we calculate the differential thermopower ΔS of SND configuration near T_c. The main theoretical result of this paper is prediction of a rather specific dependence of ΔS on relative phase shift $\theta = \phi_s - \phi_d$, with a sharp maximum for the mixed $s \pm id$ state (near $\theta = \pm \pi/2$).

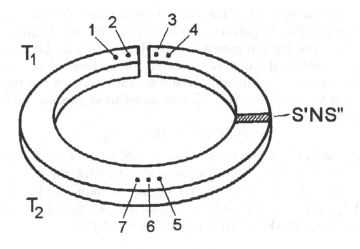

Figure 1. Schematic view of the sample geometry with $S'NS''$-junction and contacts configuration. The thermopowers S_R and S_L result from the thermal voltages detected by the contact pairs $4 - 5$ and $1 - 7$, respectively.

2. SNS configuration: a review

2.1. EXPERIMENTAL SETUP AND MAIN RESULTS

Before turning to the main subject of the present paper, let us briefly review the previous results concerning a carrier-type-dependent thermoelectric response of SNS configuration in a C-shaped $Bi_xPb_{1-x}Sr_2CaCu_2O_y$ sample (see Ref. [18] for details). The sample geometry used is sketched in Fig.1, where the contact arrangement and the position of the sample with respect to the temperature gradient $\nabla_x T$ is shown as well. Two cuts are inserted at $90°$ to each other into a ring-shaped superconducting sample. The first cut lies parallel to the applied temperature gradient serving to define a vertical symmetry axis. The second cut lies in the middle of the right wing, normal to the symmetry axis, separating an s-wave superconductor $(S' = S)$ from another s-wave superconductor $(S'' = S)$ and completely interrupting the passage of supercurrents in this wing. The passage of any normal component of current density is made possible by filling up the cut with a normal metal N. The carrier type of the normal-metal insert N was chosen to be either an electronlike N_e (silver) or holelike N_h (indium). Thermal voltages resulting from the same temperature gradient acting on both continuous and normal-metal-filled halves of the sample were detected as a function of temperature around T_c. The measured difference between the thermopowers of the two halves $\Delta S = S_R - S_L$ was found to approximately follow the linear dependence

$$\Delta S(T) \simeq S_p - B(T_c - T), \qquad (1)$$

where $S_p = \Delta S(T_c)$ is the peak value of $\Delta S(T)$ at $T = T_c$, and B is a constant. The best fit of the experimental data with the above equation yields the following values for silver (Ag) and indium (In) inserts, respectively: (i) $S_p(Ag) = -0.26 \pm 0.01 \mu V/K$, $B(Ag) = -0.16 \pm 0.1 \mu V/K^2$; (ii) $S_p(In) = 0.83 \pm 0.01 \mu V/K$, $B(In) = 0.17 \pm 0.1 \mu V/K^2$.

2.2. INTERPRETATION

Assuming that the net result of the normal-metal insert is to break up Cooper pairs that flow toward the hotter end of the sample and to produce holelike (In) or electronlike (Ag) quasiparticles, we can write the difference in the generalized GL free energy functional $\Delta \mathcal{G}$ of the right and left halves of the C-shaped sample as

$$\Delta \mathcal{G}[\psi] = \Delta \mathcal{F}[\psi] - \Delta \mu |\psi|^2, \qquad (2)$$

where

$$\Delta \mathcal{F}[\psi] \equiv \mathcal{F}_R - \mathcal{F}_L = a(T)|\psi|^2 + \frac{\beta}{2}|\psi|^4 \qquad (3)$$

and

$$\Delta \mu \equiv \mu_R - \mu_L = \mu_p - \mu_q. \qquad (4)$$

Here $\psi = |\psi|e^{i\phi}$ is the superconducting order parameter, μ_p and μ_q are the chemical potentials of quasiparticles and Cooper pairs, respectively; $a(T) = \alpha(T - T_c)$ and the GL parameters α and β are related to the critical temperature T_c, zero-temperature gap $\Delta_0 = 1.76 k_B T_c$, the Fermi energy E_F, and the total particle number density n as $\alpha = 2\Delta_0 k_B/E_F$ and $\beta = \alpha T_c/n$.

As usual, the equilibrium state of such a system is determined from the minimum energy condition $\partial \mathcal{G}/\partial|\psi| = 0$ which yields for $T < T_c$

$$|\psi_0|^2 = \frac{\alpha(T_c - T) + \Delta \mu}{\beta} \qquad (5)$$

Substituting $|\psi_0|^2$ into Eq.(2) we obtain for the generalized free energy density

$$\Delta \Omega(T) \equiv \Delta \mathcal{G}[\psi_0] = -\frac{[\alpha(T_c - T) + \Delta \mu]^2}{2\beta} \qquad (6)$$

In turn, the observed difference of thermopowers $\Delta S(T)$ can be related to the corresponding difference of transport entropies $\Delta \sigma \equiv -\partial \Delta \Omega / \partial T$ as $\Delta S(T) = \Delta \sigma(T)/nq$, where q is the charge of the quasiparticle. Thus finally the thermopower associated with a pair-breaking event reads

$$\Delta S(T) = \frac{\Delta \mu}{q T_c} - \frac{2 \Delta_0 k_B}{q \tilde{E}_F T_c}(T_c - T), \tag{7}$$

where $\tilde{E}_F = E_F - \mu_q$ accounts for the shift of the Fermi energy E_F due to the quasiparticle chemical potential μ_q. Let us discuss now separately the case of In and Ag normal-metal inserts.

2.2.1. $N = In$ (holelike metal insert)

In this case, the principal carriers are holes, therefore $q = +e$ in Eq.(7). Let the holelike quasiparticle chemical potential (measured relative to the Fermi level of the free-hole gas) be positive, then $\mu_q = +\mu$ and $\Delta \mu \equiv \mu_q - \mu_p = \mu - (-2\mu) = 3\mu$. Here $\mu_p = -2\mu$ comes from the change of the pair chemical potential of the holelike condensate with respect to the holelike quasiparticle branch. Therefore, for this case Eq.(7) takes the form

$$\Delta S^h(T) = 3 \left(\frac{k_B}{e} \right) \left(\frac{\mu}{k_B T_c} \right) - \frac{2 \Delta_0 k_B}{e \tilde{E}_F^h T_c}(T_c - T), \tag{8}$$

where $\tilde{E}_F^h = E_F - \mu$.

2.2.2. $N = Ag$ (electronlike metal insert)

The principal carriers in this case are electrons, therefore $q = -e$. The electronlike quasiparticle chemical potential (measured relative to the Fermi level of the free-hole gas) is $-\mu$. Then $\mu_q = -\mu$ and $\Delta \mu = -\mu - (-2\mu) = \mu$. For this case Eq.(7) takes the form

$$\Delta S^e(T) = - \left(\frac{k_B}{e} \right) \left(\frac{\mu}{k_B T_c} \right) + \frac{2 \Delta_0 k_B}{e \tilde{E}_F^e T_c}(T_c - T), \tag{9}$$

where $\tilde{E}_F^e = E_F + \mu$.

Using the above-mentioned experimental findings for the slope B and the peak S_p values for the two normal-metal inserts (see Eq.(1)), we can estimate the order of magnitude of the Fermi energy E_F and quasiparticle potential μ. The result is: $E_F = 0.16 eV$ and $\mu = 5 \times 10^{-3} eV$, in reasonable agreement with the other known estimates of these parameters. Besides, as it follows from Eqs.(8) and (9), the calculated ratio for peaks $|S_p(In)/S_p(Ag)| = 3$ is very close to the corresponding experimental value $|S_p^{exp}(In)/S_p^{exp}(Ag)| = 3.2 \pm 0.2$ observed by Gridin et al [18].

3. SND configuration: prediction

Since Eqs.(2)-(4) do not depend on the phase of the order parameter, they will preserve their form for a DND junction (created by two d-wave superconductors, $S' = S'' = D$, see Fig.1) bringing about the result similar to that given by Eqs.(7)-(9). It means that the experimental method under discussion (and its interpretation) can not be used to tell the difference between SNS and DND configurations, at least for temperatures close to T_c. As for low enough temperatures, the situation may change drastically due to a markedly different behavior of s-wave and d-wave order parameters at $T \ll T_c$. As we will show, this method, however, is quite sensitive to the mixed SND configuration (when $S' = S$ has an s-wave symmetry while $S'' = D$ is of a d-wave symmetry type, see Fig.1) and predicts a rather specific relative phase ($\theta = \phi_s - \phi_d$) dependences of both the slope $B(\theta)$ and peak $S_p(\theta)$ of the observable thermopower difference $\Delta S(T, \theta)$.

Following Feder et al [13], who incorporated chemical potential effects near twin boundaries into the approach suggested by Sigrist et al [10], we can represent the generalized GL free energy functional $\Delta \mathcal{G}$ for SND configuration of the C-shaped sample in the following form

$$\Delta \mathcal{G}[\psi_s, \psi_d] = \Delta \mathcal{G}[\psi_s] + \Delta \mathcal{G}[\psi_d] + \Delta \mathcal{G}_{int}, \tag{10}$$

where

$$\Delta \mathcal{G}[\psi_s] = \Delta \mathcal{F}[\psi_s] - \Delta \mu |\psi_s|^2, \tag{11}$$

$$\Delta \mathcal{G}[\psi_d] = \Delta \mathcal{F}[\psi_d] - \Delta \mu |\psi_d|^2, \tag{12}$$

and

$$\Delta \mathcal{G}_{int} = \gamma_1 |\psi_s|^2 |\psi_d|^2 + \frac{\gamma_2}{2}(\psi_s^{*2}\psi_d^2 + \psi_s^2\psi_d^{*2}) \tag{13}$$
$$-2\delta_1 |\psi_s||\psi_d| - \delta_2(\psi_s^*\psi_d + \psi_s\psi_d^*).$$

Here $\psi_n = |\psi_n|e^{i\phi_n}$ is the n-wave order parameter ($n = \{s, d\}$); $\Delta \mathcal{F}[\psi_{s,d}]$ is given by Eq.(3) with the corresponding parameters $a_s(T) = \alpha_s(T - T_{cs})$, β_s, $a_d(T) = \alpha_d(T - T_{cd})$, and β_d for s-wave and d-wave symmetry, respectively.

An equilibrium state of such a mixed system is determined from the minimum energy conditions $\partial \mathcal{G}/\partial |\psi_s| = 0$ and $\partial \mathcal{G}/\partial |\psi_d| = 0$ which result in the following system of equations for the two equilibrium order parameters ψ_{s0} and ψ_{d0}

$$[a_s(T) - \Delta \mu]|\psi_{s0}| + \beta_s|\psi_{s0}|^3 + \Gamma(\theta)|\psi_{s0}||\psi_{d0}|^2 = \Delta(\theta)|\psi_{d0}| \tag{14}$$

$$[a_d(T) - \Delta \mu]|\psi_{d0}| + \beta_d|\psi_{d0}|^3 + \Gamma(\theta)|\psi_{d0}||\psi_{s0}|^2 = \Delta(\theta)|\psi_{s0}| \tag{15}$$

where we introduced relative phase $\theta = \phi_s - \phi_d$ dependent parameters

$$\Gamma(\theta) = \gamma_1 + \gamma_2 \cos 2\theta \tag{16}$$
$$\Delta(\theta) = \delta_1 + \delta_2 \cos \theta$$

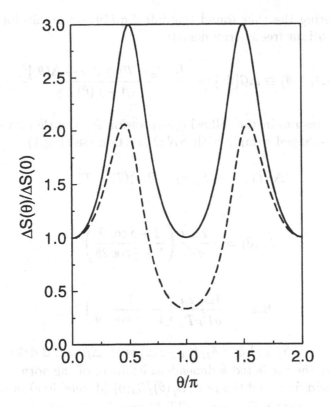

Figure 2. Predicted phase-dependent thermopower response of SND configuration in a C-shaped sample (see Fig.1). Solid and dashed lines depict, respectively, the relative phase θ dependence of the normalized slope $B(\theta)/B(0)$ and peak value $S_p(\theta)/S_p(0)$ of the induced thermopower difference, according to Eqs.(20) and (21) with $\tilde{\gamma} = \tilde{\delta} = 1/2$.

Notice that the $\Delta(\theta)$ term favors $\theta = l\pi$ (l integer), while the $\Gamma(\theta)$ term favors $\theta = n\pi/2$ ($n = 1, 3, 5\ldots$) which corresponds to a $\mathcal{T}$-violating phase [13]. In principle, we can resolve the above system (given by Eqs.(14)-(16)) and find ψ_{n0} for any set of parameters α_n, β_n, and T_{cn}. Since, however, in the present paper we are interested in describing the thermoelectric response of the mixed $s \pm id$ state only, in what follows we assume that $|\psi_{s0}| = |\psi_{d0}| \equiv |\psi_0|$, $\alpha_s = \alpha_d \equiv \alpha$, $\beta_s = \beta_d \equiv \beta$, and $T_{cs} = T_{cd} \equiv T_c$. Under these simplifying conditions, Eqs.(14) and (15) yield for $T < T_c$

$$|\psi_0|^2 = \frac{\alpha(T_c - T) + \Delta\mu + \Delta(\theta)}{\beta + \Gamma(\theta)} \tag{17}$$

After substituting the thus found $|\psi_0|$ into Eq.(10) we obtain for the generalized equilibrium free energy density

$$\Delta\Omega(T,\theta) \equiv \Delta\mathcal{G}[\psi_0] = -\frac{[\alpha(T_c - T) + \Delta\mu + \Delta(\theta)]^2}{\beta + \Gamma(\theta)} \tag{18}$$

which in turn results in the following expression for the thermopower difference in a C-shaped sample with SND junction (see Fig.1)

$$\Delta S(T,\theta) = S_p(\theta) - B(\theta)(T_c - T), \tag{19}$$

where

$$S_p(\theta) = -\frac{2\Delta\mu}{qT_c}\left(\frac{1 + \tilde{\delta}\cos\theta}{1 + \tilde{\gamma}\cos 2\theta}\right) \tag{20}$$

and

$$B(\theta) = \frac{4\Delta_0 k_B}{q\tilde{E}_F T_c}\left(\frac{1}{1 + \tilde{\gamma}\cos 2\theta}\right) \tag{21}$$

Here, $\tilde{\gamma} = \gamma_2/(\gamma_1 + \beta)$ and $\tilde{\delta} = \delta_2/(\delta_1 + \Delta\mu)$ with $\Delta\mu$ and β defined earlier.

Fig.2 shows the predicted θ-dependent behavior of the normalized slope $B(\theta)/B(0)$ (solid line) and the peak $S_p(\theta)/S_p(0)$ (dashed line) of the SND-induced thermopower difference $\Delta S(T,\theta)$ near T_c, for $\tilde{\gamma} = \tilde{\delta} = 1/2$. As is seen, both the slope and the peak exhibit a maximum for the mixed $s \pm id$ state at $\theta = n\pi/2$ ($n = 1, 3, 5\ldots$). Such a sharp dependence suggests quite an optimistic possibility to observe the above-predicted behavior of the induced thermopower, using the described in Section 2 sample geometry and experimental technique. Besides, by a controllable variation of the carrier type of the normal-metal insert, we can get a more detailed information about the mixed state and use it to estimate the phenomenological parameters $\gamma_{1,2}$ and $\delta_{1,2}$.

In summary, to probe into the mixed $s \pm id$ pairing state of high-T_c superconductors, we calculated the differential thermopower ΔS of SND junction in the presence of the strong charge imbalance effects (due to a nonzero difference between the quasiparticle μ_q and Cooper pair μ_p chemical potentials) using the generalized Ginzburg-Landau theory for a homogeneous admixture of s-wave and d-wave superconductors near T_c. The calculated thermopower was found to strongly depend on the relative phase $\theta = \phi_s - \phi_d$ between the two superconductors exhibiting a pronounced maximum near the mixed $s \pm id$ state with $\theta = \pm\pi/2$. The experimental conditions under which the predicted behavior of the induced thermopower could be observed were discussed.

References

1. van Harlingen, D.J. (1995) Phase sensitive tests of the symmetry of the pairing state in high-T_c superconductors: Evidence for $d_{x^2-y^2}$ symmetry, *Rev. Mod. Phys.* **67**, pp. 515–535
2. Sigrist, M. and Rice, T.M. (1995) Unusual paramagnetic phenomena in granular high-T_c superconductors: A consequence of d-wave pairing, *Rev. Mod. Phys.* **67**, pp. 503–513
3. Bruder, C., van Otterlo, A. and Zimanyi, G.T. (1995) Tunnel junctions of unconventional superconductors, *Phys. Rev.* **B51**, pp. 12904–12907
4. Yip, S. (1995) Josephson current-phase relationships with unconventional superconductors, *Phys. Rev.* **B52**, pp. 3087–3090
5. Tsuei, C.C., Kirtley, J.R., Rupp, M., Sun, J.Z., Gupta, A., Ketchen, M.B., Wang, C.A., Ren, Z.F., Wang, J.H. and Blushan, M. (1996) Pairing symmetry in single-layer tetragonal $Tl_2Ba_2CuO_{6+x}$ superconductors, *Science* **271**, pp. 329–332
6. Hu, C.-R. (1994) Midgap surface states as a novel signature for d-wave superconductivity, *Phys. Rev. Lett.* **72**, pp. 1526–1529
7. Tanaka, Y. Kashiwaya, S. (1996) Theory of the Josephson effect in d-wave superconductors, *Phys. Rev.* **B53**, pp. R11957–R11960
8. Walker, M.B. (1996) Orthorhombically mixed s-and $d_{x^2-y^2}$-wave superconductivity and Josephson tunneling, *Phys. Rev.* **B53**, pp. 5835–5838
9. Sigrist, M., Bailey, D.B. and Laughlin, R.B. (1995) Fractional vortices as evidence of time-reversal symmetry breaking in high-T_c superconductors, *Phys. Rev. Lett.* **74**, pp. 3249–3252 *Phys. Rev. Lett.* **78**, pp. 4841–4844
10. Sigrist, M., Kuboki, K., Lee, P.A., Millis, A.J. and Rice, T.M. (1996) Influence of twin boundaries on Josephson junctions between high-T_c and conventional superconductors, *Phys. Rev.* **B53**, pp. 2835–2849
11. Kuboki, K. and Sigrist, M. (1996) Proximity-induced time-reversal symmetry breaking of Josephson junctions between unconventional superconductors, *J.Phys. Soc. Jpn.* **65**, pp. 361–364
12. Huck, A., van Otterlo, A. and Sigrist, M. (1997) Time-reversed symmetry breaking and spontaneous currents in SND sandwiches, *Phys. Rev.* **B56**, pp. 14163–14167
13. Feder, D.L., Beardsall, A., Berlinsky, A.J. and Kallin, C. (1997) Twin boundaries in d-wave superconductors, *Phys. Rev.* **B56**, pp. R5751–R5754
14. Kuklov, A.B. (1995) Deformation-induced time-reversal symmetry breaking in d-wave superconductors, *Phys. Rev.* **B52**, pp. R7002–R7005
15. Matsumoto, M. and Shiba, H. (1995) Coexistence of different symmetry order parameters near a surface in d-wave superconductors, *J. Phys. Soc. Jpn.* **64**, pp. 3384–3396; ibid. **64**, pp. 4867–4881; ibid. **65**, pp. 2194–2203
16. Shevchenko, P.V. and Sushkov, O.P. (1997) Phase oscillations between two superconducting condensates in cuprate superconductors, *Phys. Lett.* **A236**, pp. 137–142
17. Gridin, V. and Datars, W. (1991) Potentiometric method of estimating the critical temperature of the superconducting transition in high-T_c materials, *Phys. Rev.* **B43**, pp. 3675–3677
18. Gridin, V., Sergeenkov, S., Doyle, R., de Villiers, P. and Ausloos, M. (1993) Carrier-type-dependent thermoelectric response of the SNS configuration in a C-shaped polycrystalline $Bi_xPb_{1-x}Sr_2CaCu_2O_y$ sample, *Phys. Rev.* **B47**, pp. 14591–14594
19. Sergeenkov, S., Gridin, V., de Villiers, P. and Ausloos, M. (1994) Estimation of effective electronic mass from Gaussian fluctuations of the magneto-thermopower of the polycrystalline $Bi_2Sr_2CaCu_2O_y$ sample, *Physica Scripta* **49**, pp. 637–640

GENERALIZED BCS-TYPE MODEL FOR THE ACOUSTIC PLASMON INDUCED D- WAVE SUPERCONDUCTIVITY

A. V. SEMENOV

Institute of Physics, National Academy of Sciences of Ukraine, Kiev, 252650, Ukraine

Abstract. An analytically treated model is proposed for the anisotropic Cooper pairing in the layered cuprate high-T_c compounds. This model assumes a cylindrical Fermi-surface with circular cross-section, but anisotropic Fermi-velocity (density of states), which is approximated by a step-like function on azimuth angle. The system under these assumptions is shown to support in the long wavelength limit collective charge-density excitations with acoustic spectrum (acoustic plasmons), which suppress the Coulomb interelectron repulsion for small transferred momenta. This result allows to approximate the effective interaction in the conventional BCS scheme by the well potential in the quasimomentum space. The effective constants of the Cooper pairing in anisotropic s- and d-wave channels are evaluated. It is shown, that the plasmon mechanism gives rise to the pairing in the d-wave channel, despite the fully isotropic (in two dimensions) model potential. The Umklapp processes are taken into account and are shown to increase the pairing constants.

1. Introduction.

The quasiparticle spectrum just below Fermi level in layered cuprates as well as the structure of high-T_c order parameter (at least for Bi-2212) are well established now due to the angle-resolved photoelectron spectroscopy [1, 2]. The main feature peculiar to the quasi-two-dimensional (2D) spectra of the hole-doped high-T_c compounds is the existence of extended saddle-point singularities (ESPS) with quasi-one-dimensional dispersion and square-root Van Hove singularity in the density of states (DOS) very close to the Fermi level [3, 4]. Almost cylindrical (closed in

101

M. Ausloos and S. Kruchinin (eds.), Symmetry and Pairing in Superconductors, 101–107.

two dimensions) Fermi surfaces (FS) of layered cuprates are centered round the point (π, π) of 2D Brillouin zone (BZ) corresponding to CuO_2 layers. The local DOS $d^2N/dEd\varphi \propto v_F^{-1}(\varphi)$ is essentially anisotropic with value $v_F(\pi/4)/v_F(0) \approx 5$ (where $v_F(\varphi)$ is the Fermi velocity, φ is the azimuth angle of quasimomentum in hole representation, the ESPS's are centered at $\varphi = n \cdot \pi/2$, see fig.1). The simple analytically treated model presented below is based on the mentioned experimental facts and aims the investigation of low-frequency collective charge-density excitations and their influence on the Cooper pairing. This model consists of two stages and several assumptions which will be grounded step by step in parallels with the consequences will being obtained.

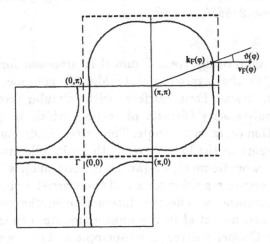

Figure 1. Schematic cross section of the FS in the original and shifted Brillouin zones.

2. Model one-particle spectrum and low-frequency collective charge-density excitations.

The first assumption of the proposed model concerns the validity of the random phase approximation (RPA) for the obtaining of low-frequency collective spectrum. The reason of such validity is the use of experimentally observed renormalized quasiparticle spectrum $E(k)$ instead of the bare one-electron band spectrum $E_0(k)$ in the conventional RPA expression for 2D polarizability. This implies leaving out of concideration the vertex corectins, while effectively retaining the self-energy ones. In the long-wave limit for zero temperature the polarization operator can be written as

$$\Pi(q, \omega) = -\frac{1}{2\pi^2} \int d^2k \, \delta(E(k) - E_F) \frac{q \cdot v(k)}{\omega - q \cdot v(k) + i\eta}, \qquad (1)$$

where $v(k) = \nabla_k E(k)$ is the group velocity of the quasi-particles. Using the cylindrical coordinates $k = \{k, \varphi\}$ and $q = \{q, \varphi_q\}$ for one-particle and transferred momenta, respectively, (1) can be rewritten as

$$\Pi(\omega, \varphi_q, q \to 0) =$$
$$-\frac{1}{2\pi^2} \int_0^{2\pi} d\varphi \, \frac{k_F(\varphi)}{|v_F(\varphi) \cos\vartheta(\varphi)|} \frac{\cos(\varphi + \vartheta(\varphi) - \varphi_q)}{U(\varphi) - \cos(\varphi + \vartheta(\varphi) - \varphi_q) + i\eta} \qquad (2)$$

where $k_F(\varphi)$ is the Fermi momentum defined by the condition $E(k_F) = E_F$, $v_F(\varphi) = v(k_F(\varphi))$ is the Fermi velocity, $U(\varphi) = \omega/qv_F(\varphi)$, and $\vartheta(\varphi)$ is the angle between $k_F(\varphi)$ and $v_F(\varphi)$.

The second step is the model approximation for the experimentally observed quasiparticle spectrum. Neglecting any violation of C_{4v} symmetry in CuO_2 layers and taking into account that the anisotropy of $k_F(\varphi)$ is much less, than that of $v_F(\varphi)$, the cylindrical FS can be modeled by the circular cross-section, so that $k_F(\varphi) = k_F$ is constant and $\vartheta(\varphi) = 0$, while the local DOS is approximated by the step function,

$$v_F(\varphi) = \begin{cases} v_1 & \text{if } 0 < \varphi \leq \varphi_0 \\ v_2 & \text{if } \varphi_0 < \varphi \leq \pi/4 \end{cases} \qquad (3)$$

where $v_1 << v_2$ and $0 < \varphi_0 < \pi/4$. The ESPSs in such model are simulated as the segments of FS with high local DOS (see fig.2). The real and imaginary parts of polarizability (2) obtained under these assumptions are given in Appendix A.

The first stage of proposed model is completed by the analysis of zeros of $\Pi(q, \omega)$, which gives the collective spectrum of charge-density excitations in the long-wave limit. The fundamental result of this stage is the existence in the system considered of low-frequency charge-density excitations with acoustic spectrum and low damping (besides the usual high-frequency plasmons). These excitations are analogous to the acoustic plasmons proposed as early as 1956 [5] for the case of multiband systems (transition metals, semimetals, multivalley semiconductors, etc.) with at least two kinds of charged carriers which essentially differ in effective masses. It is shown here that in quasi 2D systems the segments of cylindrical FS with different local DOS can play a part of such different kinds of charged carriers. In the zero-order approximation on the small parameter v_1/v_2 the acoustic plasmons are undamped and their velocity is isotropic:

$$u_{pl} = \omega_{pl}(q)/q = (v_1 \cdot v_2)^{1/2} \cdot [2\varphi_0/(\pi - 4\varphi_0)]^{1/2}. \qquad (4)$$

The peculiarities of anisotropy of acoustic plasmon velocity and damping in higher-order approximations will be given elsewhere, while it is emphasized here, that the condition of acoustic plasmon existence, namely

104

$v_1 \leq u_{pl} \leq v_2$, is doubtless fulfilled for the parameters of the model being extracted from the experiment: $\varphi_0 \approx \pi/8$, $v_2/v_1 \approx 5$.

These results suggest the second stage of the model concerning the influence of acoustic plasmons on the Cooper pairing.

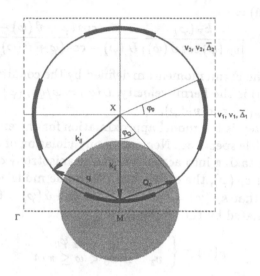

Figure 2. Model cross section of the FS in the hole representation and acoustic plasmon-mediated interparticle interaction (hatched region).

3. Influence of the acoustic plasma excitations on the Cooper pairing.

The mechanism of acoustic plasmon mediated superconductivity proposed as early as 1968 ([6, 7]) was widely debated during last decades. The main advantage of this mechanism in comparison with the conventional phonon mediated one is a higher energy of virtual boson involving in interelectron interaction, while the respective shortcoming is believed to be the restriction to relatively small values of transferred momenta q. This feature of acoustic plasmon assisted attraction which suppress the Coulomb repulsion for small q is taken into account in the present model. Besides the standard assumptions of the BCS scheme the effective potential of interelectron interaction is characterized by additional step-like dependence on the two-dimensional transferred quasimomenta:

$$ V(k, k') = \theta(\omega_c - \xi) \cdot \theta(\omega_c - \xi') \cdot \left\{ V_c - V_{pl} \cdot \theta \left[Q_0^2 - (k - k')^2 \right] \right\}, \quad (5) $$

where $\xi = v_F \cdot (k - k_F)$, $Q_0 = 2k_F \sin(\varphi_Q/2)$ is the radius (see fig.2) of plasmon assisted attraction in quasimomentum space; $\omega_c \approx u_{pl} \cdot Q_0$ is the cutoff frequency. The gap equation for cylindrical momentum coordinates takes the form

$$\Delta(\varphi) = -\int_0^{2\pi} \frac{d\varphi'}{2\pi} V(\varphi, \varphi') \nu(\varphi') \cdot \int_0^{\omega_c} d\xi' \frac{\Delta(\varphi')}{\sqrt{(\xi')^2 + \Delta^2(\varphi')}} \times$$

$$\tanh \frac{\sqrt{(\xi')^2 + \Delta^2(\varphi')}}{2T}, \qquad (6)$$

where $V(\varphi, \varphi') = V_c - V_{pl} \cdot \theta \left[\varphi_Q^2 - (\varphi - \varphi')^2\right]$, and $\nu(\varphi) = k_F/2\pi v_F(\varphi)$ is given by (3). Averaging of (6) over the segments of FS with constant local DOS leads to the set of algebraic equations for the mean values of the order parameter on these segments Δ_1 and Δ_2, which has two types of solutions, corresponding to anisotropic s- and d-wave pairing. In the latter case $\Delta_2 = 0$ and Δ_1 has opposite signs on the neighboring (matched by $\pi/2$ rotation) segments with high DOS. The values of critical temperature $T_c^{s,d}$ in both channels have the conventional BCS-type form with effective coupling constants $\Lambda_{s,d}$, which were determined in [8] through the partial coupling constants λ_{ij} on the different segments of FS. These partial constants are explicitly evaluated below in the framework of present model. In the case of $d_{x^2-y^2}$-pairing $\Lambda_d = \lambda_{11} - 2\lambda_{11}' + \lambda_{11}''$, where

$$\lambda_{11} = -\frac{\nu_1}{\varphi_0} \int_0^{\varphi_0} d\varphi \int_{-\varphi_0}^{\varphi_0} \frac{d\varphi'}{2\pi} V(\varphi, \varphi') = -\frac{\nu_1}{\pi} \left[V_c \varphi_0 - V_{pl}\left(\varphi_Q - \frac{\varphi_Q^2}{4\varphi_0}\right)\right],$$

$$\lambda_{11}' = -\frac{\nu_1}{2\varphi_0} \int_{-\varphi_0}^{\varphi_0} d\varphi \int_{\pi/2-\varphi_0}^{\pi/2+\varphi_0} \frac{d\varphi'}{2\pi} V(\varphi, \varphi') = -\frac{\nu_1}{\pi} V_c \varphi_0, \qquad (7)$$

$$\lambda_{11}'' = -\frac{\nu_1}{2\varphi_0} \int_{-\varphi_0}^{\varphi_0} d\varphi \int_{\pi-\varphi_0}^{\pi+\varphi_0} \frac{d\varphi'}{2\pi} V(\varphi, \varphi') = -\frac{\nu_1}{\pi} (V_c - U_1) \varphi_0.$$

Here U_1 is the contribution of the Umklapp processes to the retarded electron-plasmon interaction adduced in Appendix B. The effective coupling constant for anisotropic s- pairing is more complicated [8] and depends on the whole set of partial coupling constants, which will be given elsewhere. But in the case $\nu_1 \gg \nu_2$ this effective constant reduces to $\Lambda_s \approx \lambda_{11} + 2\lambda_{11}' + \lambda_{11}''$, thus $\Lambda_s < \Lambda_d$ for $\lambda_{11}' < 0$. According to (7), the effective coupling constant equals

$$\Lambda_d = \frac{\nu_1}{\pi} \cdot V_{pl} \cdot \left(\varphi_Q - \frac{\varphi_Q^2}{4\varphi_0}\right). \qquad (8)$$

The maximal value of U_1 is equal to $V_{pl}/8$ for $\zeta = 1$ and $\varphi_Q = 2\varphi_0 = \pi/4$, thus $\Lambda_d^{max} = \nu_1 \cdot V_{pl}/4$. It would be emphasized that the plasmon mechanism gives rise to the pairing in the d-wave channel, despite the fully isotropic (in two dimensions) model potential (5).

The author is grateful to Prof. E. A. Pashitskii and Dr. V. I. Pentegov for useful discussions. This work was supported by Grant 2.4/561 of the State Foundation for the Fundamental Research of Ukraine.

Appendix A

The real part of polarization operator (2) is evaluated as the principle part of integral with respect to φ:

$\Pi\left(\omega/q, \varphi_q\right) = \sum \Pi_i$, with $\Pi_i = \frac{k_F}{\pi^2 V_i} \cdot \Psi_i$, $\Psi_1 = 4\varphi_0 + \Phi_1$, $\Psi_2 = \pi - 4\varphi_0 - \Phi_2$,

$$\Phi_i = \frac{u_i}{2\sqrt{1 - u_i^2}} \cdot \ln \frac{1 + F\left(u_i\right)}{1 - F\left(u_i\right)}$$

if $u_i \leq 1$, and

$$\Phi_i = \frac{u_i}{2\sqrt{u_i^2 - 1}} \cdot \left\{\arctan F\left(u_i\right) + \pi \cdot \theta\left[-F\left(u_i\right)\right]\right\}$$

if $u_i \geq 1$;

$$u_i = \omega/qV_i, \quad F\left(u\right) = \frac{4u \cdot \left(2u^2 - 1\right) \sqrt{\left|u^2 - 1\right|} \cdot \sin 4\varphi_0}{\cos 4\varphi_q - \left[8u^2 \cdot \left(u^2 - 1\right) + 1\right] \cdot \cos 4\varphi_0}.$$

The imaginary part of the polarizability is found as the sum of residues in (2):

$$Im\Pi\left(\omega/q, \varphi_q\right) = \sum Im\Pi_i , \text{ with } Im\Pi_i = \frac{k_F}{\pi V_i} \cdot \frac{u_i}{2\sqrt{1 - u_i^2}} \cdot \Theta_i\left(u_i, \varphi_0, \varphi_q\right) ,$$

where

$\Theta_1 = 2 \cdot \theta\left(1 - u_1^2\right) \cdot \theta\left(\varphi_0 - \varphi_q\right) + sgn\left(\varphi_0 - \varphi_q\right) \cdot \left(\Theta_1^- + \Theta_1^+\right) ,$
$\Theta_2 = 2 \cdot \theta\left(1 - u_2^2\right) \cdot \theta\left(\varphi_q - \varphi_0\right) + sgn\left(\varphi_q - \varphi_0\right) \cdot \left(\Theta_2^- - \Theta_2^+\right) ,$
$\Theta_i^\pm = \theta\left[u_i^2 - \cos^2\left(\varphi_0 \pm \varphi_q\right)\right] - \theta\left[u_i^2 - \sin^2\left(\varphi_0 \pm \varphi_q\right)\right].$

Appendix B

The contribution of the Umklapp processes to the retarded electron-plasmon interaction:

$$U_1 = \frac{V_{pl}}{\pi\varphi_0} \cdot \int_0^{\Phi_0} d\varphi \cdot \left[\Phi_+\left(\varphi\right) - \Phi_-\left(\varphi\right)\right],$$

where

$$\Phi_0 = \min\left\{\varphi_0; \arccos\left[\frac{1}{\zeta} - \zeta \sin\frac{\varphi_Q}{2}\left(1 + \sin\frac{\varphi_Q}{2}\right)\right]\right\},$$

$$\Phi_+ (\varphi) = \min \left\{ \varphi_0; \arcsin \left(\frac{\alpha\gamma + \beta\sqrt{\beta^2 + \gamma^2 - \alpha^2}}{\beta^2 + \gamma^2} \right) \right\},$$

$$\Phi_- (\varphi) = \max \left\{ -\varphi_0; \arcsin \left(\frac{\alpha\gamma - \beta\sqrt{\beta^2 + \gamma^2 - \alpha^2}}{\beta^2 + \gamma^2} \right) \right\},$$

$$\alpha(\varphi) = 1 + \frac{\zeta^2}{2} \cos\varphi_Q - \zeta \cos\varphi, \ \beta(\varphi) = \zeta \left(1 - \frac{\zeta}{2} \cos\varphi \right), \ \gamma(\varphi) = \frac{\zeta^2}{2} \sin\varphi,$$

where $\zeta = k_F a / \pi$.

References

1. D.S.Dessau, Z.-X.Shen, D.M.King, D.S.Marshall, L.M.Lombardo, P.H.Dickinson, A.G.Loeser, J.Di Carlo, C.-H.Park, A.Kapitulnik, W.E.Spicer. Key Features in the Measured Band Structure of $Bi_2 Sr_2 CaCu_2 O_{8+\delta}$: Flat Bands at E_F and Fermi Surface Nesting. - Phys.Rev.Lett., 1993, v.71, N17, pp.2781-2784.
2. J.G.Tobin, C.G.Olson, C.Gu, J.Z.Liu, F.R.Solal, M.J.Fluss, R.H.Howell, J.C.O'Brien, H.B.Radovsky, P.A.Sterne. Valence bands and Fermi-surface topology of untwinned single-crystal $YBa_2 Cu_3 O_{6.9}$. Phys.Rev.B, 1992, v.45, N10, pp.5563-5576.
3. K.Gofron, J.C.Campuzano, A.A.Abrikosov, M.Lindroos, A.Bansil, H.Ding, D.Koelling, B.Dobrowski. Observation of an Extended Van Hove Singularity in $YBa_2 Cu_4 O_8$ by Ultrahigh Energy Resolution Angle-Resolved Photoemission. Phys.Rev.Lett., 1994, v.73, N24, pp.3302-3305.
4. D.M.King, Z.-X.Shen, D.S.Dessau, D.S.Marshall, C.-H.Park, W.E.Spicer, J.L.Peng, Z.Y.Li, R.L.Greene. Observation of a Saddle-Point Singularity in $Bi_2 (Sr_{0.97} Pr_{0.03})_2 CuO_{6+\delta}$ and its Implications for Normal and Superconducting State Properties. - Phys.Rev.Lett., 1994, v.73, N24, pp.3298-3301.
5. D.Pines. Electron interaction in solids. Canad. J.Phys.,1956, v.34, N12 A, pp.1379-1394.
6. E.A.Pashitskii. Plasmon mechanism of superconductivity in degenerate semiconductors and semimetals. Zh.Eksp.Teor.Fiz., 1968, v.55, N6, pp.2387-2394. (Sov.Ph. JETP, 1969, v.28, N6, p.1267).
7. H.Fröhlich. Superconductivity in metals with incomplete inner shells. J.Phys.C, 1968, v.1, N2, pp.544-548.
8. E.A.Pashitskii, V.I.Pentegov, A.V.Semenov. On competition between s- and d-symmetry types of the order parameter in high-temperature superconductors. Fiz. Nizk. Temp., 1997, v.23, N2, pp.140-145; On the Nature of the Anisotropic Gap Structure in the High Temperature Superconductors: Competition between s- and d-Symmetry Types. Physica C, 1997, v.282-287, pp.1843-1844.

SPIN FLUCTUATION PAIRING IN THE t−J MODEL

N.M.PLAKIDA AND V.S. OUDOVENKO

Joint Institute for Nuclear Research, 141980 Dubna, Russia

Abstract. Superconductivity due to exchange and kinematical interactions in the t−J model in a paramagnetic state is studied by numerical solution of the linearized Eliashberg equations. It is shown that coherent quasiparticle excitations exist near the Fermi energy at moderate doping and the d-wave-like superconducting pairing with high T_c is observed.

1. INTRODUCTION

Recent experimental evidence of the d-wave superconducting pairing in high-T_c cuprates strongly supports the spin-exchange pairing mechanism for high-temperature superconductivity (see, for example, [1, 2]). Earlier this mechanism was proposed by several groups on the basis of phenomenological models (for references see, e.g., [2]). It should be also pointed out that the superconducting pairing due to the kinematic interaction in the Hubbard model in the limit of strong electron correlations ($U \rightarrow \infty$) was first obtained by Zaitsev and Ivanov [3]. However, they have considered only the mean field approximation which results in the s-wave pairing irrelevant for strongly correlated systems (for a discussion see [4]). Later on the theory in the mean field approximation was considered for the t − J model within the Green function approach in [4] where the d-wave spin-fluctuation superconducting pairing due to exchange interaction J was studied .

A number of numerical methods for finite clusters in the strong coupling limit for the Hubbard model and the t − J model has also been developed (for a review see [5]). These studies show strong antiferromagnetic correlations which produce the $d_{x^2-y^2}$ pairing correlations. However, the finite cluster calculations due to known limitations can give only restricted information, and to prove the superconducting pairing in the strong coupling limit, an analytical treatment is highly demanded.

109

M. Ausloos and S. Kruchinin (eds.), Symmetry and Pairing in Superconductors, 109–119.

The most important analytic results for the $t - J$ model were obtained in the limit of small concentrations of holes when one can consider the motion of holes on the antiferromagnetic background within the spin-polaron model. A number of studies of this model predicts that a doped hole dressed by strong antiferromagnetic spin fluctuations can propagate coherently as a quasiparticle (QP), spin-polaron, in a narrow band of order J even at finite doping (see, e.g. [6]). It is quite natural to suggest that the same spin fluctuations could mediate a superconducting pairing of spin-polarons. This problem was treated in the framework of the weak coupling BCS formalism by considering a simple phenomenological model of quasiparticles with an effective pairing interaction in the atomic limit [7] or with that mediated by antiferromagnetic magnon exchange [8]. However, since the pairing spin-fluctuation energy J is of the same order as a QP bandwidth, the weak coupling BCS equation is inadequate to treat the problem.

A self-consistent numerical treatment of the strong coupling Eliashberg equations for spin-polarons in the $t - J$ model was given in [9]. A strong renormalization of the QP hole spectrum due to spin-fluctuations was obtained and the d- wave pairing with maximum $T_c \simeq 0.01t$ at optimal concentration of doped holes $\delta \simeq 0.2$ was proved. However, a two-sublattice representation used in [9] can rigorously be proved only for a small doping. At a moderate doping one has to consider a paramagnetic (spin-rotationally invariant) state in the $t - J$ model.

The opposite limit of low electron densities in the $t - J$ model was studied by M. Kagan and Rice [10]. They observed various forms of electron pairing at low temperatures including the d-wave instability at values $J/t > 1$. A special diagram technique for the Hubbard operators was applied by Izyumov et al. [11] to consider spin fluctuations and superconducting pairing in the $t - J$ model. However, no numerical results were presented.

In the present paper we investigate a paramagnetic state only with short-range dynamic spin fluctuations at a moderate doping. We develop the theory of superconductivity for the $t - J$ model by applying the equation-of-motion method for the Green functions (GF) [12] in terms of the Hubbard operators. The self-energy operator due to kinematic and exchange interactions in the noncrossing approximation is calculated. A numerical solution of the linearized Eliashberg equations proves the existence of narrow QP peaks for the one-electron spectral density near the Fermi surface (FS) and the d-wave like solution for the superconducting gap function with high T_c. These results are in accord with the calculations for the $t - J$ model in the polaron representation [9]. QP excitations near the FS were also observed by Prelovšek [13] who considered the equation-of-motion method for the GF at $T = 0$.

In Sect. 2, the Dyson equation for the matrix Green function is obtained by the projection technique. In Sec.3, the results of numerical solution of the self-consistent system of Eliashberg equations are presented for the one-electron spectral density, Fermi surface and the superconducting gap function and T_c.

2. Dyson equation for the matrix Green function

The Hamiltonian of the $t - J$ model in terms of the Hubbard operators (HO's) reads

$$H_{t-J} = - \sum_{i \neq j, \sigma} t_{ij} X_i^{\sigma 0} X_j^{0\sigma} - \mu \sum_{i\sigma} X_i^{\sigma\sigma} + \frac{1}{2} \sum_{i \neq j, \sigma} J_{ij} \left(X_i^{\sigma\bar{\sigma}} X_j^{\bar{\sigma}\sigma} - X_i^{\sigma\sigma} X_j^{\bar{\sigma}\bar{\sigma}} \right) \tag{1}$$

where $t_{ij} = t, t'$ is the electron hopping energy for the nearest and the second neighbors on the 2D square lattice, respectively, and J_{ij} is the exchange interaction. The HO's are defined as $X_i^{\alpha\beta} = |i, \alpha\rangle\langle i, \beta|$ for three possible states at a lattice site i: $|i, \alpha\rangle = |i, 0\rangle$, $|i, \sigma\rangle$ for an empty site and for a singly occupied site by an electron with spin $\sigma/2$ ($\sigma = \pm 1$). They obey the completeness relation $X_i^{00} + \sum_\sigma X_i^{\sigma\sigma} = 1$ which rigorously preserves the constraint of no double occupancy.

To discuss the superconducting pairing within the model (1), we consider the matrix Green function (GF)

$$\hat{G}_{ij,\sigma}(t - t') = \langle\langle \Psi_{i\sigma}(t) | \Psi_{j\sigma}^+(t') \rangle\rangle \tag{2}$$

in terms of the Nambu operators:

$$\Psi_{i\sigma} = \begin{pmatrix} X_i^{0\sigma} \\ X_i^{\bar{\sigma}0} \end{pmatrix}, \qquad \Psi_{i\sigma}^+ = \begin{pmatrix} X_i^{\sigma 0} & X_i^{0\bar{\sigma}} \end{pmatrix} \tag{3}$$

where the Zubarev notation for the anticommutator Green function (2) is used [12].

To calculate the GF (2), we use the equation of motion for the HO's

$$\left(i\frac{d}{dt} + \mu \right) X_i^{0\sigma} = -\sum_l t_{il} B_{i\sigma\sigma'} X_l^{0\sigma'} + \sum_l J_{il}(B_{l\sigma\sigma'} - \delta_{\sigma\sigma'}) X_i^{0\sigma'} \tag{4}$$

where $B_{i\sigma\sigma'} = (X_i^{00} + X_i^{\sigma\sigma})\delta_{\sigma'\sigma} + X_i^{\bar{\sigma}\sigma}\delta_{\sigma'\bar{\sigma}} = (1 - \frac{1}{2}n_i + \sigma S_i^z)\delta_{\sigma'\sigma} + S_i^{\bar{\sigma}\sigma}\delta_{\sigma'\bar{\sigma}}$. The boson-like operator $B_{i\sigma\sigma'}$ describes electron scattering on spin and charge fluctuations caused by the nonfermionic commutation relations for the HO's (the first term in (4) – the so-called kinematical interaction) and by the exchange spin-spin interaction (the second term in (4)).

By differentiating the GF (2) with respect to times t and t' and employing the projection technique (see, e.g. [14]) we get the following Dyson equation

$$\hat{G}_{ij\sigma}(\omega) = \hat{G}^0_{ij\sigma}(\omega) + \sum_{kl} \hat{G}^0_{ik\sigma}(\omega)\, \hat{\Sigma}_{kl\sigma}(\omega)\, \hat{G}_{lj\sigma}(\omega) \tag{5}$$

for the Fourier component. Here the zero–order GF is calculated in the mean-field approximation

$$\hat{G}^0_{ij\sigma}(\omega) = Q\, \{\omega\hat{\tau}_0\delta_{ij} - \hat{E}_{ij\sigma}\}^{-1} \tag{6}$$

with the frequency matrix $\hat{E}_{ij\sigma} = \langle\{[\Psi_{i\sigma}, H], \Psi^+_{j\sigma}\}\rangle\, Q^{-1}$ and the correlation function $Q = \langle X^{00}_i + X^{\sigma\sigma}_i\rangle = 1 - n/2$. In a paramagnetic state it depends only on the average number of electrons $n = \langle n_i\rangle = \sum_\sigma\langle X^{\sigma\sigma}_i\rangle$. The self-energy operator $\hat{\Sigma}_{kl\sigma}(\omega)$ is defined by the equation:

$$\hat{\Sigma}_{ij\sigma}(\omega) = Q^{-1}\, \tilde{\Sigma}_{ij\sigma}(\omega) = Q^{-1}\, \langle\langle \hat{Z}^{(irr)}_{i\sigma} \mid \hat{Z}^{(irr)+}_{j\sigma} \rangle\rangle^{(irr)}_\omega\, Q^{-1} \tag{7}$$

where the irreducible part of the operator $\hat{Z}_{i\sigma} = [\Psi_{i\sigma}, H]$ is defined by the projection equation $\hat{Z}^{(irr)}_{i\sigma} = \hat{Z}_{i\sigma} - \sum_l \hat{E}_{il\sigma}\Psi_{l\sigma}$, $\langle\{\hat{Z}^{(irr)}_{i\sigma}, \Psi^+_{j\sigma}\}\rangle = 0$.

Equations (5) - (7) give an exact representation for the one-electron GF (2). To calculate it, however, one has to apply approximations for the many-particle GF in the self-energy matrix (7) which describes inelastic scattering of electrons on spin and charge fluctuations. Here we employ the noncrossing approximation (or the self-consistent Born approximation) for the irreducible part of the many-particle Green functions in (7). It neglects vertex corrections and is given by the following two-time decoupling for the correlation functions:

$$\langle X^{\sigma'0}_{j'} B^+_{j\sigma\sigma'} X^{0\sigma'}_i(t) B_{i\sigma\sigma'}(t)\rangle_{(j\neq j',\, i\neq i')} \simeq \langle X^{\sigma'0}_{j'} X^{0\sigma'}_i(t)\rangle\langle B^+_{j\sigma\sigma'} B_{i\sigma\sigma'}(t)\rangle. \tag{8}$$

Using the spectral representation for the GF we obtain the following results for the self-energy matrix elements in the k-representation:

$$\tilde{\Sigma}^\sigma_{11(12)}(k,\omega) = \frac{1}{N}\sum_q \iint\limits_{-\infty}^{+\infty} dz\,d\Omega N(\omega, z, \Omega)\lambda_{11(12)}(q, k-q \mid \Omega)A^\sigma_{11(12)}(q, z), \tag{9}$$

with $N(\omega, z, \Omega) = (1/2)(\tanh(z/2T) + \coth(\Omega/2T))/(\omega - z - \Omega)$. Here we introduced the spectral density for the normal (G_{11}) and anomalous (G_{12}) GF:

$$A^\sigma_{11}(q, z) = -\frac{1}{Q\pi}\text{Im}\, \langle\langle X^{0\sigma}_q \mid X^{\sigma0}_q\rangle\rangle_{z+i\delta},$$

$$A_{12}^{\sigma}(q, z) = -\frac{1}{Q\pi}\text{Im}\,\langle\langle X_q^{0\sigma} \mid X_{-q}^{0\bar{\sigma}}\rangle\rangle_{z+i\delta} \tag{10}$$

and the electron - electron interaction functions caused by spin and charge fluctuations

$$\lambda_{11(12)}(q, k - q \mid \Omega) = g^2(q, k - q)\, D^{+(-)}(k - q, \Omega), \tag{11}$$

where $g(q, k - q) = t(q) - J(k - q)$ and the spectral density of bosonic excitations is given by the imaginary part of the spin and charge susceptibilities:

$$D^{\pm}(q, \Omega) = -(1/\pi)\text{Im}\,\left\{\langle\langle \mathbf{S}_q \mid \mathbf{S}_{-q}\rangle\rangle_{\Omega+i\delta} \pm (1/4)\langle\langle n_q \mid n_q^+\rangle\rangle_{\Omega+i\delta}\right\}.$$

3. Numerical results and discussion

A linearized system of the Eliashberg equations close to T_c can be written as the self-consistent equations for the normal GF and its self-energy operator

$$\tilde{G}_{11}^{\sigma}(k, i\omega_n) = \{i\omega_n - E_k + \tilde{\mu} - \tilde{\Sigma}_{11}^{\sigma}(k, i\omega_n)\}^{-1},$$

$$\tilde{\Sigma}_{11}^{\sigma}(k, i\omega_n) = -\frac{T}{N}\sum_q\sum_m \tilde{G}_{11}^{\sigma}(q, i\omega_m)\lambda_{11}(q, k - q \mid i\omega_n - i\omega_m) \tag{12}$$

where we have used the imaginary frequency representation, $\omega = i\omega_n = i\pi T(2n+1)$. The energy of quasiparticles E_k^{σ} and the renormalized chemical potential $\tilde{\mu} = \mu - \delta\mu$ in the MFA (6) are given by

$$E_k^{\sigma} = -\epsilon(k)Q - \epsilon_s(k)/Q - \frac{4J}{N}\sum_q \gamma(k - q)N_{q\sigma} \tag{13}$$

where $\epsilon(k) = t(k) = 4t\gamma(k) + 4t'\gamma'(k)$, $\epsilon_s(k) = 4t\gamma(k)\chi_{1s} + 4t'\gamma'(k)\chi_{2s}$ with $\gamma(k) = (1/2)(\cos a_x q_x + \cos a_y q_y)$, $\gamma'(k) = \cos a_x q_x \cos a_y q_y$.

$$\delta\mu = \frac{1}{N}\sum_q \epsilon(q)N_{q\sigma} - 4J(n/2 - \chi_{1s}/Q). \tag{14}$$

The average number of electrons in the k-representation is written in the form:

$$n = \frac{1}{N}\sum_{k,\sigma}\langle X_k^{\sigma 0}X_k^{0\sigma}\rangle = \frac{Q}{N}\sum_{k,\sigma}N_{k\sigma} = (1 - \frac{n}{2})\{1 + \frac{2T}{N}\sum_k\sum_{n=-\infty}^{\infty}\tilde{G}_{11}^{\sigma}(k, i\omega_n)\} \tag{15}$$

which defines the function $N_{q\sigma}$ in Eqs. (13), (14). In calculating the normal part of the frequency matrix (13) we neglected the charge fluctuations and introduced the spin correlation functions for the nearest, $a_1 =$

$(\pm a_x, \pm a_y)$, and the second, $a_2 = \pm(a_x \pm a_y)$, neighbor lattice sites :
$\chi_{1s} = \langle S_i S_{i+a_1} \rangle$, $\chi_{2s} = \langle S_i S_{i+a_2} \rangle$.

The linearized gap equation reads:

$$\Phi^\sigma(k, i\omega_n) = \Delta_k^\sigma + \tilde{\Sigma}_{12}^\sigma(k, i\omega_n) = \frac{T}{N} \sum_q \sum_m \{2J(k - q) +$$

$$+ \lambda_{12}(q, k - q \mid i\omega_n - i\omega_m)\}\tilde{G}_{11}^\sigma(q, i\omega_m)\tilde{G}_{11}^{\bar{\sigma}}(q, -i\omega_m)\Phi^\sigma(q, i\omega_m) \qquad (16)$$

where for the gap function Δ_k^σ in the MFA (6) we omitted the k-independent part caused by the kinematic interaction [3] since it gives no contribution to the d-wave pairing ([4]).

In the present numerical calculations we took into account only the spin-fluctuation contribution modelled by the spin-fluctuation susceptibility (see, e. g., [13, 15]):

$$\chi_s''(q, \omega) = \chi_s(q) \, \chi_s''(\omega) = \frac{\chi_0}{1 + \xi^2(\mathbf{q} - \mathbf{Q}_{AF})^2} \tanh \frac{\omega}{2T} \frac{1}{1 + (\omega/\omega_s)^2} \qquad (17)$$

with the characteristic AFM correlation length ξ and spin-fluctuation energy $\omega_s \simeq J$. To fix the constant χ_0 in (17), we use the following normalization condition

$$\frac{1}{N} \sum_i \langle S_i S_i \rangle = \frac{1}{N} \sum_q \chi_s(q) \int_{-\infty}^{+\infty} \frac{dz}{\exp \frac{z}{T} - 1} \chi_s''(z) = \frac{\pi \omega_s}{2N} \sum_q \chi_s(q) = \frac{3}{4} n \qquad (18)$$

In this approximation we get for the interaction functions (11)

$$\lambda_{11}(q, k - q \mid i\omega_\nu) = -g^2(q, k - q)\chi_s(k - q) \int_0^{+\infty} \frac{2z \, dz}{z^2 + \omega_\nu^2} \frac{\tanh(z/2T)}{1 + (z/\omega_s)^2} \qquad (19)$$

and $\lambda_{12}(q, k - q \mid i\omega_\nu) = \lambda_{11}(q, k - q \mid i\omega_\nu)$. For the model (17) we can calculate the static spin correlation functions $\chi_{1s} = \langle S_i S_{i+a_1} \rangle$ and $\chi_{2s} = \langle S_i S_{i+a_2} \rangle$ with the equation $\langle S_q S_{-q} \rangle = \chi_s(q)\pi/2\omega_s$.

The numerical calculations were performed using the fast Fourier transformation [16] for a mesh of 32×32 k_x, k_y-points in the full Brillouin zone. In the summation over the Matsubara frequencies we used up to 700 points with the constant cut-off $\omega_{max} = 20$ t. Usually 10 – 30 iterations were needed to obtain a solution for the self-energy with an accuracy of order 0.001. The Padé approximation was used to calculate the one-electron spectral function $A_{11}(k, \omega)$ (10) on the real frequency axis. At first the system of equations for the normal GF (12) was solved numerically for a given concentration of electrons (15). Then the eigenvalues and eigenfunctions of the linear equation for the gap function (16) was calculated to obtain the superconducting transition temperature T_c and the gap function $\Phi(k, \omega)$.

The calculations were performed for various parameters of the $t - J$ model (J, t'), the AFM correlation length ξ in the model function (17), $\omega_s = J$, and the hole concentration $\delta = 1 - n$. Below we present several results for parameters $J = 0.4$, $t' = 0$, $\xi = 1$ and $T = 0.0125$ if other values are not indicated. All the energies and temperature are measured in units of t.

In Fig.1(a-c) results for the electron spectral density in the normal state, $A(k, \omega) = A_{11}(k, \omega)$ in (10), are shown along three directions in the Brillouin zone (BZ): $\Gamma(0, 0) \to X(\pi, 0) \to M(\pi, \pi) \to \Gamma$. We see quite narrow quasiparticle (QP) peaks at the wave vectors crossing the Fermi surface (FS) (in $M \to X$ and $M \to \Gamma$ directions, the $X \to \Gamma$ wave vectors are below the FS, see Fig. 3.). In addition to the QP dispersion we see also the incoherent excitations with large dispersion under the FS which appear due to the self-energy contributions peaked at the AFM wave vector ("shadow bands"). The same behavior are observed for higher hole concentrations and other values of ξ. However, the QP band width strongly increases with doping as it is demonstrated in Fig. 2(b) where the QP dispersion for $\delta = 0.1$, 0.4 is calculated from the maxima of spectral density. The next nearest hopping energy changes the QP spectrum as shown in Fig. 2(a) for $t' = \pm 0.1$ though it does not influence other physical properties very much.

The most interesting results were obtained for the electron occupation numbers $N(k) = N_{k\sigma}$ given by Eq. (15). $N(k)$ strongly depends on the hole concentrations as shown in Fig.3(a)-(d) for $\delta = 0.1 - 0.4$. The shape of the FS changes from the hole-like at a small doping to the electron-like for a large doping. However, the drop of $N(k)$ at the FS is quite small, especially at small doping, which is specific for strongly correlated electronic systems. Large occupation numbers throughout the hole BZ are due to the incoherent contribution in the spectral density $A(k, \omega)$ under the FS. As a result, the FS is large even for a small doping in agreement with the Luttinger theorem. It is clearly seen in Fig.4 (a-d) where the FS is shown for different hole concentrations $\delta = 0.1 - 0.4$ for $\xi = 1$. In the figure the contour plots for the gap function $\Phi(k, \omega = 0)$ obtained from a direct numerical solution of the linear equation for the gap function (16) are also given. At a small doping, $\delta = 0.1$, it has a more complicated k-dependence, with two positive and two negative maxima, while at $\delta \geq 0.2$ only one positive and one negative maxima survive.

However, in all cases the gap function obeys the condition of the d-wave pairing: $\Phi(k_x, k_y) = -\Phi(k_y, k_x)$ which violates the 4-fold symmetry of the

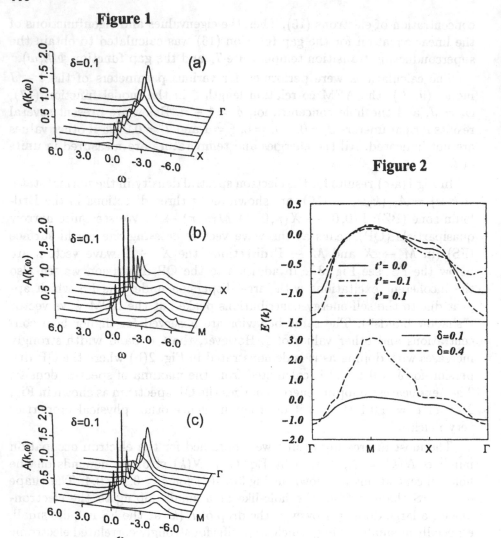

Figure 1

Figure 2

Figure 1. The electron spectral density $A(k,\omega)$ for the hole concentration $\delta = 0.1$ and AFM correlation length $\xi = 1$.

Figure 2. The electron quasiparticle spectrum $E(k)$ for (a) $\delta = 0.1$ and $t' = 0, \pm 0.1t$ and (b) $\delta = 0.1, 0.4$.

FS in the k-space.

In Fig.5 we present the superconducting T_c versus δ where the results obtained from solution of Eq.(16) are shown by the solid line. For a comparison, we give also the results of $T_c(\delta)$ calculations in the mean field

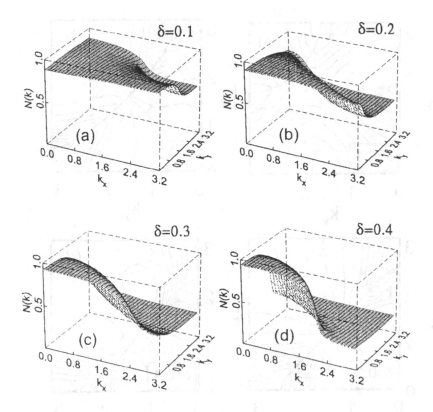

Figure 3. The electron occupation numbers $N(k)$, $(0 \leq (k_x, k_y) \leq \pi)$ for the hole concentrations: $\delta = 0.1$ (a), 0.2 (b), 0.3 (c), 0.4 (d).

approximation (MFA) for $\xi = 1$ and 3 when the self-energy operator contributions are omitted in Eqs. (12), (16). The MFA solutions give higher T_c as it should be expected.

The maximum value of T_c is quite high, of the order $0.06t \simeq 300$ K. It weakly depends on the value of J for small $J \leq t$ (at fixed ξ) since in that case the main contribution in Eq.(16) comes from $t(q)$ in the vertex $g(k, k - q)$. Therefore, we can argue that the kinematic pairing mechanism for a strongly correlated system proposed in [3] really works and provides d-wave symmetry due to the AFM dynamic spin fluctuation susceptibility.

To summarize, we have solved numerically self-consistent Eliashberg equations, (12), (16), for the model dynamic spin susceptibility, Eq.(17). The results for the electron spectral density (see Fig.1) prove the existence

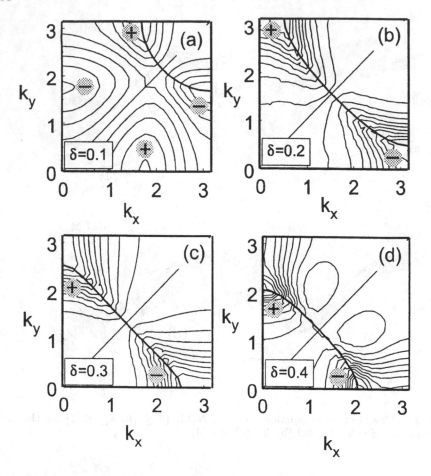

Figure 4. The FS (solid line) and the gap function $\Phi(\mathbf{k}, \omega = 0)$ versus k for $\delta = 0.1$ (a), 0.2 (b), 0.3 (c), 0.4 (d).

of quasiparticles near the FS. The occupation numbers $N(\mathbf{k})$ have the characteristic behavior for strongly correlated systems, Fig.4, which results in the large FS. The superconducting pairing due to the exchange, J, and kinematic, t^2, interactions has d-wave symmetry and occurs at high T_c as shown in Fig.5. We believe that vertex corrections disregarded here do not change the main conclusions of our calculations.

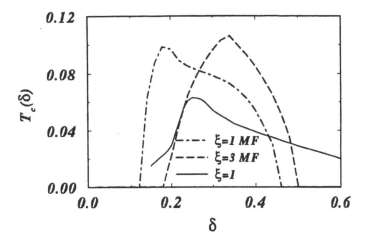

Figure 5. The superconducting temperature T_c versus hole concentration δ.

Acknowledgments

We gratefully acknowledge stimulating discussions with P. Horsch and A. Liechtenstein. We are also indebted to G. Jackely and V. Yushankhai for valuable discussions and remarks. We acknowledge usage of computational facilities of the Max-Planck-Institute in Stuttgart kindly provided for us by Prof. L. Hedin. Partial financial support by the INTAS–RFBR Program, Grant No 95–591, and by NREL, Subcontract AAX-6-16763-01, are acknowledged.

References

1. Tsuei, C.C., Kirtley, J.R., Rupp, M., et al., 1996, *Science*, **271**, 329.
2. Scalapino, D.J., 1995, *Phys. Reports*, **250**, 329.
3. Zaitsev R.O., Ivanov V.A., 1987, *Fiz. Tver. Tela*, **29**, 2554, 3111 (in Russian).
4. Plakida N.M., Yushankhai V.Yu., Stasyuk I.V., 1989, *Physica C*, **160**, 80; Yushankhai V.Yu., Plakida N.M., Kalinay P., 1991, *Physica C*, **174**, 401.
5. Dagotto, E., 1994, *Rev. Mod. Phys.*, **66**, 763.
6. Plakida, N.M., Oudovenko V.S., and Yushankai V.Yu., 1994, *Phys. Rev. B* **50**, 6431.
7. Dagotto, E., Nazarenko, A., and Moreo, A., 1995, *Phys. Rev. Lett.* **74**, 310.
8. Belinicher, V.I., Chernyshov, A,L., et al., 1995, *Phys. Rev. B*, **51**, 6076.
9. Plakida N.M., Oudovenko V.S., Horsch P., Liechtenstein A.I., 1997, *Phys. Rev. B*, **55**, R11997.
10. Kagan M.Yu., Rice T.M., 1994, *J. Phys.: Condens. Matter*, **3**, 5373.
11. Izyumov Yu.A., Letfulov B.M., 1991, *J. Phys.: Condens. Matter*, **3**, 5373.
12. Zubarev D.N., 1960, *Sov. Phys. Uspekhi*, **3**, 320.
13. Prelovśek P., 1997, *Z. Phys. B*, **103**, 363.
14. Plakida, N.M., Hayn, R., Richard, J-L., 1995, *Phys. Rev. B* **51**, 16599.
15. Jaklič J., Prelovśek P., 1995, *Phys. Rev. Lett.*, **74**, 3411; *ibid.* **75**, 1340.
16. Serene J.W., et al., 1991, *Phys. Rev. B* **44**, 3391.

CHARGE DENSITY FLUCTUATIONS AND GAP SYMMETRY IN HIGH-T_C SUPERCONDUCTORS WITH EXTENDED SADDLE-POINT FEATURES IN ELECTRON SPECTRUM

E. A. PASHITSKII, V. I. PENTEGOV, A. V. SEMENOV

Institute of Physics, National Academy of Sciences of Ukraine, Kiev, 252650, Ukraine

Abstract. It is shown that the strong anisotropy of the one-particle electron spectrum, due to the existence of extended saddle-point features (ESPF) close to the Fermi level in the hole-type cuprates *YBCO* and *BSCCO*, leads to the appearance of a low-frequency peak in the spectral function of the charge density fluctuations due to the presence of acoustic plasmon branch in the collective electron spectrum. The retarded anisotropic electron-plasmon interaction leads to the suppression of the static screened Coulomb repulsion for small transferred momenta and, consequently, to the effective attraction between electrons in the $d_{x^2-y^2}$-wave channel of the Cooper pairing of current carriers. Breaking of C_{4v} symmetry in and *YBCO* crystals leads to a possibility of a change of $d_{x^2-y^2}$-wave symmetry of the gap to mixed $s-d$ gap symmetry for singlet Cooper pairs or to p-wave gap symmetry for triplet pairs.

1. Introduction.

Any theoretical model intended for an adequate description of the nature of the HTS in cuprates should account for the d-wave superconducting gap symmetry, manifested in spontaneous Josephson currents [1], in generation of the half-integer quanta of the magnetic flux [2, 3], and also in the strong anisotropy of the gap in the plane of CuO_2 layers [4].

One of the Cooper pairing mechanisms, producing the d-wave symmetry of the superconducting gap in high-T_c superconductors, is described by the model of the electron-magnon interaction in an almost antiferro-

M. Ausloos and S. Kruchinin (eds.), Symmetry and Pairing in Superconductors, 121–130.

magnetic quasi-2D Fermi liquid [5, 6]. This model leads to an anisotropic repulsion between electrons (or holes) in the entire 2D momentum space with peaks in the corners of the Brillouin zone (BZ). Such interaction results in an effective attraction responsible for $d_{x^2-y^2}$-wave singlet Cooper pairing. However, the main question about the sufficiently large value of the coupling constant of electron-magnon interaction for obtaining a high value (~ 100 K) of the critical temperature remains open.

New important information about the structure of the electronic spectrum of high-T_c superconductors was recently obtained from angle-resolved photoemission spectroscopy with high energy resolution [7, 8]. These experiments exhibited the existence of extended saddle point features (ESPF) near the Fermi level in cuprates with hole-type conductivity. According to Refs. [9, 10], the ESPF in the band spectrum can be the result of a strong hybridization of the overlapping broad and narrow 2D bands in the layered cuprate crystals.

In this paper we present results of our theoretical and numerical investigations of the influence of the ESPF in the band spectrum on the HTS mechanism in the layered crystals of cuprates. We consider the ESPF effects on the spectrum of the collective charge-density excitations, the screened Coulomb interaction and the superconducting gap symmetry.

In §2 it is shown that the strong anisotropy of the one-particle electron spectrum in the CuO_2 layers due to the existence of ESPF leads to the appearance of the low frequency branch with an acoustic dispersion in the collective spectrum of the electron density excitations. These excitations are similar to acoustic plasmons (AP) in metals having a multiply-connected Fermi surface (FS) with essentially different effective masses of the current carriers in different extrema of the band spectra [11, 12]. The spectral function of the charge-density excitations is strongly anisotropic in this case and is peaked at frequencies corresponding to the AP branch. Due to the Kramers-Kronig relations for the reciprocal dielectric function, the static screened Coulomb interaction has an anisotropic structure with the pronounced minimum in the region of small transferred momenta $|\mathbf{q}| \ll \pi/a$ (where a is the lattice constant). This suppression of the Coulomb repulsion is a consequence of the retarded interaction between electrons due to the exchange of virtual AP.

In §3 we argue that the deep minimum of the screened Coulomb repulsion due to the electron-plasmon interaction leads to the effective attraction between electrons in the $d_{x^2-y^2}$-wave channel of the singlet Cooper pairing with the gap $\Delta_d(\varphi) \propto \cos 2\varphi$. This mechanism of the d-wave pairing differs from those proposed by the authors of [5, 6], who accentuated the important role of the sharp repulsion peak at $\mathbf{q} = (\pi, \pi)$ and believed the peculiarities of the interaction at small $\mathbf{q}$ to be irrelevant. Notice that the AP-mediated

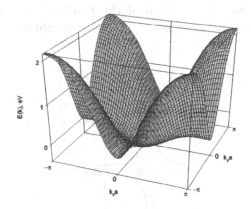

Figure 1. The conductivity band $E(k_x, k_y)$ of the 2D electron hybrid spectrum, calculated in [9].

interaction was considered earlier as a possible mechanism of superconductivity in quasi-isotropic transition metals [13, 14], in multi-valley semiconductors and semi-metals [15], in layered semiconducting heterostructures [16], and in high-T_c cuprate superconductors [17].

Breaking of the C_{4v} symmetry occurs also in $YBCO$ crystals due to the existence of ordered 1D CuO chains along the b-axis, leading to the mixed $s + d_{x^2-y^2}$ or $s + d_{xy}$ symmetry of the gap parameter for singlet Cooper pairing, and to the possibility of the p-wave gap symmetry with $\Delta_p(\varphi) \propto \sin \varphi$ or $\Delta_p(\varphi) \propto \cos \varphi$ in the triplet pairing channel.

2. Acoustic plasmons and screened Coulomb interaction in crystals with ESPF

It is known [11, 12] that a low-frequency AP branch in the collective electron spectrum can exist in multi-band (multi-valley) crystals having a multiply-connected FS and several groups of current carriers (electrons, holes) with significantly different effective masses. We show that a similar collective electron branch with acoustic dispersion relation can exist in layered crystals having a singly-connected FS but strong anisotropy in the electron density of states (DOS) and the Fermi velocity due to the existence of the ESPF [7, 8]. The experimental values of the Fermi energy and Fermi momentum for the quasi-1D spectrum near the bottom of ESPF are respectively $\mu_1 \approx 20$meV and $k_{F1} \approx 0.15$Å^{-1}. In the parabolic spectrum approximation $\mu_1 = k_{F1}^2 / 2m_1^*$ we obtain an effective electronic mass $m_1^* \approx 4.3 m_0$ (where m_0 is the bare electron mass).

In the present paper when choosing parameters for the hybridized conductivity band we are using the results of the multiple band calculations of Ref. [9]. This conductivity band $E(k_x, k_y)$ is shown in Fig. 1. When the

Fermi level lies above the bottom of the ESPF, it is convenient to use a shifted BZ centered in (π, π) point, in order to obtain a closed hole-type FS (Fig. 2).

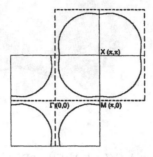

Figure 2. Cross section of the Fermi surface in the original and shifted Brillouin zones.

We proceed to show that the strong anisotropy of the DOS on the FS (and consequently that of the Fermi velocity of the quasi-particles) leads to the appearance of the low frequency AP branch in the collective electron spectrum, despite the fact that the FS is singly connected.

The dispersion relation $\omega_{pl}(q)$ of the AP branch is determined by zero of the real part of the longitudinal complex dielectric function

$$\varepsilon\left(\mathbf{q}, \omega\right) = 1 + V_c\left(\mathbf{q}\right) \Pi\left(\mathbf{q}_{\|}, \omega\right) \qquad (1)$$

where V_c is the matrix element of the Coulomb interaction in the layered crystals,

$$V_c\left(\mathbf{q}\right) = \frac{2\pi e^2}{q_{\|}} \cdot \frac{\sinh q_{\|} d}{\cosh q_{\|} d - \cos q_{\perp} d} \qquad (2)$$

and Π is the polarization operator, corresponding to the 2D band $E(\mathbf{k}_{\|})$, which is crossed by the Fermi level. Here $\mathbf{q}_{\|}$ and $\mathbf{k}_{\|}$ are the longitudinal momenta in the $a - b$ plane, $q_{\perp}$ is the transverse momentum along the c-axis, d is the distance between layers.

In Fig. 3 the real and imaginary parts of the polarization operator are shown as functions of $\omega/q_{\|}$ for $q_{\|} \rightarrow 0$.

The spectral function of the electronic charge density fluctuations (virtual plasmons) given by

$$S_{pl}\left(\mathbf{q}, \omega\right) = -\frac{1}{\pi} \operatorname{Im} \varepsilon^{-1}\left(\mathbf{q}, \omega\right) = \frac{1}{\pi} \frac{\operatorname{Im} \varepsilon\left(\mathbf{q}, \omega\right)}{\left[\operatorname{Re} \varepsilon\left(\mathbf{q}, \omega\right)\right]^2 + \left[\operatorname{Im} \varepsilon\left(\mathbf{q}, \omega\right)\right]^2} \qquad (3)$$

and calculated with the electron spectrum of Fig. 1, is plotted in Fig. 4 for several $\mathbf{q}_{\|} \neq 0$ along one of the two BZ diagonals and for $q_{\perp} = 0$. The

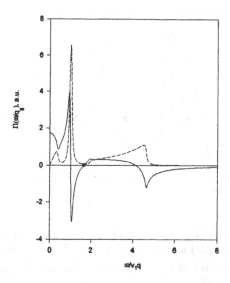

Figure 3. The real (solid curve) and imaginary (dashed curve) parts of the polarization operator in the long wave limit $(q \to 0)$, when q is parallel to one of the main crystallographic axes, in a function of $\frac{\omega}{v_1 q_\parallel}$ for the case of strong anisotropy of the Fermi velocity.

function $S_{pl}(\mathbf{q}, \omega)$ has maximum at the frequency $\omega_{pl}(q)$ of the AP since $\operatorname{Re}\varepsilon(\mathbf{q}, \omega_{pl}) = 0$, while for $\omega \to 0$ according to (3) $S_{pl}(\mathbf{q}, \omega) \sim \omega$ since $\operatorname{Im}\varepsilon(\mathbf{q}, \omega) \sim \omega$.

By virtue of the Kramers-Kronig relation for the reciprocal dielectric function $\varepsilon^{-1}(\mathbf{q}, \omega)$ the matrix element of the static (for $\omega = 0$) screened Coulomb repulsion between electrons can be written as

$$\tilde{V}_c(\mathbf{q}) \equiv \frac{V_c(\mathbf{q})}{\varepsilon(\mathbf{q}, 0)} = V_c(\mathbf{q})\left[1 - 2\int_0^\infty \frac{d\omega'}{\omega'} S_{pl}(\mathbf{q}, \omega')\right]. \tag{4}$$

The plot of $\tilde{V}_c\left(\mathbf{q}_\parallel\right)$ calculated for the band of Fig. 1 is presented in Fig. 5. As we see, $\tilde{V}_c(\mathbf{q})$ has a deep minimum in the region of small transferred momenta $\mathbf{q}_\parallel$ due to the low frequency maximum of the charge density spectral function $S(\mathbf{q}, \omega)$ in the region of the AP branch existence. Such a suppression of the static Coulomb repulsion is a consequence of the effective electron-electron attraction through the exchange of the virtual AP.

In the general case of C_{4v} symmetry of the electron spectrum the Fourier series expansion of the screened Coulomb matrix element $\tilde{V}_c\left(\mathbf{k}_\parallel - \mathbf{k}'_\parallel\right)$ with

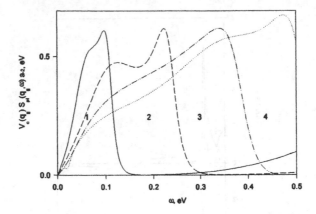

Figure 4. The frequency dependence of the spectral function of the electronic charge density fluctuations $S_{pl}(\mathbf{q}, \omega)$ multiplied by the bare Coulomb matrix element $V_c(\mathbf{q})$ for $q_\perp = 0$ and for several values of $\mathbf{q}_\parallel$ along the BZ diagonal: $1 - q_\parallel = \frac{\sqrt{2}\pi}{16a}$, $2 - q_\parallel = \frac{\sqrt{2}\pi}{8a}$, $3 - q_\parallel = \frac{3\sqrt{2}\pi}{16a}$, $4 - q_\parallel = \frac{\sqrt{2}\pi}{4a}$.

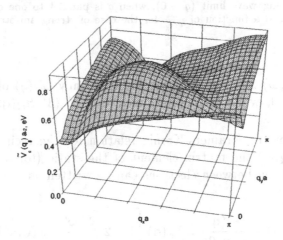

Figure 5. The plot of the static screened Coulomb repulsion $\tilde{V}_c(\mathbf{q}_\parallel)$ calculated for the electron spectrum of Fig. 1.

respect to the angles φ and φ'

$$\tilde{V}_c(\varphi, \varphi') = \sum_{n,m} V_{nm} e^{in\varphi + im\varphi'}. \tag{5}$$

will contain harmonics V_{nm} with indices satisfying the condition $n + m = 4l$, where l is an integer.

3. Influence of the electron spectrum anisotropy on the gap symmetry

In what follows, we show that the marked momentum dependence of the screened Coulomb interaction (4), connected to the low frequency peak of the plasmon spectral function (3), determine the symmetry of the superconducting gap.

We will use the Eliashberg equations [18] for the gap in superconductors with strong coupling, taking into account the retarded interaction between electrons due to the exchange of virtual phonons and acoustic plasmons, and also the screened Coulomb repulsion. Near the critical temperature, $T \to T_c$, the linearized equation for the anisotropic gap $\Delta(\mathbf{k}_{\parallel}, \omega)$ on the FS, given the quasi-2D character of the electron spectrum, using the static Kramers-Kronig relation (4), can be written on the FS ($\left|\mathbf{k}_{\parallel}\right| = k_F$) as

$$(1 + \lambda) \, \Delta(\varphi, 0) = \frac{1}{2} \int_0^{2\pi} \frac{d\varphi'}{2\pi} \int_{-\tilde{\Omega}}^{\tilde{\Omega}} \frac{d\omega}{\omega} \Delta(\varphi', \omega) \, \nu(\varphi', \omega) \times \\ \left[W_{ph} \, \theta\left(\tilde{\Omega}_{ph} - |\omega|\right) - \tilde{V}_c(\varphi, \varphi') \right] \tanh \frac{\omega}{2T_c}, \tag{6}$$

where λ is the renormalization constant, connected to the normal self-energy, $\theta(x)$ is the unit step function, the electron-phonon interaction W_{ph} is taken to be quasi isotropic, $\tilde{\Omega}$ is a cut-off energy of the Coulomb interaction ($\tilde{\Omega} \approx E_F$), and $\nu(\varphi', \omega)$ is the anisotropic DOS.

3.1. UNBROKEN C_{4v} SYMMETRY

For the layered crystals, having the unbroken C_{4v} symmetry of the electron spectrum (for instance, $BiSrCaCuO$, $TlBaCaCuO$, $HgBaCaCuO$), the anisotropic DOS can be approximated by

$$\nu(\varphi, \omega) = \nu_+(\omega) + \nu_-(\omega) \cos 4\varphi \tag{7}$$

where

$$\nu_{\pm}(\omega) = \frac{1}{2} \left[\nu_1 \, \mathrm{Re} \, \sqrt{\frac{\mu_1}{\mu_{1+\omega}}} \pm \nu_2 \right]. \tag{8}$$

In the same time it is possible to retain in the Fourier expansion (5) only the main terms, corresponding to the A_1 and B_1 representations of the C_{4v} group:

$$\tilde{V}_c(\varphi, \varphi') \approx V_{00} + V_{22} \left(\cos 2\varphi \cos 2\varphi' - \sin 2\varphi \sin 2\varphi' \right) + \\ V_{44} \left(\cos 4\varphi \cos 4\varphi' - \sin 4\varphi \sin 4\varphi' \right), \tag{9}$$

where the isotropic part of the Coulomb repulsion $V_{00} > 0$, while V_{22} and V_{44}, containing the anisotropic contribution of the electron-plasmon interaction, may be either positive or negative (see below). Substituting (7) and (9) in Eq. (6) we conclude that the singlet Cooper pairing is possible either in d- or s-channels. For the s-wave pairing with the anisotropic gap

$$\Delta_s(\varphi) = \Delta_0 + \Delta_4 \cos(4\varphi) \tag{10}$$

we obtain the following coupled equations determining the critical temperature T_c^s

$$(1+\lambda)\Delta_0 = \frac{W_{ph} - V_{00}^*}{2} \int_{-\tilde{\Omega}_{ph}}^{\tilde{\Omega}_{ph}} \frac{d\omega}{\omega} \left[\nu_+(\omega)\Delta_0 + \frac{\nu_-(\omega)\Delta_4}{2} \right] \tanh \frac{\omega}{2T_c^s} \tag{11}$$

$$(1+\lambda)\Delta_4 = -\frac{1}{4}V_{44} \int_{-\tilde{\Omega}}^{\tilde{\Omega}} \frac{d\omega}{\omega} [\nu_+(\omega)\Delta_4 + \nu_-(\omega)\Delta_0] \tanh \frac{\omega}{2T_c^s} \tag{12}$$

where

$$V_{00}^* = \frac{V_{00}}{1 + \nu_2 V_{00} \ln(E_F/\tilde{\Omega}_{ph})} \tag{13}$$

For the d-wave pairing the critical temperature T_c^d is given by equation

$$(1+\lambda) \cdot \Delta_d(\varphi) = -\frac{V_{22}}{2} \int_0^{2\pi} \frac{d\varphi'}{2\pi} \int_{-\tilde{\Omega}}^{\tilde{\Omega}} \frac{d\omega}{\omega} \nu(\varphi',\omega) \Delta_d(\varphi') \times$$
$$(\cos 2\varphi \cos 2\varphi' - \sin 2\varphi \sin 2\varphi') \tanh \frac{\omega}{2T_c^d} \tag{14}$$

The sign of the coefficient V_{22} determines the type of d-wave symmetry of the gap. For the negative value of V_{22} the gap has $d_{x^2-y^2}$ symmetry, $\Delta_d(\varphi) = \Delta_2 \cos 2\varphi$, and the equation (14) is reduced to

$$(1+\lambda) = \frac{|V_{22}|}{4} \int_{-\tilde{\Omega}}^{\tilde{\Omega}} \frac{d\omega}{\omega} \left(\nu_+(\omega) + \frac{1}{2}\nu_-(\omega) \right) \tanh \frac{\omega}{2T_c^d}. \tag{15}$$

For the positive value of V_{22} from (14) the solution of the form $\Delta_d(\varphi) = \Delta_2 \sin 2\varphi$ follows, corresponding to the d_{xy}-wave gap symmetry.

3.2. $YBCO$-TYPE CRYSTALS WITH CUO CHAINS

For the $YBa_2Cu_3O_7$ and $YBa_2Cu_4O_8$ crystals, where C_{4v} symmetry is broken due to the existence of the ordered 1D chains CuO, the anisotropic

DOS has approximately the following angular dependence

$$\nu\left(\varphi,\omega\right) = \tilde{\nu}_0\left(\omega\right) - \frac{\tilde{\nu}_c}{2}\cos 2\varphi + \nu_-\left(\omega\right)\cos 4\varphi \qquad (16)$$

where $\tilde{\nu}_0(\omega) = \nu_+(\omega) + \tilde{\nu}_c/2$, and $\tilde{\nu}_c$ is the electron DOS in the 1D chains.

The asymmetric terms in the electron-electron interaction in this case are

$$U\left(\varphi,\varphi'\right) = \tilde{U}_{11}\left(1 + \cos\varphi\cos\varphi' - \sin\varphi\cdot\sin\varphi'\right) \qquad (17)$$

Substitution of (16) and (17) in (6) leads, for the singlet Cooper pairing, to the system of coupled gap equations

$$(1+\lambda)\,\Delta_0 = \frac{W_{ph} - \tilde{U}_{11} - V_{00}^*}{2} \int\limits_{-\tilde{\Omega}_{ph}}^{\tilde{\Omega}_{ph}} \frac{d\omega}{\omega}\left[\Delta_0\tilde{\nu}_0\left(\omega\right) + \frac{\Delta_2\tilde{\nu}_c}{2}\right]\tanh\frac{\omega}{2T_c} \qquad (18)$$

$$(1+\lambda)\,\Delta_2 = \frac{|V_{22}|}{4} \int\limits_{-\tilde{\Omega}}^{\tilde{\Omega}} \frac{d\omega}{\omega}\left\{\Delta_0\tilde{\nu}_c + \Delta_2\left[\tilde{\nu}_0\left(\omega\right) + \frac{\nu_-\left(\omega\right)}{2}\right]\right\}\tanh\frac{\omega}{2T_c} \qquad (19)$$

for s- and d-wave components of the anisotropic superconducting gap which is given by,

$$\Delta_{sd}\left(\varphi\right) = \begin{cases} \Delta_0 + \Delta_2\cos 2\varphi & \text{if } V_{22} < 0 \\ \Delta_0 + \Delta_2\sin 2\varphi & \text{if } V_{22} > 0 \end{cases} \qquad (20)$$

In other words, the presence of the 1D chains of CuO in $YBCO$ should bring about the mixed s-d symmetry of the gap.

Furthermore, a sufficiently strong interaction $\tilde{U}_{11}$ in (17), violating the C_{4v} symmetry, could give rise to triplet Cooper pairing of electrons in 1D chains with p-type order parameter symmetry

$$\Delta_p\left(\varphi\right) = \begin{cases} \Delta_1\cos\varphi & \text{if } \tilde{U}_{11} < 0 \\ \Delta_1\sin\varphi & \text{if } \tilde{U}_{11} > 0 \end{cases}. \qquad (21)$$

In this case the critical temperature T_c^p will be defined by the equation

$$(1+\lambda) = \frac{1}{4}|\tilde{U}_{11}| \int\limits_{-\tilde{\Omega}}^{\tilde{\Omega}} \frac{d\omega}{\omega}\tilde{\nu}_0\left(\omega\right)\tanh\frac{\omega}{2T_c^p} \qquad (22)$$

Notice that the negative value of $\tilde{U}_{11}$ can result from the strong electron-phonon interaction in the CuO chains.

130

4. Conclusions

We have shown that strong anisotropy of the one-particle electron spectrum, associated, in particular, with the presence of the ESPF near the Fermi level, may lead to the appearance of the acoustic plasmon branch in the collective electron spectrum. These electronic excitations cause the low frequency peak in the spectral function of the charge density fluctuations $S_{pl}(\mathbf{q}, \omega) = -\frac{1}{\pi} \operatorname{Im} \varepsilon^{-1}(\mathbf{q}, \omega)$ and, through the Kramers-Kronig relation for the reciprocal dielectric function $\varepsilon^{-1}(\mathbf{q}, \omega)$, lead to the deep minimum in the static screened Coulomb repulsion for the small transferred momenta $\mathbf{q}$. Such suppression of the Coulomb repulsion, which is the result of the effective electron-electron attraction due to the exchange of the virtual acoustic plasmons, favors the $d_{x^2-y^2}$-wave Cooper pairing of the current carriers with the superconducting gap structure $\Delta_d(\varphi) \sim \cos 2\varphi$ in the layered crystals of the cuprate metal-oxide compounds, having C_{4v} symmetry of the CuO_2 layers. Breaking of C_{4v} symmetry leads to the mixed $s - d$ wave singlet Cooper pairing, or to the p-wave triplet pairing of current carriers.

References

1. D. A. Wollman et al. Phys.Rev.Lett. **71**, 2134(1993); **74**, 797 (1995).
2. S. S. Tsuei, J. R. Kirtley, C. C. Chi, Lock See Yu-Jahnes, A. Gupta, T. Shaw, J. Z. Sem, M. B. Ketchen. Phys.Rev.Lett **73**, 593 (1994).
3. J. R. Kirtley, S. S. Tsuei, M. Rupp et al. Phys.Rev.Lett. **76**, 1336 (1996).
4. H. Ding, M. R. Norman, J. C. Campuzano M. Rnaderia, A. F. Bellman, T. Yokoya, T. Takahashi, T. Mochiku, K. Kadowaki. Phys.Rev. *B* **54**, R9678 (1996).
5. A. J. Millis, H. Monien, D. Pines. Phys.Rev *B* **42**, 167 (1990).
6. D. Pines, Physica *C* **235**, 113(1994); Zeit.Phys. *B* **103**, 129 (1997).
7. D. M. King, Z.-X. Shen, D. S. Dessau. Phys.Rev.Lett. **73** , 3298 (1994).
8. K. Gofron, J. C. Campuzano, A. A. Abrikosov et al. Phys.Rev.Lett. **73**, 3302 (1994).
9. O. K. Andersen, O. Jensen, A. I. Liechtenstein, I. I. Mazin. Phys.Rev. *B* **49**, 4145 (1994).
10. E. A. Pashitskii, V. I. Pentegov, JETP **84**, 164 (1997).
11. D. Pines. Canad.J.Phys. **34**, 1379 (1956); D. Pines, J. R. Shrieffer. Phys.Rev. **124**, 1387 (1961).
12. J. Ruvalds, Adv.Phys. **30**, 677 (1981).
13. J. W. Garland Jr., Phys.Rev. **153**, 460 (1967).
14. H. Frölich, Phys.Lett. *A* **26**, 169 (1968); J.Phys. *C* **1**, 544 (1968).
15. E. A. Pashitskii, Sov.Phys.JETP **28**, 1267(1968); E. A. Pashitskii and V.M. Chernousenko, Sov.Phys.JETP **33**, 802 (1971).
16. E. A. Pashitskii, Sov.Phys.JETP **28**, 362 (1969); E. A. Pashitskii and Yu. A. Romanov, Ukr.Phys.J. **15**, 1594 (1970).
17. E. A. Pashitskii, JETP Lett. **55**, 336 (1992); *ibid* **56**, 364 (1992); JETP **76**, 425 (1993); Low Temp.Phys. **21**, 763 (1995); *ibid* 837 (1995).
18. G. M. Eliashberg. Sov. Phys. JETP **11**, 696 (1960); **12**, 1000 (1961).

VORTEX LATTICE MELTING THEORIES AS AN EXAMPLE OF SCIENCE FICTION

A.V.NIKULOV
Institute of Microelectronics Technology and
High Purity Materials,
Russian Academy of Sciences,
142432 Chernogolovka, Moscow District, Russia.

Abstract. It is shown that the popular concept of vortex lattice melting has appeared as a consequence of incorrect notions about the Abrikosov state and an incorrect definition of the phase coherence. The famous Abrikosov solution gives qualitatively incorrect results. The transition to the Abrikosov state must be first order in an ideal (without disorder) superconductor. Such a sharp transition is observed in bulk superconductors with weak disorder below H_{c2}. No experimental evidence of the vortex lattice melting exists now. The absence of a sharp transition in thin films with weak disorder is interpreted as due to the absence of long-rang phase coherence down to very low magnetic field. The observed smooth phase coherence appearance in superconductors with strong disorder is explained by an increase of the effective dimensionality. It is proposed to return to the Mendelssohn model for the explanation of the resistive properties of superconductors with strong disorder. It is conjectured that the Abrikosov state is not a vortex lattice with crystalline long-range order.

1. Introduction

I think nobody will contest that some theoretical works is science fiction only. It can be easily proven. Some different points of view exist about many problems, but only one of them corresponds to reality. Consequently, other points of view are science fiction only. I understand this situation is inevitable. The theorist's fantasy is much wider than a reality. It is very difficult sometimes to distinguish fiction from the reality. But we must

M. Ausloos and S. Kruchinin (eds.), Symmetry and Pairing in Superconductors, 131–140.

strive for this. And this is a very important place for a fair discussion. In the present work I try to convince readers that the conception of the vortex lattice melting in the mixed state of type II superconductors, which was very popular for the last ten years and continues to be popular now, is science fiction only.

2. Why has the concept of vortex lattice melting appeared and become popular?

The concept of vortex lattice melting has appeared and became popular because the point of view prevailed that the Abrikosov state is a flux line (or vortex) lattice (FLL) [1] like an atomic lattice, or a lattice of long molecules [2]. And now, almost nobody doubts that it is so. This opinion was caused by direct observation of the vortex lattice [3]. The Abrikosov vortex was considered as a magnetic flux. The transport properties were connected with the motion of the magnetic flux structures (or FLL) [4]. According to all textbooks (see for example [4]), the steady-state motion of the magnetic flux structure induces a time-averaged macroscopic electric field following Faraday's law. Motion of the magnetic flux structure can be induced by the Lorentz force. The resistivity in the Abrikosov state is called "flux flow resistivity" [4]. The Lorentz force is compensated by the damping force and the pinning force. The pinning force is caused because the vortices are pinned by superconductor disorder. The vortex pinning is a very important effect. Application of type II superconductors in high magnetic fields is possible owing to the pinning effect only. Therefore it was a strong disappointment [5] when soon after the high-Tc superconductor discovery it was detected that the pinning effect is absent in a wide region below the second critical field, H_{c2}. This was interpreted as a consequence of the vortex lattice melting because the Abrikosov state was considered as the flux line (or vortex) lattice.

According to a popular point of view [5], the process of vortex pinning is very different for vortex liquids or solids. In the case of a vortex solid, a few pinning centers can hold the entire lattice because it is stiff. But it is impossible to hold in place a vortex liquid with a few pinning centers.

3. The notions on which the concept of vortex lattice melting is founded are not quite right.

3.1. THE ABRIKOSOV VORTEX IS NOT A FLUX LINE BUT IS A SINGULARITY IN THE MIXED STATE WITH THE PHASE COHERENCE.

The Abrikosov vortices appear because a magnetic flux cannot exist within a superconducting region with long-rang phase coherence and without singularities. This follows from the relation [4]

$$\frac{\Phi_0}{2\pi} \int_l dR \frac{d\phi}{dR} = \int_l dR \lambda_L^2 j_s + \Phi \qquad (1)$$

for the superconducting current where l is a closed path of integration; $\lambda_L = (mc/e^2 n_s)^{0.5}$ is the London penetration depth; j_s is the superconducting current density; Φ is the magnetic flux contained within the closed path of integration l. If the singularity is absent $\int_l dR d\phi/dR = 0$. In this case the relation (1) is the equation postulated by F. and H.London [6] for explaning the Meissner effect [7] (see [4]).

According to Eq.(1) a magnetic field can penetrate a superconductor only if: 1) superconductivity is destroyed, or 2) singularities appear, or 3) the long-rang phase coherence is absent. The first case is observed in type I superconductors. The second situation corresponds to the Abrikosov state. The existence of vortices is an evidence of the existence of the phase coherence. Consequently, the Abrikosov state is the mixed state with long-rang phase coherence. $(\int_l dR d\phi/dR)/2\pi = n$ is the number of the vortices contained within l. If the l-radius $\gg \lambda_L$ then $n = \Phi/\Phi_0$. The third case is a mixed state without long range phase coherence.

3.2. THE RESISTIVITY IN THE ABRIKOSOV STATE IS CAUSED BY THE VORTEX MOTION BUT NOT THE MOTION OF THE MAGNETIC FLUX.

The time-averaged macroscopic electric field appears in the Abrikosov state in accordance not with Faraday's law but with the Josephson's relation

$$V = \frac{\hbar}{2e} \frac{d\phi}{dt} \qquad (2)$$

The vortex flow causes a change of the phase difference ϕ in time and as consequence of Eq.(2) it induces a voltage. The Faraday's law and the Josephson relation give the same result. Therefore this has led to the situation where the resistivity in the Abrikosov state is considered as a flux flow resistivity in all textbooks and in a majority of papers, although it is obvious that the magnetic flux does not flow in a superconductor. This mistaken notion is one of the causes of the wide popularity of the concept of vortex lattice melting. I will use a more correct term "vortex flow resistivity" instead of "flux flow resistivity" in the following.

3.3. THE VORTEX PINNING IS A CONSEQUENCE OF THE LONG-RANGE PHASE COHERENCE.

The vortex pinning is a consequence of the long-range phase coherence, because the Abrikosov vortex cannot exist without phase coherence. Consequently, the pinning disappearance can be interpreted as the phase coherence disappearance. It is important to note this because most authors interpret the pinning disappearance as the vortex lattice melting itself.

The vortex pinning has an influence on resistive properties first of all. The resistivity has a different nature in the states with and without the phase coherence. Consequently, resistive properties change first of all at the long-range phase coherence appearance transition.

3.4. NO EVIDENCE EXISTS NOW THAT THE ABRIKOSOV STATE IS THE VORTEX LATTICE WITH THE CRYSTALLINE LONG-RANGE ORDER.

Many scientists think that the direct observation in [3] is an evidence that the Abrikosov state is the vortex lattice. But it is not right. If we lay along a fishing net with the help of stakes it will look like a lattice. But from this direct observation we can not draw a conclusion that the fishing net is a lattice which can melt.

A real vortex lattice is a structure in an inhomogeneous space, because disorders exist in any real superconductor sample. Larkin [8] has shown that the crystalline long-range order of the vortex lattice is unstable against the introduction of random pinning. Consequently we cannot contend ourselves on the basis of the "direct observation" [3] that the Abrikosov state is a vortex lattice which can melt because it can be a structure like a fishing net.

Some theorists base their vortex lattice melting theories on the Abrikosov solution [9] and subsequent results [10]. According to the mean field approximation the Abrikosov state is a triangular vortex lattice with crystalline long-range order [11]. But according to [12] the mean field approximation cannot be used for the description of the mixed state in the thermodynamic limit.

Maki and Takayama [12] have shown that the fluctuation correction $\Delta n_{s,fl}$ to the Abrikosov solution calculated in the linear approximation depends on the superconductor size L across the magnetic field direction: $\Delta n_{s,fl}$ is proportional to $\ln(L/\xi)$ in a three-dimensional superconductor and $\Delta n_{s,fl}$ is proportional to $(L/\xi)^2$ in a two-dimensional superconductor. This result seems very queer for most scientists, because they think that it contradicts experimental results. Therefore almost nobody has believed in the reality of this result, not even the authors themselves. In order to "amend" this result Maki (with Thompson) [13] have even made more incorrect work .

In spite of the opinion of most scientists, I claim that the Maki- Takayama result [12] is right in fact. It does not contradict any direct observation of the Abrikosov state. But according to this result the thermal fluctuation in the mixed state cannot be considered as perturbations in the thermodynamic limit. Therefore the Abrikosov solution cannot be used for a the starting work for vortex lattice melting theories.

4. Theories of the vortex lattice melting and theories of the vortex liquid solidification.

Soon after the HTSC discovery the elastic theories of the vortex lattice melting appeared [14]. In these theories the vortex lattice melting is considered as a consequence of increased thermal displacement of the vortex position. In analogy to a crystal lattice the melting temperature T_m was estimated from the Lindeman criterion [15].

Dislocation-mediated melting was discussed in some works [16]. Numerous other theories on a possible melting transition were published (see review [1, 17]). All these theories are based on the assumption that the Abrikosov state is the vortex (or flux line) lattice with crystalline long-range order.

The elastic theories of vortex lattice melting raised doubts soon after they appeared, because the description in these theories is unsatisfactory in principle : they start from the state in which the translation symmetry has been broken "at first" [18]. Therefore some theorists consider not the vortex lattice melting but the solidification transition of vortices. The solidification theories are a revision of the Abrikosov solution [9]. The theorists try to find and to describe the transition into the vortex lattice state, taking into account the thermal fluctuation.

The first attempt to find the solidification transition was made before the HTSC discovery [19]. No transition was found in these approaches [19, 20] based on perturbation theory. Thus, the transition into the Abrikosov state caanot be derived from perturbation fluctuation theory. But because it is well known from direct observation that the Abrikosov state exists attempts to find the solidification transition were continued in many works along different approaches. Most authors found the solidification transition [21]. And only few authors [22] ventured to state that the solidification transition is absent.

I agree with M.A. Moore that the solidification transition is absent. But this does not mean that the transition into the Abrikosov state is absent, because the Abrikosov state is the mixed state with long range phase coherence. The solidification theories do not consider he long-range phase coherence appearance but the destruction of the translation symmetry because a wrong definition of the phase coherence is used in these theories.

5. Definition of the phase coherence.

The phase coherence is defined in solidification theory through the correlation function. According to this definition the long-range phase coherence cannot exist without the crystalline long-range order of the vortex lattice. It is claimed in some work [23] that the phase coherence can remain short-ranged even in the vortex solid phase.

The phase coherence definition used in the solidification theory is logically contradictory because the existence of the vortices is evidence of the long-rang phase coherence. And this must follow from the right definition. We can use the relation (1) for the definition of the phase coherence: the phase coherence exists in some region if the relation (1) is valid for any closed path in this region.

According to the correct definition the long-range phase coherence must be both in the vortex lattice and in the vortex liquid. Consequently the phase coherence disappearance transition must be observed above the vortex lattice melting one.

6. Why I think that the vortex lattice melting is science fiction.

Only one transition is observed. This transition exists for sure in bulk superconductors with weak disorder. It is observed first of all by investigating the resistivity in a perpendicular magnetic field. First this transition was observed in conventional superconductors [24] before the HTSC discovery. Later, this result was repeated in many works [25] by investigations of $YBa_2Cu_3O_{7-x}$ single crystals with weak disorder. This transition is observed below H_{c2} at a magnetic field denoted as H_{c4} in [26]. A difference of the $H_{c2} - H_{c4}$ values observed in conventional superconductors [24] and in $YBa_2Cu_3O_{7-x}$ [25] conforms to the scaling law [27].

This transition was interpreted in our work [24] as a phase transition from a "one-dimensional" state (the mixed state without the long-range phase coherence) into the Abrikosov state. But most authors interpret this transition as the vortex lattice melting [25]. This interpretation cannot be correct because no transition is observed above that one. According to this interpretation the vortex liquid exists above H_{c4}. But no experimental evidence of the vortex liquid exists nowadays. Moreover, our investigation of bulk conventional superconductors [28] and thin films [29, 30] show that the vortex liquid does not exist.

Some properties of the vortex liquid must differ from the one of the mixed state without the phase coherence. For example, a nonlocal resistivity must be observed in the vortex liquid [31] and be not observed in the mixed state without the phase coherence. Our investigations of nonlocal conductivity in thin films of conventional superconductor [30] show that the magnetic field destroys the phase coherence in fact. In a wide region below H_{c2} the phase coherence is absent. Therefore the opinion of most authors that the observed transition is the vortex lattice melting one and the state between H_{c4} and H_{c2} is the vortex liquid cannot be right.

7. The famous Abrikosov solution gives qualitatively incorrect results.

According to the Abrikosov solution [9] a second order phase transition from the normal state into the superconducting state with the long-range phase coherence takes place in H_{c2}. It is right in the mean field approximation. But it is no correct in a fluctuation theory.

The long-range phase coherence appearance may be considered as a consequence of the increasing up to infinity of the coherence length. The correlation length becomes anisotropic in a high magnetic field [32]. The longitudinal (along a magnetic field) coherence length calculated in the linear approximation $\xi_l = (\Phi_0/2\pi(H - H_{c2})^{0.5}$ increases up to infinity at H_{c2}, whereas the transversal coherence length $\xi_t = (2\Phi_0/\pi H)^{0.5}$ changes little near H_{c2}. The longitudinal coherence must be renormalized near H_{c2}, whereas transversal coherence length changes little in the critical region. This means that the correlation function of bulk superconductors near H_{c2} (in the lowest Landau level (LLL) approximation region) is similar to that of the one-dimensional superconductor [33]. Consequently, if we define the phase coherence in the mixed state by the correlation function we can conclude that the long-range coherence cannot exist in the mixed state of type II superconductors.

On the other hand we know that the Abrikosov state is a mixed state with long-range phase coherence (with the length of the phase coherence equal to the sample size L). A mixed state with phase coherence length perpendicular to the magnetic field direction $(\Phi_0/H)^{0.5}$ exists also. I call this state the mixed state without the phase coherence (or "one-dimensional" state). There are only two characteristic lengths, $(\Phi_0/H)^{0.5}$ and sample size L in a superconductor without disorder. Consequently, the length for the phase coherence in an ideal superconductor can change only by a jump from $(\Phi_0/H)^{0.5}$ to L, i.e the transition into the Abrikosov state must be a first order (sharp) phase transition in the ideal case.

8. The position of the transition into the Abrikosov state depends on the amount of disorder in thin films.

The sharp transition is observed at H_{c4} in bulk superconductors with weak disorder [24, 25]. The results of work [34] show that it may be a first order phase transition indeed. But no sharp transition is observed in thin films [29, 35] nor in bulk superconductors with strong disorder [36].

The absence of any features of the resistive properties up to low magnetic field in amorphous $Nb_{1-x}O_x$ films was interpreted in our paper [29] as the absence of the transition into the Abrikosov state. This interpretation was criticized by Theunessen and Kes [35]. They contend that we do not observe any feature because our measuring current is extremely high in comparison to the critical current. But sharp features of the vortex flow resistivity must be observed also at the transition into the Abrikosov state in superconductors with weak disorder according to fluctuation theory [12].

Such sharp features are observed in all "enough homogeneous" bulk superconductors [28, 37]. And only in not enough homogeneous samples the "classical" flux flow resistivity dependencies [38] are observed. The observed features differ qualitatively from the mean-field vortex flow resistivity dependence [39].

Smooth features of the vortex flow resistivity are observed in moderate fields in a-NbGe films with intermediate strength of disorder [35] (see the inset of Fig.7). This feature is a consequence of the phase coherence appearance. But no feature is observed down to very low fields in the $Nb_{1-x}O_x$ films with extremely small pinning [29]. The features of the vortex flow resistivity can be observed at high measuring current. Therefore I can claim that the phase coherence appearance in thin films depends on the amount of disorder. This claim is confirmed by my theoretical results [40].

9. Transition into the Abrikosov state in superconductors with strong disorder. The return to the Mendelson model.

The transition into the Abrikosov state in superconductors with strong disorder [35, 36] is smooth. The length of the phase coherence does not change by a jump but increases gradually with the magnetic field (or the temperature) decreasing [41].

To explain this difference from the ideal case I propose in [41] to return to the Mendelssohn model [42]. The Mendelssohn's sponge [42] can be considered as a limiting case of strong disorder. Real superconductors can be considered as intermediate cases between the Mendelssohn's [42] and Abrikosov's [9] models. The Mendelssohn sponge is a one-dimensional system. Therefore a thin film with strong disorder can be considered as a system like the one-dimensional superconductor. In a one-dimensional superconductor [43] the length of the phase coherence increases smoothly with decreasing temperature. In consequence of this the resistive transition is also smooth.

10. Conclusion

The Abrikosov state is the mixed state with long-range phase coherence. Consequently two long-range orders exist in the Abrikosov state if it is the vortex lattice with crystalline long-range order. But only one transition is observed on the way from the Abrikosov state into the normal state. Consequently, either both orders disappear at this transition, or the Abrikosov state is not the vortex lattice. I conjecture that the latter corresponds to reality. Observed correlation between the vortex lattice and the crystal lattice of superconductor [4] confirms my opinion.

The second critical field H_{c2} is not a critical point neither in superconductors with weak disorder nor in superconductors with strong disorder. The phase coherence appears below H_{c2} in superconductors with weak disorder [24, 25] and above H_{c2} in superconductors with strong disorder [41].

The sharp transition into the Abrikosov state predicted by the fluctuation theory in ideal cases is observed in bulk superconductors with weak disorder [24, 25] only. No sharp transition is observed in thin films with weak disorder [29]. This difference can be explained by differences of the fluctuation value in three- and two-dimensional superconductors [12].

The smooth phase coherence appearance in superconductors with strong disorder can be explained qualitatively by the increase of the effective dimensionality of the fluctuations.

Acknowledgements
I thank for financial support the International Association for the Promotion of Co-operation with Scientists from the New Independent States (Project INTAS-96-0452) and the National Scientific Council on Superconductivity of SSTP "ADPCM" (Project 98013).

References

1. E.H.Brandt, Rep. Progr. Phys. **58**, 1465 (1995).
2. D.R.Nelson, Nature **375**, 356 (1995).
3. D.Cribier, B.Jacrot, L.M.Rao, and B.Farnoux, Phys. Lett. **9**, 106 (1964); U.Essmann and H.Trauble, Phys. Lett. A **24**, 526 (1967).
4. R.P.Huebener, Magnetic Flux Structures in Superconductors (Springer-Verlag, Berlin Heidelberg New York, 1919).
5. D.Bishop, Nature **382**, 760 (1996).
6. F.London and H.London, Proc.Roy.Soc. (London) A **149**, 71 (1935).
7. W.Meissner and R.Ochsenfeld, Naturwiss. **21**, 787 (1933).
8. A.I.Larkin, Zh.Eksp.Teor.Fiz. **58**, 1466 (1970) (Sov. Phys.-JETP **31**, 784 (1970)).
9. A.A.Abrikosov, Zh.Eksp.Teor.Fiz.**32**,1442 (1957) (Sov.Phys.-JETP **5**, 1174 (1957)).
10. A.S.Fetter and P.C.Hohenberg, Phys.Rev. **147**, 140 (1966); **159**, 330 (1967); E.Cohen and A.Schmid, J.Low Temp.Phys. **17**, 331 (1974).
11. W.H.Kleiner, L.M.Roth, and S.H.Autler, Phys. Rev. A **133**, 1226 (1964).
12. K.Maki and H.Takayama, Prog. Theor. Phys. **46**, 1651 (1971).
13. K.Maki and R.S.Thompson, Physica C 162-164, 275 (1989).

14. D.R.Nelson, Phys.Rev.Lett. **60**, 1973 (1988); D.R.Nelson and H.S.Seung, Phys.Rev.
    B **39**, 9153 (1989); A.Houghton, R.A.Pelcovits, and A.Sudbo, Phys.Rev. B **40**,
    6763 (1989); E.H.Brandt, Phys.Rev.Lett. **63**, 1106 (1989); Physica C **165-166**, 1129
    (1990); Physica B **169**, 91 (1991); S.Sengupta et al., Phys.Rev.Lett. **67**, 3444 (1991);
    G.I.Menon and C.Dasgupta, Phys.Rev.Lett. **73**, 1023 (1994).
15. F.Lindemann, Phys. Z. (Leipzig) **11**, 69 (1910).
16. B.A.Hubermann and S.Doniach, Phys.Rev.Lett. **43**, 950 (1979); D.S.Fisher, Phys.
    Rev. B **22**, 1190 (1980); M.V.Feigel'man, V.B.Geshkenbein, and A.I.Larkin, Phys-
    ica C **167**,177 (1990); H.Ma and S.T.Chui, Phys.Rev.Lett.**67**, 505(1991); **68**,
    2528(1992).
17. G.Blatter, M.V.Feigel'man, V.B.Geshkenbein, A.I.Larkin, and V.M.Vinokur, Rev.
    Mod. Phys. **66**, 1125 (1994).
18. Z.Tesanovic, Phys. Rev. B **44**, 12635 (1991).
19. G.J.Ruggeri and D.J.Thouless, J. Phys. F **6**, 2063 (1976)
20. E.Brezin, A.Fujita, and S.Hikami, Phys. Rev. Lett. **65**, 1949 (1990).
21. Z.Tesanovic and L.Xing, Phys.Rev.Lett. **67**, 2729 (1991); Y.Kato and N.Nagaosa,
    Phys.Rev. B **47**, 2932 (1993); Phys.Rev. B **48**, 7383 (1993); J.Hu and A.H. MacDon-
    ald, Phys.Rev.Lett **71**, 432 (1993); J.Hu and A.H. MacDonald, Phys.Rev. B **52**, 1286
    (1995); R.Sasik and D.Stroud, Phys.Rev.Lett **72**, 2462 (1994); Phys.Rev.Lett **75**,
    2582 (1995); Phys.Rev. B **48**, 9938 (1993); Phys.Rev. B **49**, 16074 (1994); Phys.Rev.
    B **52**, 3696 (1995).
22. M.A.Moore, Phys.Rev. B **45**, 7336 (1992); N.Wilkin and M.A.Moore, Phys.Rev.
    B **48**, 3464 (1993); J.A.O'Neill and M.A.Moore, Phys.Rev.Lett. **69**, 2582 (1992);
    J.A.O'Neill and M.A.Moore, Phys.Rev. B **48**, 374 (1993); H.H.Lee and M.A.Moore,
    Phys.Rev. B **49**, 9240 (1994).
23. R.Ikeda, J. Phys. Soc. Jpn. **65**, 3998 (1996); R.Sasik, D.Stroud and Z.Tesanovic,
    Phys. Rev. B **51**, 3041 (1995).
24. V.A.Marchenko and A.V.Nikulov, Pisma Zh.Eksp.Teor.Fiz. **34**, 19 (1981) (JETP
    Lett. **34**, 17 (1981)).
25. W.K.Kwok et al., Phys.Rev.Lett. **64**, 966 (1990); H.Safar et al., Phys.Rev.Lett.
    69, 824 (1992); W.K.Kwok et al., Phys.Rev.Lett. **69**, 3370 (1992); W.Jiang et al.,
    Phys.Rev.Lett. **74**, 1438 (1995).
26. A.V.Nikulov, Supercond. Sci. Technol. **3**, 377 (1990).
27. A.V.Nikulov, in *Fluctuation Phenomena in High Temperature Superconductors* (Ed.
    M. Ausloos and A.A. Varlamov) (Kluwer,Dordrecht,1997) p.271.
28. A.V.Nikulov, Thesis, Institute of Solid State Physics, Chernogolovka, 1985.
29. A.V.Nikulov, D.Yu.Remisov, and V.A.Oboznov, Phys.Rev.Lett. **75**, 2586 (1995).
30. A.V.Nikulov, S.V.Dubonos, and Y.I.Koval, J.Low Temp.Phys. **109**, 643 (1997)
31. D.A.Huse and S.N.Majumdar, Phys.Rev.Lett. **71**, 2473 (1993); Chung-Yu Mou,
    R.Wortis, A.T.Dorsey, and D.A.Huse, Phys.Rev.B **51**, 6575 (1995).
32. M.Tinkham, *Introduction to Superconductivity* (McGraw-Hill, New York, 1975)
33. P.A.Lee and S.R.Shenoy, Phys. Rev. Lett. **28**, 1025 (1972).
34. A.Schilling et al., Nature **382**, 791 (1996).
35. M.H.Theunissen and P.H.Kes, Phys.Rev.B **55**, 15183 (1997).
36. J.A.Fendrich et al., Phys. Rev. Lett. **74**, 1210 (1995).
37. V.A.Marchenko and A.V. Nikulov, Zh. Eksp. Teor.Fiz. **80**, 745 (1981) (Sov.Phys.-
    JETP **53**, 377 (1981)).
38. Y.B.Kim, C.F.Hempsted, and A.R.Strnad, Phys. Rev. **131**, 2486 (1963); Phys. Rev.
    139, A1163 (1965).
39. L.P.Gor'kov and N.B.Kopnin, Usp.Fiz.Nauk **116**, 413 (1975) (Sov.Phys. - Uspeki
    18, 496 (1976)).
40. A.V.Nikulov, Phys.Rev. B **52**, 10429 (1995) .
41. A.V.Nikulov,http://publish.aps.org/ eprint/gateway/epget/aps1998mar20-002.
42. K.Mendelssohn, Proc. Roy. Soc. **152A**, 34 (1935).
43. L.W.Grunberg and L.Gunther, Phys. Lett. A **38**, 463 (1972).

NUMERICAL SIMULATION OF FLUX-CREEP IN PRESENCE OF A PINNING STRENGTH DISTRIBUTION FOR THE WASHBOARD-TYPE POTENTIAL

A.N.LYKOV

P.N.Lebedev Physical Institute, RAS, 117924 Moscow, Russia

Abstract. Anderson theory is modified to explain features of flux creep in high-T_c superconductors. The approach is based on the consideration of an uncorrelated interaction of isolated vortices with a washboard potential in the presence of a normal pinning strength distribution. The method provides the possibility of estimating the main pinning parameters of superconductors.

1. Introduction

Considerable effort has gone into the study of flux creep in high-temperature superconductors, which show many unexpected features. The most interesting experimental results in this field are the following. The first is the logarithmic shape of the current-dependent potential barrier $U(J) \sim log(J_o/J)$ derived from resistive characteristics of $YBa_2Cu_3O_{7-x}$ films [1], where J_o is constant, and confirmed by magnetic relaxation experiments [2]. The second is the scaling behavior of the electric field as a function of current density (the $E - J$ curve), which is accompanied by the collapse of the $logE$-$logJ$ curves into two curves with different signs of curvature [3, 4]. This cannot be explained in the framework of the conventional Anderson theory for flux creep [5], which assumes a thermal activation mechanism and uncorrelated motion of vortex bundles. In this theory, the curves always have positive curvature, and the potential barrier depends linearly on the applied current. The exact nature of these features is still under discussion. All models can partially explain some experimental data, but all have some problems. For example, a model for the phase transition be-

M. Ausloos and S. Kruchinin (eds.), Symmetry and Pairing in Superconductors, 141–150.
© 1999 *Kluwer Academic Publishers. Printed in the Netherlands.*

142

tween vortex liquid and vortex glass states was proposed [3, 4] to explain the behavior of the $E - J$ curves. This approach complicates the study of pinning in superconductors, and cannot explain the logarithmic shape of the current-dependent potential barrier. Both this model and the theory of collective flux creep [6] predict a power-law dependence for the activation barrier $U(J) \sim J^{\alpha}$, where $\alpha < 0$. On the other hand, in spite of the large effort put into attempting to modify the Anderson model [7, 8, 9, 10], this approach also cannot explain these effects. The change of the sign of the curvature of the $\log E$-$\log J$ curves was explained in different models, but the values of the calculated critical exponents significantly differ from the experimental ones. Moreover, these models cannot explain the logarithmic shape of the potential barrier and other results of magnetic studies of flux creep in high-T_c superconductors. In this letter, we have modified Anderson theory in order to explain these features of flux creep in high-T_c superconductors using well-known flux-dynamics equations and traditional ideas of mixed-state theory.

2. MODELING

Following the classical ideas of Anderson theory, our approach is based on a consideration of the uncorrelated interaction of isolated vortices with a modified washboard potential. This approach is possible in the case of a strong pinning potential and a sufficiently weak magnetic field. We restrict our consideration to thin films in a transverse magnetic field. The transport current flows parallel to the grooves of the harmonic pinning potential. Unlike a usual harmonic one-dimensional potential, we assume that the period of every single sinusoidal pinning potential is equal to the coherence length $\xi(T)$, and that their amplitudes U_i are distributed. This is close to the experimental situation, because the smallest transverse length scale that can be resolved by the vortex core is the coherence length. Moreover, the dimension of the most effective pinning center should also equal $\xi(T)$. Each amplitude of the potential can be associated with its own critical current density (J_{ci}) since $F_{pi} \sim \nabla U_i$, where F_{pi} is the pinning force per unit length arising from the interaction of the vortex lattice with the washboard potential. Thus (J_{ci}) is the virtual critical current density of the film with a usual single sinusoidal pinning potential (U_i) in the creep-free case. In our case, we can divide the pinning centers into two types for every transport current density J. These are strong pinning centers, when (J_{ci}) is greater then J, and weak pinning centers, when (J_{ci}) is smaller then J. In the first case, the vortex system enters the flux creep regime. It is obvious that the transport current suppresses (U_i) [11, 12], so that for the sinusoidal pinning potential, its current dependence is:

$$U_i(j) = U_i(0)[(1 - j^2)^{0.5} - j \cos^{-1}(j)], \qquad (1)$$

where $j = J/(J_{ci})$. Thermal activation results in a hopping motion of the vortices, which leads to a electric field. The flux creep gives in the case of a uniform washboard potential (without amplitude distribution) the following expression for the induced electric field:

$$E = B\xi(T)\Omega exp[-\frac{U_i(j)}{k_B T}][1 - \exp(-\frac{\pi U_i(0)J}{k_B T})] \qquad (2)$$

where k_B is Boltzmann's constant, B is the magnetic induction, and Ω is the depinning attempt frequency with which vortices try to escape from the pinning well. At the present Ω is unknown in details. For example, Ω was estimated by Brandt as the characteristic frequency of the thermal fluctuations of a vortex lattice [9]. Another mechanism for vortex excitation in pins is based on the following consideration. Vortices located at strong pinning centers interact with neighboring fast-moving vortices. The flux motion in that part of the superconductor with weak pinning potential creates an oscillating electromagnetic field that excites the pinned vortices. In real superconductors, every vortex interacts with many other vortices, due to their wandering motion[13]. Thus, the motion of even a small number of vortices is of great importance for the excitation of the vortex system. Since $E = Bv$, where v is the vortex velocity, equation (2) leads to the time spent by a vortex in one strong pinning center

$$\tau_{ci} = \Omega^{-1} exp[\frac{U_i(j)}{k_B T}][1 - \exp(-\frac{\pi U_i(0)J}{k_B T})]^{-1}. \qquad (3)$$

For large current or weak pinning centers, when $J > J_{ci}$, the vortex system enters a flux flow regime. The equation of motion of every vortex line in the washboard potential can be written

$$\eta v = F_L - F_{pi}, \qquad (4)$$

where η is the viscous damping coefficient of the flux motion and F_L is the Lorenz driving force corresponding to a uniform transport current density. As usual, we do not take into account the inertial component in this equation, that component which is significantly smaller than the viscous drag force ηv. For a uniform washboard pinning potential, this equation is very similar to that for the time-dependent phase in a resistively shunted Josephson junction [14]. As a result, equation (4) yields an oscillating electric field, and its time average is given by

$$E = \rho_f(J^2 - J_{ci}^2)^{0.5}, \qquad (5)$$

144

where ρ_f is the flux flow resistivity. In this case, the time for a vortex to move over one pinning center with dimensions equal to $\xi(T)$ is given by

$$\tau_{fi} = \xi(T)\eta(J^2 - J_{ci}^2)^{-0.5}/\Phi_0, \tag{6}$$

where Φ_0 is the magnetic flux quantum. The total time τ spent by a vortex in the film is determined by the time spent in the strong pinning centers τ_c and by the time for the viscous flux motion τ_f in the remaining part of the sample. In order to find τ, a normal distribution for the amplitudes of the single sinusoidal pinning potential was assumed,

$$N_i = N_0 exp[-\frac{(U_i(0) - U_0)^2}{2\sigma^2}], \tag{7}$$

where $U_0(T)$ and σ are the distribution parameters chosen to obtain the best agreement with experiment. The value N_0 is defined so that the total number of pinning centers is equal $w/\xi(T)$. As a result, the electrical field can be written $E = Bw/\tau$, where w is the width of the film. However this equation is applicable only to uniform samples in which different fragments of the sample along its length have identical superconducting properties. In real superconductors, e.g.in superconducting films this is not so because there are many spatial inhomogeneities. Thus, first, an electric field arises in the fragment with the smallest critical current, and the size of superconductor in a resistive state increases with increasing the transport current. To take into account this phenomenon, we believe that the fragments or channels for the transport current are distinguished by their parameter U_0. A normal distribution for U_0 was adopted for the calculation. Thus we should sum the electric field of all fragments to find the $E - J$ curves of the sample.

This method for calculating the $E - J$ curves has difficulties in taking into account pinning centers with J_{ci} near J, since equation (6) yields $\tau_{fi} \rightarrow \infty$. That is not realistic, since the maximum time spent at a pinning center is equal to Ω^{-1}, from (6). To overcome this problem we exchanged the τ_f resulting from (6) when $\tau_{fi} > \Omega^{-1}$ with Ω^{-1} in our program. Thus, flux motion is transformed into hopping motion for small electric fields. The pinning strength distribution makes it possible to decrease the calculation error even by using this simple technique, thinking that the time spent by the vortices at these centers is a small fraction the τ. In our case, the additional error in the calculation of the $E - J$ curves does not exceed one percent.

3. RESULTS OF CALCULATIONS AND DISCUSSION

The calculated $E - J$ curves for the sample's parameters reported in [3,4] are plotted in a double logarithmic scale in Fig.1a. If, in accordance with the experiment, we restrict the voltage range $-1 < log_{10} E < 2$, the qualitative agreement between the series of curves presented in [3] and in Fig.1a becomes evident. The best agreement with experiments [3, 4] was achieved for $\Omega = 1.510^9$ Hz. The curvature of the $E - J$ curves changes sign at $T = 77.5K$, which corresponds to the melting point T_g in the model of the liquid-glass vortex state transition. In our case, this is the transition from flux flow to flux creep in the investigated voltage region, and no change of states occurs. Moreover, the model gives a power-law $E - J$ curve for $T = T_g$ in this voltage range. Finally, our model can also explain the scaling of the experimental $E - J$ curves presented in [4], where, in accordance with this work, the scaled resistance $(E|T - T_g|^{\gamma(1-z)}/J)$ is plotted against the scaled current $(J/|T - T_g|^{2\gamma})$ where z and γ are the critical exponents. The collapsed $E - J$ curves calculated using our model are shown in Fig.1b.

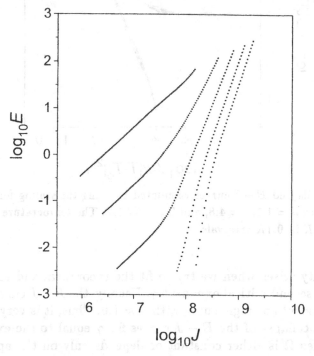

Fig.1a $E - J$ curves calculated with $\Omega = 1.5 * 10^9$ Hz for the sample reported in ref.(3). The temperature ranges from $75.5K$ to $79.5K$ in $2K$ intervals and $B = 4T$.

Note that the model makes it possible to change the slope of an $E -$

146

J curve plotted in a double logarithmic scale at $T = 77.5K$ in a wide range by varying the parameters of the sample and spatial distribution of the pinning potential. As a result, agreement with the experimental slope can be achieved. This makes it possible to achieve equality between the experimental and calculated values of z, which is a linear function of the slope of the $\log E$-$\log J$ curve at T_g. In Fig.1b, $z = 4.8$ in agreement with the experiment [3, 4].

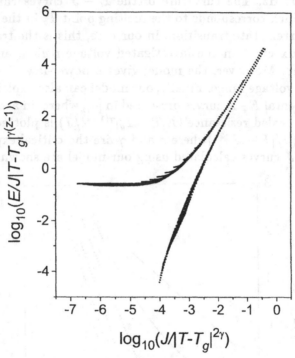

Fig.1b The collapsed $E - J$ curves calculated by using the scaling forms reported in the text with $\nu = 1.1$, $z = 4.8$, and $T_g = 77.5K$. The temperature ranges from $74.5K$ to $79.5K$ in $0.1K$ intervals.

A difficulty arises when we try to fit the theoretical and experimental values of the second critical exponent γ. Usually, the $E - J$ curves from the model collapse into a single curve with $\gamma \approx 1.0$. Thus, it is very difficult to get a precise collapse of the $E - J$ curves for γ equal to the experimental value 1.7 when Ω is either constant or depends only on the applied magnetic field and temperature, as in [9]. For example, the best collapse of the $E - J$ curves calculated with parameters typical for $YBa_2Cu_3O_{7-x}$ films occurs with $\gamma = 1.1$, as shown in Fig.1b. Similar collapsed $E - J$ curves were also obtained for Brandt's dependence of $\Omega(T, B)$. While γ values that were approximately equal to 1 were found in some experiments [15, 16, 17] the problem of increasing γ remains. The agreement can be achieved taking into account the excitation of the vortices located at strong pinning centers by neighboring fast-moving vortices. In this case, the characteristic or maximum frequency of the excitation,i.e. the depinning attempt frequency, is proportional to the transport current in agreement with equation (4):

$$\Omega = \Phi_0 J/\eta\xi. \tag{8}$$

This equation offers the possibility to take into account the mutual interaction of the vortices in the model. The $E - J$ curves and collapsed curves calculated using this relation for Ω are plotted in Fig.2(a,b). In this case, we believed $U_0(T) = U_0(0) * (1 - (T/T_c)^2)(1 - (T/T_c)^4)^{0.5}$, $U_0/k_B=8000$, and $\sigma = 0.5U_0$. The associated parameter J_{c0} of the J_{ci} distribution is assumed to have the following temperature dependence $J_{c0}(T) = J_{c0}(0)(1 - (T/T_c)^2)^{1.5}$, where $J_{c0}(0) = 5 * 10^9 A/m^2$. Evidently, the quality of our collapse is good enough. Here, in agreement with experiments [3, 4], $z = 4.8$ and $\gamma = 1.7$. Thus, the model explains not only the scaling behavior of the $E - J$ curves but also the $\log E$-$\log J$ collapse.

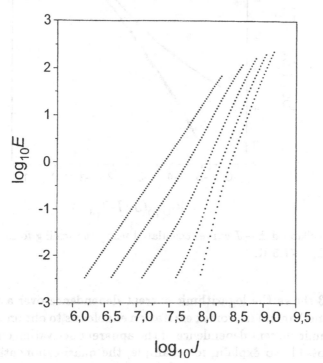

Fig.2a The $E - J$ curves calculated for the same temperatures as for the curves in Fig. 1(a) using the current dependence of Ω

148

On the other hand, the influence of $\Omega(J)$ on the $E - J$ curves is equivalent at small current to a logarithmic current dependence of the activation energy in models with constant Ω, since $(J_0/J)exp[-U_0/k_BT] = exp[ln(J_0/J) - (U_0/k_BT)]$. This is supported by our calculations. Figure 3 shows the current dependence of the apparent activation energy derived from the model. This dependence was obtained using the following method. First, $E(J)$ was calculated for a current J using relation (8) for the attempt frequency. Then by calculating the $E - J$ curves with a constant attempt frequency, a new U_0 was selected to give an $E(J)$ equal to that first obtained. In other words, the $E - J$ curves, calculated using relation (8), are analyzed in the usual framework of the flux-creep equations with constant Ω [1, 2, 18].

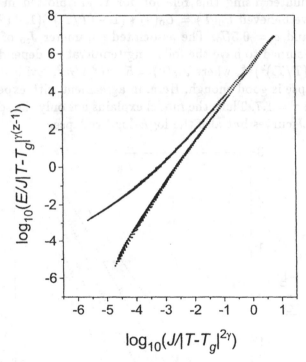

Fig.2b The collapsed $E - J$ curves calculated using the scaling forms with $\nu=1.7$, $z=4.8$, and $T_g =77.5$ K.

Figure 3 shows the logarithmic current dependence over a wide range. Thus, this mechanism for vortex excitation enables us to obtain the origin of the logarithmic current dependence of the apparent activation energy, which makes it possible to explain, for example, the quasi-exponential behavior of the measured $J_c(T)$ dependence [18, 19]. The role of this mechanism for the excitation of pinned vortices reduces with decreasing transport current, since the vortices spend more time at the pinning center and do not move. This increases the role of vortex lattice vibration. At some J, the first mechanism is replaced by the second, and $U_0(J)$ will approach a constant at small transport currents. A similar current dependence for the activation energy was found in $YBa_2Cu_3O_{7-x}$ films [20] giving unambiguous support for this model.

Using this approach, it is possible to explain other dependences, for example, the decrease of T_g with increasing B [3] or with decreasing film thickness [21], which cannot be explained by a model with a phase transition between vortex liquid and vortex glass states.

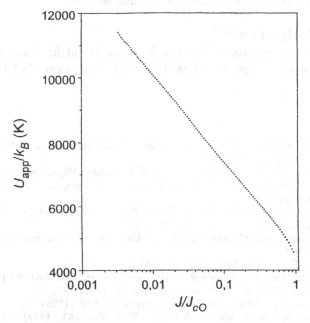

Fig.3 Logarithmic current dependence of the apparent activation energy obtained using our calculations.

In our model, T_g depends on U_0. The larger U_0, the larger T_g, since the temperature interval where the role of the hopping motion is large in comparison with the flux flow motion increases. As a result, the transition from flux flow to flux creep should occur at a higher temperature in some electric field window. For the most effective pinning centers, such as crystal grain boundaries, twin planes, dislocations, and intergrowth of second phases displaced perpendicular to the film surface, the pinning potential is proportional to the film thickness for thin films in a perpendicular magnetic

150

field. On the other hand, the pinning energy is suppressed by an external magnetic field. Thus, T_g increases with thickness for thin films and decreases with B. The calculated dependences are similar to the experimental ones.

4. CONCLUSIONS

Thus, both the transport current and magnetic field features of high-T_c superconductors can be explained by thermally activated flux creep of isolated vortices for a modified washboard pinning potential. The key features of the model that resolve controversial results are the pinning strength distribution and the current dependence of the attempt frequency. The method makes it possible to estimate the main pinning parameters of superconductors by fitting the calculated and experimental data.

Acknowledgements

This work was supported by the Russian Scientific Council NSTP on Condensed Matter (grant No 96041), and RFBR (grant No 97-02-17545).

References

1. E.Zeldov, N.M.Amer, G.Koren, A.Gupta, R.J.Gambino, and M.W.McElfresh, *Phys.Rev.Lett.* **62** (1989) 3093.
2. M.P.Maley, J.O.Willis, H.Lessure, and M.E.McHenry, *Phys.Rev. B* **42** (1990) 2639.
3. R.H.Koch, V.Foglietti, W.J.Gallager, G.Koren, A.Gupta, and M.P.A.Fisher, *Phys. Rev.Lett.* **63** (1989) 1151.
4. R.H.Koch, V.Foglietti, and M.P.A.Fisher, *Phys. Rev. Lett.* **64** (1990) 2586.
5. P.W.Anderson, *Phys. Rev. Lett.* **9** (1962) 309.
6. M.V.Feigel'man, V.B.Geshkebein, A.I.Larkin, and V.M.Vinokur, *Phys. Rev. Lett.* **63** (1989) 2303.
7. R.Griessen, *Phys. Rev. Lett.* **64** (1990) 1674.
8. S.N.Coppersmith, M.Inue, and P.B.Littlewood, *Phys. Rev. Lett.* **64** (1990) 2585..
9. E.H.Brandt, *Z.Phys. B - Condensed Matter* **80** (1990) 167.
10. T.Matsushita, T.Tohdoh, and N.Ihara, *Physica C* **259** (1996) 321
11. M.R.Beasley, R.Labush, and W.W.Webb, *Phys. Rev.* **181** (1969) 682.
12. T.Matsushita and E.S.Otabe, *Jpn. J. Appl. Phys.* **31** (1992) L33.
13. D.R.Nelson and H.S.Seung, *Phys. Rev. B* **39** (1989) 9153.
14. P.Martinoli, *Phys. Rev. B* **17** (1978) 1175.
15. Qiang Li, H.J.Wiesmann, M.Suenaga, L.Motowidlow, P.Haldar, *Phys. Rev. B* **50** (1994) 4256.
16. W.Jiang, N.-C.Yeh, CX.-C.Fu, M.Konczykkowski, and F.Holtzberg, *Physica C* **282 - 287** (1997) 1947.
17. W.J.Yeh and Z.Q.Yu, *Physica C* **282- 287** (1997) 2005.
18. A.P.Malozemoff, *Physica C* **185-189** (1991) 264.
19. S.Senoussi, M.Oussena, G.Collin, and I.A.Campbell, *Phys. Rev. B* **37** (1988) 9792.
20. Z.L.Xiao, J.Häring, Ch.Heinzel, and P.Ziemann, *Solid State Commun.* **95** (1995) 153.
21. A.Sawa, H.Yamasaki, Y.Mawatary, H.Obara, M.Umeda, and S.Kosaka. *Physica C* **282 - 287** (1997) 2071.

ELECTRIC FIELD EFFECTS IN HIGH-T_C CUPRATES WITH DIFFERENT BULK AND SURFACE CONDUCTIVITIES

P. KONSIN, B. SORKIN

Institute of Physics, University of Tartu, Riia 142, EE-2400 Tartu, Estonia

AND

M. AUSLOOS

University of Liege, Institute of Physics-B5 SUPRAS, Sart-Tilman, B-4000, Liege, Belgium

Abstract. Dependences of T_c on hole concentration in $YBa_2Cu_3O_{7-y}$ and $Bi_2Sr_2CaCu_2O_{8+x}$ cuprates have been calculated in the transverse electric field $\vec{E}$ in the framework of the two-band model. The electric field effects in cuprates are considered with the use of calculated $T_c(n_h)$. In metal-insulator-superconductor structures the electric field induces changes in the carrier concentration n_h and in the position of chemical potential. The field-induced shifts of T_c depend on the strength and polarity of electric field. For maximum field effect the superconductor has to be away from optimal doping, i.e. maximum T_c, $dT_c/dn_h = 0$, and has to be closer to the maximum of the slope dT_c/dn_h. The asymmetry $|\Delta T_c(\vec{E})| \neq |\Delta T_c(-\vec{E})|$ changes with $\vec{E}$. In agreement with experiment the shifts $|\Delta T_c(\vec{E})|$ decrease when the film thickness increases. The field effects in both systems are similar but in Bi-2212 they are stronger. The theory agrees qualitatively and in some cases quantitatively with experiment. The dependence of the ratio of surface and bulk conductivities on the film thickness (d) allows us to obtain quantitative agreement between the calculated d-dependence of $|\Delta T_c(\vec{E})|$ and the experimental data in Y-123 and Bi-2212.

M. Ausloos and S. Kruchinin (eds.), Symmetry and Pairing in Superconductors, 151–160.

1. Introduction

Since the beginning of the 1960's there has been an ongoing effort to use the transverse static electric fields to modulate the superconducting properties of the films [1]. These activities were extended to the high-T_c cuprates shortly after the discovery of these superconductors [2-9]. For example, the superconducting transition temperature T_c of 5–10 nm thick Y-123 and Bi-2212 films is shifted by several kelvin [2-9]. The effect is large (25–30 K) in a Y-123 film with weak links [4, 5] for applied polarizations $P = \epsilon E \approx 2 \cdot 10^8$ V/cm. Under these conditions the critical current density is changed by ≈30–50% at 4.2 K [2, 8]. Here ϵ designates the dielectric constant of the gate insulator in the metal-insulator-superconductor structures.

Currently it is still being debated which mechanism induces these large electric field effects. Two different mechanisms [7,9-12] have been proposed and it is not yet clear which one, or possibly the combination of the two, leads to the effects observed.

According to the first mechanism (see, e.g., [1, 7, 9, 11, 12]), the electric field interacts with the charge carriers via the Coulomb forces and directly influences their concentration n_h. The altered mobile carrier concentration in turn changes the superconducting transition temperature and order parameter(s) and thus leads to the observed field effects.

The second mechanism [10] is based on an electric-field-induced oxygen rearrangement. As in any electronic device, the temporal response is determined by the RC time constant, the product of the channel resistance and the gate capacitance. The model based on field-induced ordering of oxygen atoms predicts characteristic RC times of the order of minutes. In [13] the measured value of the RC time constant of the device on the base of Y-123 film equals 0.5 μs. This result is consistent with the interpretation of the field effects in the high-T_c cuprates as a direct charge transfer by the electric field, but inconsistent with a model based on electric field induced rearrangement of oxygen atoms. One counterargument to second mechanism contends that the oxygen binding energies are too high to be overcome by the suggested processes in this mechanism.

In [7] the electric-field effects in $Bi_2Sr_2CaCu_2O_{8+x}$ and $YBa_2Cu_3O_{7-y}$ films have been measured. It was shown that the field effects attained are rather similar. This provides direct evidence that in Y-123 and Bi-2212 superconductors the field effects are based on direct field-induced changes of carrier concentration n_h which implies that the large electric field effects are a generic property of the high-T_c superconductors.

In this paper we consider the electric field effects in high-temperature superconducting films in metal-insulator-superconductor structures. We shall describe the situation microscopically in the framework of the two-band model [14-17]. We shall calculate electric-field-induced effects in superconducting films connected with the change of the chemical potential and, consequently, with the change of the carrier concentration in the electric field. The electric field effects in Bi-2212 and Y-123 are considered. The comparative analysis of field effects in these model superconductors Bi-2212 and Y-123 are made.

2. Model

The equation determining T_c ($\zeta = -\mu$ is the chemical potential of holes) has the form in the two-band model [14]:

$$\kappa F(\Gamma_1 - \zeta, \Gamma_2 - \zeta) F(\Gamma_3 - \zeta, \Gamma_4 - \zeta) = 1, \tag{1}$$

with

$$F(\Gamma_\sigma - \zeta, \Gamma_{\sigma'} - \zeta) = \int_{\Gamma_\sigma - \zeta}^{\Gamma_{\sigma'} - \zeta} \frac{dE}{E} \tanh \frac{E}{2k_B T_c}, \tag{2}$$

$$\kappa = \frac{1}{4} W^2 \rho_1 \rho_2. \tag{3}$$

It is supposed that the limits of integration in Eq.(2) for the higher band ($\sigma = 1$) are $\{-\Gamma_1, -\Gamma_2\}$ and for the lower one ($\sigma = 2$), $\{-\Gamma_3, -\Gamma_4\}$. The densities of states ρ_σ per spin are supposed to be constant.

We consider the case when $\Gamma_1 = 0$ fixes the top energy of the higher band of the width E_1, $\Gamma_3 = E_0$, where $-E_0$ is the top energy of the lower band and the cut-off energy $-E_c$ determines $\Gamma_2 = \Gamma_4 = E_c$.

The necessary relations between ζ and n_h look like [15] $\zeta = \frac{n_h}{\rho_1} + \frac{1}{2}E_1$, if $\zeta < E_0$; $\zeta = (\rho_1 + \rho_2)^{-1}(n_h + \rho_2 E_0 + (1/2)\rho_1 E_1)$, if $\zeta > E_0$; $\zeta = (\frac{n_h}{\rho_1} + \frac{1}{2}E_1 - \frac{\rho_2}{\rho_1}k_B T \ln 2)$, if $\zeta = E_0$. Here n_h is treated as the number of holes per cell, which are added by doping to CuO_2 planes in $YBa_2Cu_3O_{7-y}$ and $Bi_2Sr_2CaCu_2O_{8+x}$.

Using Eq.(1) the dependences of $T_c(n_h)$ for pure Bi-2212 and impurity doped Bi-2212 are calculated [9]. We use the following densities of states $\rho_1 = 1.0$ $(eV)^{-1}$, $\rho_2 = 2.2$ $(eV)^{-1}$, the width of the higher band $E_1 = 4$ eV and fitting parameters $W = 0.22$ eV, $E_c = 2.33$ eV, $E_0 = 2.18$ eV for clean Bi-2212 and $\rho_1 = 1.0$ $(eV)^{-1}$, $\rho_2 = 2.2$ $(eV)^{-1}$, $W = 0.19$ eV, $E_c = 2.39$ eV, $E_1 = 4.0$ eV, $E_0 = 2.15$ eV for Bi-2212-Y,Tm and Bi-2212-Na,K.

For $YBa_2Cu_3O_{7-y}$ superconductor we use the parameter values $\rho_1 = 0.9$ $(eV)^{-1}$, $\rho_2 = 2.2$ $(eV)^{-1}$, $W = 0.23$ eV, $E_c = 2.33$ eV, $E_1 = 4.0$ eV and $E_0 = 2.18$ eV. This set of parameters differs slightly from that used in [16]. The calculated dependence of T_c on Δp for Y-123 in comparison with the experimental data [18] is depicted in Fig.1a. As follows from fitting for these cases, the lower band is narrow.

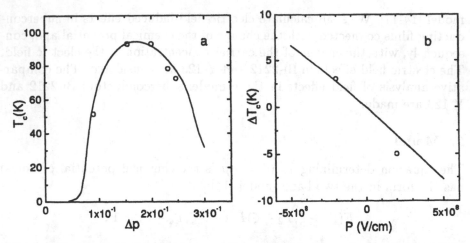

Fig.1a. The theoretical dependence of T_c on the hole concentration Δp for Y-123 (solid line) in comparison with the experiment (circles) [18].

Fig.1b. The theoretical dependence of ΔT_c on P for Y-123 at $d = 100$ nm, $T_c(0) = 31$ K (solid line) in comparison with the experiment (circles) [6].

Electrostatic screening, as described in the Thomas-Fermi model, counteracts the penetration of the electric field into a superconducting film and thus reduces the field effects. The Thomas-Fermi charge screening length equals

$$l_{TF} = \left(\frac{\hbar^2 \pi \epsilon_s}{4 m_e k_F e^2} \right)^{1/2} = \left(\frac{\epsilon_s E_{F0}}{4 \pi^2 e^2 n_0} \right)^{1/2}, \tag{4}$$

where ϵ_s is the dielectric constant of superconductor; E_{F0} and n_0 are the Fermi energy and the carrier concentration in the sample in the absence of the external electric field ($\vec{E} = 0$), respectively, and e is the electron charge. For Y-123 the carrier concentration $n_0 \sim 5 \cdot 10^{21}$ cm^{-3}, $\epsilon_s = 26$ and thus $l_{TF} = 5 \mathring{A}$ [8]. In [8] estimates of $l_{TF} = 5 - 10 \mathring{A}$ have been obtained. In superconducting films of conventional metals $l_{TF} < 1 \mathring{A}$ which leads to relatively small $\Delta T_c(\vec{E})$ in these metals. In high-T_c superconductors the low density of charge carriers is advantageous, leading to relatively large screening lengths. Furthermore, the small coherence lengths of high-T_c superconductors allow the fabrication of ultrathin films in which the total carrier density can be changed to a substantial extent. Correspondingly, the superconducting order parameter(s) can rapidly change (due to the small correlation length) to probe the field-penetrated area. An additional reduction in screening, attained e.g. by using samples in which weak links [4, 5] have been incorporated, can allow one to achieve even larger field effects. If a transverse electric field penetrates into a superconductor, its response to the field will depend sensitively on the value of the parameter $\tau = l_{TF}/\xi_z$, where ξ_z is the coherence length in the field direction.

In our model the electric-field-induced effects are connected with the changes of the carrier concentration in the electric field and as a consequence with the change of the chemical potential $\zeta(\vec{E})$.

In the framework of the Thomas-Fermi approximation it was obtained in [19] for the carrier concentration of superconductor $n_s(z)$ in the field E (the coordinate $\vec{z} \| \vec{E}$)

$$n_s(z) = n_0 \frac{E}{E^*} \exp(-z/l_{TF}) , \qquad (5)$$

where $E^* = e l_{TF} n_0/\epsilon\epsilon_0$, ϵ is the dielectric constant of the gate insulator, $e < 0$, $E = V_g/l$, V_g is the applied voltage and l is the thickness of the gate insulator.

Further we suppose that the total charge induced by the electric field participates in conductivity and the surface and bulk mobilities of carriers are equal. This assumption is more justified in the case $\tau = l_{TF}/\xi_z \geq 1$. Then using Eq.(5), the field induced changes of the carrier concentration Δn averaged over the thickness d of a superconducting film is given by the expression

$$\Delta n = \frac{1}{d} \int_0^d n_s(z) dz = \frac{\epsilon\epsilon_0 E}{ed}(1 - e^{-d/l_{TF}}) . \qquad (6)$$

It is clear that for $d \gg l_{TF}$ $\Delta n = \Delta Q/d$ (see also [11, 12]), where $\Delta Q = e^{-1}\epsilon_0\epsilon E$ is the field induced surface carrier density (for the surface layer).

The expression for the chemical potential $\zeta(\vec{E})$ follow from the equations for the chemical potential for ζ by replacing $n_h \to n_h + n_{\vec{E}}$ with $n_{\vec{E}} = v\,\Delta n$ equal to the number of holes (electrons) per cell of the volume v, i.e. the electric field changes the average carrier concentration n_h in CuO_2 planes (electric field doping). The thickness of superconducting films influences essentially $n_{\vec{E}}$ and the magnitude of the field-induced effects.

Using Eq.(1), we obtain the equation for the superconducting temperature in the electric field by replacing $\zeta = \zeta(0)$ by $\zeta(\vec{E})$,

$$\kappa F(-\zeta(\vec{E}), E_c - \zeta(\vec{E})) F(E_0 - \zeta(\vec{E}), E_c - \zeta(\vec{E})) = 1 . \qquad (7)$$

The shift of the superconducting transition temperature of films in an electric field is defined as $\Delta T_c(\vec{E}) = T_c(\vec{E}) - T_c(0)$. In the positive electric field E the carrier concentration $n_{\vec{E}} < 0$ and, if $(dT_c/dn_h) > 0$, it can be shown that $\Delta T_c(\vec{E}) < 0$. For the negative voltage V_g $n_{\vec{E}} > 0$ and $\Delta T_c(-\vec{E}) > 0$. In the case $(dT_c/dn_h) < 0$, the shifts mentioned are opposite. This behavior correlates with experiment [2, 3, 4, 5, 6, 7, 8, 13] which shows that the electric field effects are indeed the bulk effects.

3. Calculations for $Bi_2Sr_2CaCu_2O_{8+x}$ and $YBa_2Cu_3O_{7-y}$

Next on the base of Eqs.(6),(1),(7) and the expression for $n_{\vec{E}}$ we calculate the dependences of the shifts $\Delta T_c(\vec{E})$ in $Bi_2Sr_2CaCu_2O_{8+x}$ and $YBa_2Cu_3O_{7-y}$ on the applied polarizations $P = \epsilon E$.

Figure 1b shows the dependence of $\Delta T_c(\vec{E})$ on the applied polarization P for relatively thick $YBa_2Cu_3O_{7-y}$ film ($d = 100$ nm). As seen from Fig.1b, the calculated $\Delta T_c(\vec{E})$ in Y-123 agree satisfactorily (practically without fitting parameters) with the experimental results [6] and reveal rather small asymmetry in comparison with the strong asymmetry of the experimental points (at $\pm E$ only two experimental points are available). Notice that $T_c(0) = 31$ K corresponds to the experimental value [6].

Figure 2a shows the dependence $\Delta T_c(\vec{E})$ in Y-123 with the carrier concentration $n_h = 0.135$ and $T_c(0) = 8$ K on the film thickness d for the positive and negative voltages. The shifts $\Delta T_c(\vec{E})$ depend strongly on thickness d because the field-induced carrier concentration (Eq.(6)) decreases with d.

We suppose that $T_c(0)$ does not depend on the film thickness d, i.e. $d \geq 5$ nm [20]. As follows from our theory , the calculated dependences $\Delta T_c(\vec{E})$ on thickness d show similar tendencies as in the experiment [3].

We present the electric field effects not only for Y-123 (Figs.1b,2a,2b) but also for Bi-2212 (Fig.2b). As follows from our calculations, the electric field effects are similar in both compounds supporting the conclusions of Frey et al. [7]. We have also calculated the dependences of $\Delta T_c(\pm\vec{E})$ on the carrier concentration Δp and ΔT_c on film thickness d for Bi-2212. The main difference between the electric field effects in these materials is in larger values of $|\Delta T_c(\pm\vec{E})|$ and $\Delta T_c(d)$ for Bi-2212 than for Y-123.

Figure 2b shows for the positive voltages the dependences of field induced shifts $|\Delta T_c(\vec{E})|$ normalized to charge density $\sigma = \epsilon_0 \epsilon E$ (added to gate electrode), i.e. $|\Delta T_c(\vec{E})|/\sigma$, on the film thickness d for Y-123 and Bi-2212 cuprates. The agreement with the experimental data on ultrathin films by Mannhart et al. [2, 4, 5, 7] and by Xi et al. [3] and for thick film [6] is good. The agreement with the experiment is good especially for ultrathin film with weak links [4, 5] (the larger electric field penetration lengths). The normalized quantities $|\Delta T_c(\vec{E})|/\sigma$ in films with intermediate thickness d is in satisfactory agreement with the experiment.

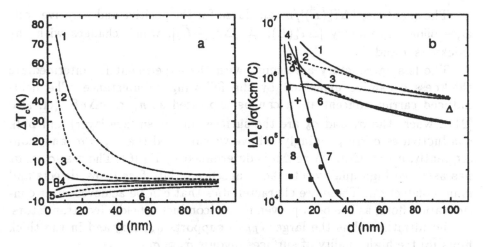

Fig.2a. The theoretical dependences of ΔT_c on thickness d for Y-123 at $T_c(0) = 8$ K; $P = -5 \cdot 10^8$ V/cm (curve 1), $-2 \cdot 10^8$ V/cm (2), $-5 \cdot 10^7$ V/cm (3), $+5 \cdot 10^7$ V/cm (4), $+2 \cdot 10^8$ V/cm (5) and $+5 \cdot 10^8$ V/cm (6) in comparison with the experimental data (circle for $P = -2 \cdot 10^7$ V/cm and square for $P = +2 \cdot 10^8$ V/cm) [3].

Fig.2b. The theoretical dependences of $|\Delta T_c|/\sigma$ on thickness d for pure Bi-2212 at $T_c(0) = 32$ K, $\sigma_s = \sigma_b$ ($P = +5 \cdot 10^7$ V/cm (curve 1), $+2 \cdot 10^8$ V/cm (2) and $+5 \cdot 10^8$ V/cm (3)); for Y-123 at $T_c(0) = 31$ K, $\sigma_s = \sigma_b$ ($P = +5 \cdot 10^7$ V/cm (4), $+2 \cdot 10^8$ V/cm (5) and $+5 \cdot 10^8$ V/cm (6)); for pure Bi-2212, $\sigma_s \neq \sigma_b$ ($P = +2 \cdot 10^8$ V/cm (7)) and for Y-123, $\sigma_s \neq \sigma_b$ ($P = +2 \cdot 10^8$ V/cm (8)) in comparison with the experimental data for pure Bi-2212 (solid circles) [7] and Y-123 (solid squares) [7], star, open circle, cross [3, 5] and open triangle [6].

A good qualitative agreement with the experiment can be achieved introducing the surface and bulk conductivities (see below).

From our calculations it follows that the thickness d of the superconducting films of Y-123 cuprates has to be by 1.5–2 times smaller than in the Bi-2212 cuprates in order to achieve the same effects, in agreement with the experiment with thin films [7]. Stronger electric field effects in Bi-2212 in comparison with Y-123 are connected with the fact that the $n_{\vec{E}}$ is proportional to the unit cell volume v, which is in Bi-2212 2.6 times larger than in Y-123. When $T_c(0)$ is lower (this case is not given in Fig.2b), the agreement with the experiment is better.

In Bi-2212-Y,Tm and Bi-2212-Na,K the electric field effects are smaller than in the pure Bi-2212. This is connected with the fact that the averaged $|dT_c/dn_h|$ decreases with the impurity doping.

158

The quantities $|\Delta T_c(\vec{E})|/\sigma$ calculated for the positive and negative voltages show asymmetry $|\Delta T_c(\vec{E})| \neq |\Delta T_c(-\vec{E})|$, which changes with the thickness d and $\vec{E}$.

The less satisfactory agreement with the experiment for intermediate thickness d can be explained by the following circumstance. The field-induced carrier concentration can be expressed as $n_{\vec{E}} \simeq v\Delta n\frac{\sigma_s}{\sigma_b}$ (cf. also [21]), where the σ_s and σ_b are the surface (in the surface layer) and bulk conductivities correspondingly. Above we supposed that $\sigma_s = \sigma_b$ and, consequently, $n_{\vec{E}} = v\Delta n$, where Δn is determined by Eq.(6). The case $\sigma_s \simeq \sigma_b$ has assumed high quality interfaces, absence of interface traps, defects and ionic conduction. These are characteristics that can be achieved after considerable effort, as has been proven in semiconductor field-effect transistors. In the ultrathin films the larger l_{TF}/d supports $\sigma_s \approx \sigma_b$ and in the thick films [6] the high quality of surfaces favours $\sigma_s \approx \sigma_b$.

We use the formula $\sigma_s/\sigma_b = B(d) = \frac{A}{d^n}$ where $B(d) < 1$ (cf., e.g. [21]). The defects and the inhomogeneities of film surfaces lead to the decrease of σ_s and more rapid change of $T_c(\vec{E})$ on thickness d. The dependence of the ratio of surface and bulk conductivities on the film thickness allows us to obtain quantitative agreement between the calculated d-dependence of $|\Delta T_c(\vec{E})|/\sigma$ and the experimental data in Y-123 and Bi-2212 (see Fig.2b). We use the following fitting parameters in σ_s/σ_b: $A = 8.5$ (nm)^2 and $n = 2$ for Y-123 in field $E = 2 \cdot 10^8$ V/cm and $A = 6.5 \cdot 10^2$ (nm)^3, $n = 3$ for Bi-2212, $E = 2 \cdot 10^8$ V/cm. For the high quality surface $\sigma_s \simeq \sigma_b$ and $A = 1$, $n = 0$. As follows from our analysis, in experiments [7] Y-123 films were higher quality than Bi-2212 ones. The level of the film surface and the sample quality introduces some uncertainty in the experimental results and their reproducibility. However, the experimental data of [2, 3, 4, 5, 6, 7, 8] are in qualitative agreement.

The calculated $|\Delta T_c(\pm\vec{E})|$ for the case $\sigma_s \neq \sigma_b$ are smaller (at $A = 8.5$ (nm)^2 and $n = 2$ for Y-123 by order of magnitude) than electric field shifts of T_c presented in Fig.2a (the case $\sigma_s = \sigma_b$) but the shapes of the curves are retained.

4. Summary

In this paper the electric field effects in cuprates are investigated. The electric field effects themselves are calculated practically without any fitting parameters if $\sigma_s = \sigma_b$. As follows from our analysis, the number of fitting parameters of the theory in which the different surface and bulk conductivities are introduced is equal to 2 (A and n). The other fitting parameters are used to fit the carrier concentration dependence of T_c in Y-123 and Bi-2212 at $\vec{E} = 0$. The agreement of theoretical results with the experiment is good and in some cases satisfactory. Some new electric field effects are predicted: $\Delta T_c(\vec{E})$ dependence on carrier concentration, the saturation effect, etc. The theory strongly supports the mechanism of the influence of the electric field on the superconducting temperature via inducing the change in the carrier concentration directly by Coulomb forces. In nonequilibrium conditions the contribution from the field-induced rearrangement of oxygen in Y-123 into $\Delta T_c(\vec{E})$ is not excluded.

Acknowledgements

We thank N. Kristoffel, A. Sherman and P. Rubin for discussions. This work was partially supported by the Estonian Science Foundation grant No. 1929 and is part of the Action de Recherche Concertée (ARC 94/99-174) of the University of Liége in contract with the Ministry of Higher Education and Research of the Communauté Francaise de Belgique.

References

1. Glover, R.E. and Sherrill, M.D. (1960) Changes in superconducting critical temperature produced by electrostatic charging, *Phys. Rev. Lett.* **5**, 248–250.
2. Mannhart, J., Bednorz, J.G., Müller, K.A., and Schlom, D.G. (1991) Electric field effect on superconducting $YBa_2Cu_3O_{7-y}$ films, *Z. Phys. B* **83**, 307–311.
3. Xi, X.X., Doughty, C., Walkenhorst, A., Kwon, L., Li, Q., and Venkatesan, T. (1992) Effects of field-induced hole-density modulation on normal-state and superconducting transport in $YBa_2Cu_3O_{7-x}$, *Phys. Rev. Lett.* **68**, 1240–1243.
4. Mannhart, J.,Ströbel, J., Bednorz, J.G., and Gerber, Ch. (1993) Large electric field effects in $Y_1Ba_2Cu_3O_{7-\delta}$ films containing weak links, *Appl. Phys. Lett.* **62**, 630–632.
5. Mannhart, J., Bednorz, J.G., Müller, K.A., Schlom, D.G., and Ströbel, J. (1993) Electric field effect in high-T_c superconductors *J. Alloys and Compounds* **195**, 519–526.
6. Taheri, E.H., Cochrane, J.W., and Russell, G.J. (1995) Electric field effects in superconducting $YBa_2Cu_3O_{7-x}$ thin films using field-effect transistor structure, *J. Appl. Phys.* **77**, 761–764.
7. Frey, T., Mannhart, J., Bednorz, J.G., and Williams, E.J. (1995) Mechanism of the electric field effect in the high-T_c cuprates, *Phys. Rev. B* **51**, 3257–3260.
8. Lemanov, V.V. and Kholkin, A.L. (1994) Effect of electric field in superconductors, *Fiz. Tverd. Tela* **36**, 1537–1587.
9. Konsin, P., Sorkin, B., and Ausloos, M. (1997) Electric field effects in high-T_c superconductors, in: Fluctuation Phenomena in High Temperature Superconductors, *Kluwer Series on High Technology*, The Netherlands, 91–100; Konsin, P. and Sorkin, B. (1998) Electric field effects in high-T_c cuprates, *Phys. Rev. B*, in press.
10. Chandrasekhar, N., Valls, O.T., and Goldman, A.M. (1994) Charging effects observed in $YBa_2Cu_3O_{7-x}$ films: Influence of oxygen ordering, *Phys. Rev. B* **49**, 6220–6227.
11. Konsin, P. (1995) Influence of electric field on $La_{2-x}Sr_xCuO_4$ superconducting films, *Physica C* **252**, 183–187.
12. Konsin, P. (1994) Influence of electric field on the properties of high-T_c superconducting films, *Physica C* **235-240**, 1437–1438.
13. Schneider, R. and Auer, R. (1995) Temporal response of a high-T_c superconducting field effect transistor, *Appl. Phys. Lett.* **67**, 2075–2077.

160

14. Konsin, P., Kristoffel, N., and Örd, T. (1988) The interband interaction as a possible cause of high-temperature superconductivity, *Phys. Lett. A* **129**, 339–342.
15. Konsin, P., Kristoffel, N., and Örd, T. (1989) The role of the Fermi level position in high-T_c superconductivity, *Phys. Lett. A* **137**, 420–422.
16. Konsin, P., Kristoffel, N., and Örd, T. (1990) Application of the interband model to the dependence of T_c on oxygen in $Y_1Ba_2Cu_3O_{7-y}$, *Phys. Lett. A* **143**, 83–84.
17. Kristoffel, N., Konsin, P., and Örd, T. (1994) Two-band model for high-temperature superconductivity, *Rivista Nuovo Cimento* **17**, 1–41.
18. Rao, C.N.R. and Ganguli, A.K. (1994) Relation between superconducting properties and structural features of cuprate superconductors, *Physica C* **235–240**, 9–12.
19. Burlachkov, L., Khalfin, I.B., and Shapiro, B.Ya. (1993) Increase of the critical current by an external electric field in high-temperature superconductors, *Phys. Rev. B* **48**, 1156–1159.
20. Li, Q., Xi, X.X., Wu, X.D., Inam, A., Vadlamannati, S., McLean, W.L., Venkatesan, T., Ramesh, R., Hwang, D.M., Martinez, J.A., and Nazar, L. (1990) Interlayer coupling effect in high-T_c superconductors probed by $YBa_2Cu_3O_{7-y}/PrBa_2Cu_3O_{7-x}$, *Phys. Rev. Lett.* **64**, 3086–3089.
21. Tulina, N.A., Emelchenko, G.A., and Kulakov, A.B. (1995) Electric field and current-induced effects on tunnel spectra of $Bi_2Sr_2CaCu_2O_{8+\delta}$ single crystal junctions, *Phys. Lett. A* **204**, 74–78.

THE ORBITAL SYMMETRY OF THE ORDER PARAMETER IN HTSC: THE CONTROVERSY CONTINUES

R. A. KLEMM

Materials Science Division, Argonne National Laboratory
9700 South Cass Ave., Argonne, Illinois 60439 USA
email: richard_klemm@qmgate.anl.gov

1. Introduction

Recently, there have been a large number of experiments purporting to give evidence with regard to the orbital symmetry of the superconducting order parameter in the high temperature superconductors. Generally, the material that has been studied the most is $YBa_2Cu_3O_{7-\delta}$ (YBCO), but a few experiments have been performed on $Tl_2Ba_2CuO_{4+\delta}$ (Tl2201), $Nd_{1.85}Ce_{0.15}CuO_4$ (NCCO), $HgBa_2CuO_{4+\delta}$ (Hg1201), and on $Bi_2Sr_2CaCu_2$ $O_{8+\delta}$ (Bi2212), and on related compounds.

These experiments can be placed into three classes. The first class includes transport and thermodynamic measurements. Angle resolved photoemission spectroscopy (ARPES) makes up the second class. The third and most interesting class involves a variety of Josephson tunneling experiments. Unfortunately, the results of these experiments are inconsistent with one another.

2. Thermodynamic and Transport Measurements

The most important experiments in this class are the non-linear Meissner (or penetration depth) effect, the specific heat, the penetration depth, the magnetoconductivity, the thermal conductivity, the paramagnetic Meissner effect (PME), nuclear magnetic resonance (NMR), and point contact and/or scanning tunneling microscopy (PC/STM). These experiments have led to confusing results, as discussed in the following.

M. Ausloos and S. Kruchinin (eds.), Symmetry and Pairing in Superconductors, 161–172.

2.1. PARAMAGNETIC MEISSNER EFFECT

The first of these to draw widespread attention was the PME, observed in some ceramic samples of Bi2212.[1] In the PME, the field-cooled magnetization below T_c exceeds that for $T > T_c$. In those samples of Bi2212 exhibiting the PME, the width of T_c was found to be particularly large, indicating that they were particularly inhomogeneous. Although ceramic samples with nearly identical texture of YBCO and of other cuprates did not show the effect, some inhomogeneous samples of Nb did show the effect.[2] Thus, we suspected that the PME was a dirt effect.

To test this hypothesis, we set out to examine another property of the PME present in some of the Bi2212 samples which exhibited the PME: the anomalous microwave absorption (AMWA). In most superconductors, the field dependence of the microwave absorption power is U-shaped, exhibiting a minimum at zero applied field. In the AMWA, this field dependence became W-shaped, with a local maximum at zero field. We found one PME-exhibing sample which also exhibited the AMWA,[3] with properties essentially identical to those of inhomogeneous Bi2212.[1] In addition, the PME in Nb is *reversible*, as in inhomogeneous Bi2212.[3, 4] Hence, the origin of the PME in Nb and Bi2212 is almost certainly the same, and unrelated to the OPsymmetry.

2.2. PENETRATION DEPTH

One important finding is that the penetration depth measured in NCCO appears to fit the standard BCS theory quantitatively, exhibiting a very flat T-dependence at low temperatures.[5] However, different results were obtained on untwinned YBCO.[6] In those measurements, it was possible by cutting the sample into sections to measure $\lambda(T)$ in all three directions on the same untwinned single crystal. Along both the a and b directions, $\lambda_{a,b}(T)$ exhibited a linear-in-T low T behavior, suggestive of line nodes of the OP. However, $\lambda_c(T)$ fit the s-wave model involving proximity coupling to normal chains better. [7] Hence, the simultaneous measurement of all three penetration depth components was not completely compatible with either OP class.

2.3. NON-LINEAR MEISSNER EFFECT

When the magnetic field is directed into the ab-plane and the resulting in-plane, transverse magnetization measured as the sample is rotated about the c-axis, any nodes of the purported d-wave variety present should appear in the magnetization non-linear in the field strength, leading to a periodicity of $\pi/2$ as the sample is rotated. In an early experiment on a rectangularly shaped sample, a substantial π-periodicity and a small $\pi/2$-periodicity were found. However, an untwinned disk-shaped sample of YBCO exhibited a greatly reduced π- and no discernible $\pi/2$-periodicity. More recently, the sample was cut into a disk with a laser, an oscillatory field was applied parallel to the static field, and the static penetration depths $\lambda_{a,b,c}(T)$ were measured on the same samples. Although $\lambda_{a,b,c}(T)$ agreed with published data purporting to give evidence for OP nodes,[6] no evidence for any OP nodes in untwinned YBCO was detected in this, the most sensitive of all thermodynamic measurements.[8, 9]

2.4. SPECIFIC HEAT

A fundamental thermodynamic property of a solid is its specific heat C_V. In C_V measurements on single crystals of YBCO grown in $BaZrO_3$ crucibles, the linear temperature dependence of C_V remains at low T, but that in the presence of a strong magnetic field, the measured specific heat appears to scale with $TH^{1/2}$.[12] This scaling behavior is suggestive of line nodes, as is a d-wave OP. However, such behavior was also seen in V_3Si, [13] which might have been due to spin fluctuations. In addition, such behavior has long been known to arise from vortex-vortex interactions, provided that $\lambda(T) \propto T$ at low T.[14] Thus, any mechanism for a linear-in-T $\lambda_{ab}(T)$ can also give rise to the observed $TH^{1/2}$ C_V behavior.

2.5. THERMAL CONDUCTIVITY

Perhaps the least reliable OP symmetry probes are thermal conductivity $\kappa(T)$ measurments. Magnetothermal conductivity experiments on YBCO samples with square and/or rectangular geometries are inherently flawed by exceedingly strong sample shape effects.[8] Such effects were quite possibly observed in magnetothermal rotation experiments on untwinned YBCO. In these experiments, an apparent $\pi/2$-periodicity was observed at high temperature, but this gave way to a low temperature behavior that was π-periodic, in a manner remarkably reminiscent of the behavior in Nb.[10] Another possible explanation could be that the actual $\pi/2$-periodicity observed at high temperatures arises from the $\pi/2$-periodicity present in the normal state electronic structure, which was observed in the in-plane angular dependence of the c-axis normal state magnetoconductivity in Tl2201.[11]

More recently, low-T $\kappa(T)$ measurements of Bi2212 in strong c-axis magnetic fields showed evidence for a phase transition.[15] This phase transition is almost certainly a vortex lattice transition, which is unrelated to the OP symmetry.[16] Also very low-T $\kappa(T)$ measurements of Ni-doped Bi2212 were performed.[17] Curiously, for undoped and rather heavily doped samples, no low-T anomalies were observed, but for two intermediate dopings, $\kappa(T)$ became linear in T below $0.2K$. Adding $H = 200$ G to these samples removed the anomaly. The authors claimed proof of unconventional superconductivity. [17] However, the H-dependence is consistent with a vortex lattice transition.[16] The $H = 0$ behavior of the two intermediate Ni dopings is most likely a dirt effect, since it is completely inconsistent with high-quality Pb/Bi2212 Josephson junctions, proving an s-wave component to the OP is present at 2K.[18]

2.6. POINT CONTACT AND SCANNING TUNNELING MICROSCOPY

Some of the more controversial measurements have been those obtained from PC/STM. These methods nominally measure the density of states, with changes observed above and below T_c normally being associated with the superconducting 'gap' or OP. Unfortunately, the results obtained by PC/STM are often inconsistent. In Hg1201, the densities of states were in excellent agreement with an s-wave OP.[19] In Tl2201, the tunneling conductance appeared d-wave-like,[20] and in YBCO and Bi2212, it varied from s-wave-like to d-wave like, even at different places on the same sample.[21, 22] However, PC/STM measurements also measure the gap associated with a charge-density wave (CDW), as observed in the CDW system $2H$-TaSe$_2$.[23] Thus, it can be confusing to determine the origin of the apparent states in the 'gap', if both superconductivity and CDWs and/or spin-density waves (SDWs) are present.

3. The Pseudogap

Recently, it has been seen by many workers that the underdoped cuprates have a region in their T versus hole doping concentration phase diagram characterized by a 'pseudogap'. In this doping regime, there is a characteristic temperature T^* (in most experiments) or a characteristic energy E_g (in NMR experiments) which distinguishes different types of physical behavior. Above T^* or for energies above E_g, the system behaves as an ordinary quasi-two-dimensionnal metal, with properties exceedingly like those in the normal state of other layered superconductors.[24] In particular the materials $2H$-TaS$_2$ and $2H$-TaSe$_2$ have normal state transport and thermodynamic behavior strikingly similar to the cuprates.[24, 25] However, those materials then exhibit a CDW at lower T values, below which the resistivity and magnetic susceptibility decrease.[25, 26] Such decreases are also observed in the cuprates. Such behaviors are highly suggestive of CDW/SDWs in the cuprates, especially in the underdoped ones.

In any event, the pseudogap is clearly not due to superconducting fluctuations. For example, early reports of magnetic susceptibility measurements in $2H$-TaS$_2$(pyridine)$_{1/2}$ claimed evidence for superconducting fluctuations out to 35 K, or $10T_c$.[27] These results proved later to arise from incomplete destruction of the CDW in $2H$-TaS$_2$ with incomplete intercalation with pyridine.[26] The similarities with the cuprates are indeed striking. However, the magnetic field dependence of the fluctuations in the NMR $1/T_1(T)$ above T_c in YBCO presented strong evidence for an s-wave OP at T_c.[28]

3.1. ARPES

An important new probe of changes in the electronic structure with temperature is ARPES. In the cuprates, ARPES has been used to study the electronic structure of YBCO, NCCO, the two-chain compound YBa$_2$Cu$_4$O$_8$, Bi2201, and especially various dopings of Bi2212. Although there have been a few reports of apparent measurements of a gap on the YBCO chains (which is most likely a CDW, rather than a superconducting gap), the only material in which the consistent measurement of a gap in the electronic spectrum has been reported is Bi2212.[29, 30] Until rather recently, only 'optimally doped' samples were studied. In those samples, the 'gap', or shift in the quasiparticle states at the Fermi energy, was found to be very anisotropic, consistent with that expected for a $d_{x^2-y^2}$-wave state.[29, 30]

More recently, underdoped samples have been studied. In these materials, it is much easier to separate out the various OPs, if indeed more than one is contributing to the observations. We now know that the gap in the spectrum first opens up on the *saddle bands*, which are very flat bands with saddle-like dispersion, centered at the $\overline{M}$ points in the first Brillouin zone.[24, 30] In Bi2212, these saddle bands touch the Fermi surface along an extended line, so the gap formation on the saddle bands affects the shifts in the quasiparticle peaks on the Fermi surface, especially on those parts nearest to the $\overline{M}$ points.

The gap formation on these saddle bands is strikingly similar to that observed recently of the CDW formation on the saddle bands in $2H$-TaSe$_2$.[31] In that material, however, the saddle bands only touch the Fermi surface at isolated points. The CDW forming on the saddle bands had a node at the Fermi surface, so that the interaction of the CDW with the superconducting gap at much lower T is negligible.[23, 31]

3.2. NEUTRON SCATTERING

From neutron scattering on optimally doped Bi2212 and on underdoped YBCO, direct evidence of the CDW/SDW formation below the pseudogap temperature T^* is now available.[32] Above T^*, both the SDW and the CDW are found to be commensurate with the lattice, appearing respectively at (π, π) and $(2\pi, 2\pi)$ in reciprocal space. (π, π) corresponds precisely to the wavevector connecting two nearest $\overline{M}$ points. Below T^*, these wavevectors become *incommensurate* with the lattice, persisting as such into the superconducting state below $T_c < T^*$. However, the CDW/SDW is inherently dynamic in nature, which makes its direct observation with electron diffraction ordinarily difficult. There is one example of an extremely underdoped sample that appears to exhibit a static CDW, however.[33] These dynamic density waves are indistinguishable by ARPES from a static density wave gap, since the time scales of the measurement are so short.

There is now evidence that $T^* > T_c$, even in overdoped materials.[34] In any event, the particular wavevectors involved in these CDW/SDWs form a dynamic gap that mimics a $d_{x^2-y^2}$-wave superconducting gap.[32] Thus, many of the superconducting properties that had been ascribed to d-wave superconductivity may actually be due to the competing density wave OP, which persists to low T.

4. Josephson Junctions

The most interesting, and perhaps the most controversial experiments purporting to give evidence regarding the symmetry of the superconducting OP involve Josephson junctions. These experiments can give evidence regarding the phase of the superconducting OP, rather than just its magnitude, as in other experiments.

4.1. BICRYSTAL YBCO GRAIN BOUNDARIES

The first of these experiments were made by Chaudhari and Lin.[35] These workers studied [001] thin films of YBCO, with regular polygonal grains imbedded in surrounding films with their principal axes rotated by 45° about the c-axis. The critical current I_c was measured from inside the grain to a point far away, and the polygon sides were systematically removed by laser degradation. They found that I_c was directly proportional to the circumference of the polygons. This result has never been understood from a d-wave standpoint, and remains strong evidence for an isotropic superconducting OP in YBCO.

4.2. C-AXIS JUNCTIONS

Direct evidence of a substantial s-wave component to the OP was found in a number of experiments.[36, 37, 38] These experiments involved c-axis Josephson junctions between Pb and YBCO, with about 10-20Å of interstitial Ag to control the oxygen diffusion. Beautiful Fraunhofer diffraction patterns were consistently obtained in a parallel applied field.[37] It was found that $I_c R_n(0)$ values of about 1.5 mV were consistently obtained using un-twinned single crystals, but these numbers decrease with increasing YBCO twinning.[38] While it is impossible to determine if there is any amount of d-wave superconductivity using Pb, these values demonstrate that the s-wave component is large, being at least 30% of the total.[38] Very recently, similar experiments have shown beautiful Fraunhofer patterns with c-axis Pb/Bi2212 Josephson junctions.[18] However, $I_c R_n \approx 1 - 2\mu V$, which is very small. Although there can be many reasons for these small values, the s-wave OP component in Bi2212 is difficult to reconcile with a pre-dominant $d_{x^2-y^2}$ OP, since they are different representations of the crystal symmetry group. However, attempts to observe c-axis Josephson tunneling in NCCO/Pb junctions have so far been unsuccessful.

4.3. SINGLE TWIN JUNCTIONS

More recently, the same groups have performed a very interesting set of experiments involving a c-axis junction straddling a single twin in the YBCO.[39] By rotating the field with respect to the twin direction, a strong dip in the 'Fraunhofer diffraction' pattern was observed, giving strong evidence that the s-wave (the only observable) component of the OP exhibited a phase shift approximating π across the twin. While other, more mundane, explanations are indeed possible, this was taken by those authors to be consistent with the notion that the d-wave component of the OP was larger than the s-wave component, with the total being consistent with the forms $d + s$ and $d - s$ on opposite sides of the twin. However, this interpretation is inconsistent with the non-observation of OP nodes in the transverse magnetization experiments.[8, 9]

4.4. *AB*-PLANE YBCO/PB JUNCTIONS

Two other groups made experiments involving YBCO and Pb Josephson junctions, but both of these were made with the junctions in the less con-trollable ab-plane.[40, 41] By first forming a thick layer (of $\approx 10^3$Å) of Au, and then a Pb junction either on a single edge or straddling a corner, an apparent phase shift of π was obtained.[40] However, the $I_c R_n(0)$ val-ues obtained in these measurements were generally very low, being about $1\mu V$. A strong possible explanation of these experiments is that the sam-ple corners can trap flux, especially flux lying in the ab-plane.[42] Recent superconducting quantum interference device (SQUID) microscope obser-vations of flux trapped in the ab-plane have been published, and confirm this conjecture.[43] The reason the flux likes to lie in the ab-plane is just that the samples are layered, and it costs much less energy for the vortices to have their cores lying between the superconducting layers.[44, 45] With flux trapped at the corners, the junction maps onto those of SNS junctions with a trapped monopole vortex in the center, which gives an apparent phase shift of π in the center of the 'Fraunhofer' diffraction pattern.[42]

Similarly, in the SQUID-like ab-plane junctions made at Maryland, the $I_c R_n$ values obtained with thick Ag junction were comparable to those obtained along the c-axis in comparable YBCO thin film junctions with Pb.[37] In the beautiful first set of experiments, SQUIDs with relative YBCO/Pb junction angles of 0 and 90° were made, and the latter exhibited apparent phase shifts of π, which were robust upon inverting the field and current directions, indicating time-reversal invariance.[41] In addition, a SQUID microscope was used to search and eliminate trapped flux. Such experiments appeared to give strong evidence for an OP consistent with the $d_{x^2-y^2}$ symmetry.

However, in a more detailed second set of experiments with a larger variety of junction angle pairs, it was found that the new results were actually consistent with a p-wave polar state, with angle pairs that differed by 180° being out of phase by π.[41] While it was speculated that the results were actually consistent with the d-wave scenario, it was necessary to eliminate a large fraction of the data in order to come to that conclusion. Apparently, there may be a problem with the Ar milling process that was used in preparing the junctions.[41]

4.5. TRICRYSTAL YBCO GRAIN BOUNDARIES

One of the most fascinating set of experiments involved a tricrystal of three grain boundaries of YBCO (or, in one case, Tl2201) connected at a point.[46] In the absence of an applied field, a SQUID microscope was used to eliminate the presence of trapped flux emanating out of the plane, although there were a number of experiments that appeared to show that trapped flux lying in the ab-plane was often present.[43] In addition, the SQUID microscopes detected flux at the center of the tricrystal junction that integrated to one-half the flux quantum, $\Phi_0/2$. This would be expected if the tricrystal junction contained an odd number of π phase shifts, as for a d-wave OP. Similar results were found in Tl2201,[47] which is also orthorhombic.[48] Taken at face value, these experiments comprise the strongest evidence for d-wave superconductivity.

4.6. MEANDERING GRAIN BOUNDARIES

However, direct observation using a transmission electron microscope of the grain boundaries used in those tricrystal experiments proved that they meandered strongly.[47] Due to the island pattern of thin film growth across a substrate grain boundary, the meandering formed an intricate pattern of rather flat facets, which were directed almost randomly. Recently, it has become possible to improve remarkably upon the quality of the YBCO grain boundaries, by either *in situ* annealing, or even better, by bicrystal seeding.[49] Hence, the unconfirmed tricrystal experiments could be improved upon.

5. Intrinsic Josephson Junctions

There are two types of intrinsic Josephson junctions that have been studied. The first type involve the junctions inside a single crystal. The second type involves breaking a single crystal, and then forming a break junction between the two pieces.

5.1. COHERENT VERSUS INCOHERENT TUNNELING

In the usual Ambegaokar-Baratoff (AB) theory of Josephson tunneling, the propagation is assumed to be *incoherent*, or that the wavevector parallel to the junction is completely unconserved.[50] With coherent tunneling, the wavevector parallel to teh junction is completely preserved. For an s-wave superconductor, it doesn't matter whether the internal c-axis propagation is coherent or incoherent. The main difference is that coherent Josephson tunneling internal to a layered superconductor gives rise to an $I_c R_n(0)$ that is independent of the OP magnitude for small c-axis bandwidth $2J$,[45, 51] whereas for incoherent tunneling, it is proportional to the OP magnitude.[50]

For a d-wave superconductor, incoherent c-axis Josephson tunneling ordinarily does not exist. Only in the unlikely case of d-wave interlayer scattering upon incoherent tunneling can a finite amplitude of c-axis Josephson tunneling be obtained with a d-wave superconductor.[51, 52] If, on the other hand, the tunneling were coherent, then the Josephson tunneling $I_c R_n(0)$ would be independent of the OP magnitude, just as for the s-wave case.[45, 51]

5.2. INTERNAL JOSEPHSON JUNCTION EXPERIMENTS

There have been three groups which have reported such intrinsic, internal Josephson behavior.[53, 54] One group measured the effect in Bi and Tl compounds, notably both underdoped and overdoped Bi2212.[53] The second group measured the effect in underdoped YBCO.[54] Both groups found behavior that was consistent with ordinary AB Josephson tunneling. The '$I_c R_n(0)$' values were on the order of 10 mV in the case of Bi2212 and 1-2 mV in the case of YBCO, and the temperature dependences fit the AB behavior.[53, 54] A third group measured Josephon tunneling across a single intrinsic junction prepared by layer-by-layer molecular beam expitaxy, and found c-axis $I_c R_n$ values of 5-10 mV.[55] These values are too small to have arisen from coherent Josephson tunneling, but they are also way too large to have arisen from incoherent d-wave superconductivity, unless 'R_n' is anomalously large at low bias voltage. Hence, these data comprise strong evidence for an s-wave OP.

5.3. C-AXIS BREAK JUNCTIONS

There have recently been two types of c-axis break junctions involving Bi2212 that have been reported. In the first case, the Bi2212 sample is deliberately cleaved between the BiO layers. One of the resulting pieces is then twisted relative to the other, and then fused back onto the first part at elevated temperature.[56] Then, electrical leads were attached onto each of the parts away from the twist. Because the critical current was so large at low temperatures, it was necessary to apply a strong magnetic field to suppress it. Near to T_c, however, it was possible to obtain data in the absence of an applied magnetic field.[57] By comparing the critical current across one of the single crystal pieces with that across the twisted junction, it was possible to obtain a measure of the effect of the twist angle on the critical current.

If the superconducting OP were $d_{x^2-y^2}$, one would expect the critical current across the twist to vanish for a twist angle of 45°, at least at temperatures near to T_c in zero applied field.[58] At low temperatures and large applied fields, a large number of twist junctions were made, and no systematic variation of the critical current with twist angle was found.[56] At high temperatures near to T_c, only a limited number of twist junctions were measured, but again, no measureable dependence of the critical current upon twist angle was found.[57] Thus, these experiments comprise very strong evidence for s-wave superconductivity.

In the second set of break junction experiments, a point contact tip was pressed onto a Bi2212 sample, breaking it in an uncontrollable fashion.[59] The broken piece could be lifted up, examined, and then placed back down on the sample. The conductance curve was measured, and Josephson tunneling across the break junction was observed. For overdoped samples, $I_c R_n(0)$ was usually about 2 mV, but in two underdoped samples, values of 15 and 25 mV were obtained. The Josephson tunneling set in below T_c, which in the underdoped samples was well below T^*, serving as another strong bit of evidence that the pseudogap regime is not characterized by substantial superconducting fluctuations. In addition, the very large values of $I_c R_n(0)$ obtained in heavily underdoped samples are very difficult to reconcile with d-wave superconductivity. In underdoped samples, the c-

axis conduction is particularly incoherent, and breaking the sample must make the conduction even more incoherent. Thus, these experiments also comprise strong evidence for s-wave superconductivity.

6. Conclusions

A variety of experiments purporting to give evidence regarding the symmetry of the superconducting order parameter in the cuprates have been performed. Unfortunately, these experiments are not all consistent with one another. Some are best explained by d-wave superconductivity. Some are best explained by s-wave superconductivity. Some are better explained by a mixture of the two, with a substantial amount of both. It is not clear that different materials have the same order parameter. For example, NCCO appears from penetration depth measurements to be s-wave, as does Hg1201 from PC/STM, but c-axis Josephson junctions with Pb have so far been unsuccessful, probably due to materials problems. In addition, it is becoming increasingly clear that another type of order parameter may be relevant to these questions: that of a charge/spin-density wave. Such density waves might be absent in Hg1201 and NCCO. Thus, the symmetry of the order parameter in the high temperature superconductors remains unknown.

References

1. W. Braunisch *et al.*, *Phys. Rev. Lett.* **68**, 1908 (1992).
2. D. J. Thompson *et al.*, *Phys. Rev. Lett.* **75**,529 1995); P. Kostić *et al.*, *Phys. Rev.* **B 53**, 791 (1996).
3. A. Anand, R. A. Klemm, H. Claus, B. Veal, and S. V. Bhat, "Anomalous nonresonant microwave absorption in Nb disks", unpublished.
4. L. Pust and L. E. Wenger, unpublished; H. Claus, unpublished.
5. D. H. Wu *et al.*, *Phys. Rev. Lett.* **70**, 85 (1993).
6. D. A. Bonn *et al.*, *J. Phys. Chem. Solids* **56**, 1941 (1995).
7. R. A. Klemm and S. H. Liu, *Phys. Rev. Lett.* **74**, 2343 (1995).
8. J. Buan *et al.*, *Phys. Rev. Lett.* **72**, 2632 (1994); R. A. Klemm *et al.*, *ibid.* **77**, 3058 (1996); A. M. Goldman, private communication.
9. D. Xu, S. Yip, and J. A. Sauls, *Phys. Rev.* **B 51**, 16233 (1995); I. Žutić and O. T. Valls, *Phys. Rev.* **B 54**, 15500 (1996) and unpublished.
10. H. Aubin *et al.*, *Phys. Rev. Lett.* **78**, 2624 (1997).
11. N. E. Hussey *et al.*, *Phys. Rev. Lett.* **76**, 122 (1996).
12. K. A. Moler *et al.*, *Phys. Rev. Lett.* **73**, 2744 (1994).
13. A. P. Ramirez, private communication.
14. A. L. Fetter and P. C. Hohenberg in R. D. Parks, Ed., *Superconductivity* (Dekker, New York, 1969), p 919.
15. K. Krishana *et al.*, *Science* **277**, 83 (1997).
16. D. T. Fuchs *et al.*, *Phys. Rev. Lett.* (to be published) (Cond-mat/9804205).
17. R. Movshovich *et al.*, *Phys. Rev. Lett.* **80**, 1968 (1998).
18. R. Kleiner *et al.*, *Physica* **C 282-287**, 2435 (1997); M. Mößle and R. Kleiner, unpublished.
19. J. Chen *et al.*, *IEEE Trans. Appl. Supercond.* **5**, (2), 1502 (1995).

172

20. L. Ozyuzer *et al.*, *Phys. Rev.* **B 57**, R3245 (1998).
21. H. Murakami and R. Aoki, *J. Phys. Soc. Jpn.* **64**, 1287 (1995).
22. T. Hasegawa *et al.*, *J. Phys. Chem. Solids* **53**, 1643 (1992).
23. T. Kumakura *et al.*, *Proc. of the 21st Int. Conf. on Low Temp. Physics, Czech. J. Phys.* **46** (S5) 2611 (1996).
24. R. A. Klemm, "Origin of the pseudogap in high temperature superconductors", ANL preprint, September 1997.
25. J. A. Wilson, F. J. DiSalvo, and S. Mahajan, *Adv. Phys.* **24**, 117 (1974).
26. F. J. DiSalvo, *Low Temperature Physics-LT13*, K. D. Timmerhaus *et al.*, Eds. (Plenum, New York, 1974), **3**, pp. 417-427.
27. T. H. Geballe *et al.*, *Phys. Rev. Lett.* **27**, 314 (1971).
28. P. Carretta *et al.*, *Phys. Rev.* **B 54**, R9682 (1996).
29. Z.-X. Shen *et al.*, *Phys. Rev. Lett.* **70**, 1553 (1993).
30. H. Ding *et al.*, *Nature* **382**, 51 (1996); H. Ding, private communication.
31. R. Liu, C. G. Olson, and R. F. Frindt, *Phys. Rev. Lett.* (to be published); B. Dardel *et al.*, *J. Phys.: Condens. Matter* **5**, 6111 (1993).
32. H. Mook and B. C. Chakoumakos, (unpublished); P. Dai, H. A. Mook, and F. Doğan, *Phys. Rev. Lett.* **80**, 1738 (1998).
33. G. Yang *et al.*, *Physica* **C 260**, 103 (1996).
34. Ch. Renner *et al.*, *Phys. Rev. Lett.* **80**, 149 (1998).
35. P. Chaudhari and S.-Y. Lin, *Phys. Rev. Lett.* **72**, 1084 (1994).
36. A. G. Sun *et al.*, *Phys. Rev. Lett.* **72**, 2267 (1994); *Phys. Rev.* **B 54**, 6734 (1996).
37. A. S. Katz *et al.*, *Appl. Phys. Lett.* **66**, 1052 (1995).
38. R. Kleiner *et al.*, *Phys. Rev. Lett.* **76**, 2161 (1996).
39. K. A. Kouznetsov *et al.*, *Phys. Rev. Lett.* **79**, 3050 (1997).
40. D. A. Wollman *et al.*, *Phys. Rev. Lett.* **74**, 797 (1995).
41. A. Mathai *et al.*, *Phys. Rev. Lett.* **74**, 4523 (1995), Y. Gim *et al.*, *J. Phys. (France)* **I 6**, 2299 (1996)
42. O. B. Hyun, J. R. Clem, and D. K. Finnemore, *Phys. Rev.* **B 40**, 175 (1989).
43. K. A. Moler *et al.*, *Science* **279**, 1193 (1998).
44. R. A. Klemm, A. Luther, and M. R. Beasley, *Phys. Rev.* **B 12**, 877 (1975).
45. L. N. Bulaevskii, *Zh. Eksp. Teor. Fiz.* **64**, 2241 (1973) [*Sov. Phys. JETP* **37**, 1133 (1973)].
46. C. C. Tsuei *et al.*, *Phys. Rev. Lett.* **73**, 593 (1994), J. R. Kirtley *et al.*, *ibid.* **76**, 1336 (1996); J. R. Kirtley and C. C. Tsuei, *Sci. Am. (Int. Ed.)* **275**(2), 50 (1996).
47. C. C. Tsuei *et al.*, *Science* **271**, 329 (1996).
48. J. L. Wagner *et al.*, *Physica* **C 277**, 170 (1997).
49. D. J. Miller *et al.*, *Appl. Phys. Lett.* **66**, 2561 (1995); private communication.
50. V. Ambegaokar and A. Baratoff, *Phys. Rev. Lett.* **10**, 486 (1963); **11**, 104 (1963).
51. R. A. Klemm, G. Arnold, C. T. Rieck, and K. Scharnberg, "Coherent versus incoherent c-axis Josephson tunneling between layered superconductors", ANL Preprint, April 1998.
52. M. J. Graf *et al.*, *Phys. Rev.* **B 52**, 588 (1995); *ibid.* **B 47**, 12089 (1993).
53. R. Kleiner *et al.*, *Phys. Rev.* **B 50**, 3942 (1994), R. Kleiner and P. Müller, *Phys. Rev.* **B 49**, 1327 (1994).
54. M. Rapp *et al.*, *Phys. Rev. Lett.* **77**, 928 (1996).
55. I. Bozovic and J. N. Eckstein, *Appl. Surf. Science* **113-114**, 189 (1997).
56. Q. Li *et al.*, *Physica* **C 282-287**, 1495 (1997); *IEEE Trans. Appl. Supercond.* **7**, 1584 (1997); Y. Zhu *et al.*, *Microsc. Microanal.* **3**, 423 (1997).
57. Q. Li, private communication.
58. R. A. Klemm, C. T. Rieck, and K. Scharnberg, *Phys. Rev.* **B 58** (to be published).
59. N. Miyakawa *et al.*, *Phys. Rev. Lett.* **80**, 157 (1998).

DYNAMIC AXIAL CHARGE TRANSFER PROCESSES IN $La_{2-x}Sr_xCuO_4$

M. GEORGIEV*+

Institute of Solid State Physics * and
Central Laboratory for Photoprocesses,+
Bulgarian Academy of Sciences,
72 Tzarigradsko Chaussee, 1784 Sofia, Bulgaria

M. DIMITROVA-IVANOVICH AND I. POLYANSKI

Institute of Solid State Physics,
Bulgarian Academy of Sciences,
72 Tzarigradsko Chaussee, 1784 Sofia, Bulgaria

AND

P. PETROVA

Institute for Nuclear Research and Nuclear Energy,
Bulgarian Academy of Sciences,
72 Tzarigradsko Chaussee, 1784 Sofia, Bulgaria

1. Introduction

A long-standing problem of HTSC Physics relates to the coupling between conducting CuO_2 planes. In-plane holes reside mostly in b_1 linear combinations of oxygen $2p_{x,y}$ orbitals which hybridize with copper $3d(x^2 - y^2)$ orbitals giving rise to the main conduction band, though a small percentage occupy the a_1 linear combination which hybridize with copper $3d(3z^2 - r^2)$ orbitals to incite a satelite conduction band. These $3d(3z^2 - r^2)$ copper orbitals split in energy from the copper $3d(x^2 - y^2)$ orbitals as a result of a Jahn-Teller distortion of the CuO_6 octahedron in which the apical Cu(P)-O(A) bond is elongated relative to the in-plane Cu(P)-O(P) bond. The respective $\varepsilon - \Theta$ splitting has been expected to give rise to Jahn- Teller polarons and has historically been the primary reason for the search for a high-T_c superconductivity in $La_{2-x}Ba_xCuO_4$ [1]. Further, the a_1 ligand holes hybridize, by virtue of symmetry, with apex oxygen $2p_z$ orbitals giving

M. Ausloos and S. Kruchinin (eds.), Symmetry and Pairing in Superconductors, 173–186.
© *1999 Kluwer Academic Publishers. Printed in the Netherlands.*

rise to a charge leak along the c-axis which secures the interplane coupling. This $3d_{z^2} - 2p_z$ leak occurs as an electronic charge transfer (CT) believed to be mode-coupled [2].

The temperature and doping dependences of the frequencies of phonon modes coupled to electronic CT have not been understood so far and no simple solutions offered to analyze experiments. In earlier works, we considered $3d_{z^2} - 2p_z$ axial CT's that coupled dynamically to the A_{2u} odd phonon mode in the tetragonal phase of $La_{2-x}Sr_xCuO_4$: The odd-mode coupling was assumed to lead to asymmetric small polarons and bipolarons formed by virtue of vibronic Van der Waals interpolaron binding,[3] while in a subsequent model the odd-mode coupling was assumed to promote carrier pairing through the mediation of highly polarizable off-center O(A) partners. [4] The latter model described the observed $T_c - x$ phase diagrams assuming a dual role for O(A) as both a CT gate and a polarizable pairing partner, but a few important problems such as the doping and temperature dependences of renormalized A_{2u} frequencies were not explained.

It is the purpose of the present study to extend the $3d_{z^2} - 2p_z$ CT model so as to cover temperature and doping dependences. For the former we consider the free energy of a system of nonlinear vibronic oscillators coupled weakly through dipolar interactions. The resulting temperature dependences are obtained numerically by solving the corresponding extremal equations. Doping dependences are introduced on grounds of the accompanying suppression of the Janh-Teller distortion of CuO_6 octahedra. Our results are to be considered against a background of experimental data [5].

2. Hamiltonian

We define the following Pseudo-Jahn-Teller (PJT) Hamiltonian for the events along the c-axis:

$$
\begin{aligned}
H_{PJT} &= t_{pd} \sum_n (a_{pn}^\dagger a_{dn} + a_{dn}^\dagger a_{pn}) + (\varepsilon_{pd}/2) \sum_n (a_{pn}^\dagger a_{pn} - a_{dn}^\dagger a_{dn}) + \\
&\quad \hbar\Omega \sum_n b_n^\dagger b_n + G\sqrt{(\hbar/2NM\Omega)} \sum_n (a_{pn}^\dagger a_{dn} + a_{dn}^\dagger a_{pn})(b_n^\dagger + b_n) \\
&= \sum_{\alpha\beta k} t_{\alpha\beta}(k) a_{\alpha k}^\dagger a_{\beta k} + \sum_{\alpha k} \varepsilon_\alpha(k) a_{\alpha k}^\dagger a_{\alpha k} + \\
&\quad \sum_q \hbar\Omega_q b_q^\dagger b_q + \sum_{k\alpha\beta q} \sqrt{(\hbar/2NM\Omega_q)} G_q(k) a_{\alpha k}^\dagger a_{\beta k}(b_q^\dagger + b_{-q}) \quad (1)
\end{aligned}
$$

where the subscripts p and d stand for the $2p_z$ and $3d_{z^2}$ orbitals, respectively, n is the site label, t_{pd} is the hopping energy, ε_{pd} is the $2p_z$ to $3d_{z^2}$ energy levels gap, Ω is the bare A_{2u}-mode frequency and G is the electron-mode coupling constant: $a_{ni}^\dagger$ (a_{ni}) for i=d,p are the creation (annihilation) operators for hole carriers upon the respective orbitals. The energy reference level is set in the middle between ε_d and ε_p and we introduce $\varepsilon_{pd} = |\varepsilon_d - \varepsilon_p|$. The first equation gives H_{PJT} in a site representation (creation operators $a_{in}^\dagger$), the second one does so in the band representation (creation operators $a_{\alpha k}^\dagger$). The electron-phonon mixing term is of the band off-diagonal type concomitant with the experimental evidence for double-well potentials sensed by apex oxygens in $La_{2-x}Sr_xCuO_4$ [6]. Double-well potentials are generated by H_{PJT} for $4E_{JT} > \varepsilon_{pd}$, where $E_{JT} = G^2/2K$ is the Jahn-Teller energy.

3. CT energy gap

We first consider the electronic Hamiltonian in the absence of phonons and the electron-phonon interaction. In a purely ionic picture, the CT gap is ε_{dp}. However, due to the partial covalent bonding through t_{dp}, the first two terms of H_{PJT} diagonalize in a basis by the eigenstates of the triatomic O(A)-Cu(P)-O(A) molecule. The eigenvalue spectrum is [4] :

$$E_{(AB-NB)} = (1/2)\{\varepsilon_{dp} + \sqrt{[\varepsilon_{dp}^2 + 8t_{dp}^2]}\} \qquad (2)$$

$$E_{NB} \equiv \varepsilon_p (= 0) \qquad (3)$$

$$E_{(NB-B)} = (1/2)\{\varepsilon_{dp} - \sqrt{[\varepsilon_{dp}^2 + 8t_{dp}^2]}\} \qquad (4)$$

corresponding to the antibonding, nonbonding, and bonding levels in their decreasing order, while the reference is set at the unperturbed oxygen energy. Another gap appears associated with the molecule: the splitting $E_{(NB-AB)}$ between the non-bonding (NB) and antibonding (AB) levels of the molecular energy spectrum.

4. Carrier hopping along c-axis

4.1. ELECTRON HOPPING

We shall next concentrate on the electron hopping from an O(A) $2p_z$ orbital to an underlying Cu(P) d_{z2} orbital corresponding to the hole transfer the opposite way. This transfer cannot occur spontaneously due to the finite energy gap $\delta_{CT} = |\varepsilon_d - \varepsilon_p| \equiv \varepsilon_{dp}$ and requires an external boost. Apart from the optical intergap excitation, there may be another effective way of promoting the electron transfer in a nondegenerate electronic system, provided its energy levels couple to an appropriate phonon mode Q. The phonon coupling may bring two electronic energies close enough to enable the electron transfer through interlevel tunneling.

We consider the coupling of the O(A)-Cu(P)-O(A) entity to the A_{2u} axial phonon mode in a tetragonal lattice symmetry. During half of a vibrational period, one of the O(A) comes closer to Cu(P) as the other O(A) moves away. These displacements may shift the oxygen $2p_z$ energies from values close to the copper $3d_{z^2}$ energy to values close to what is characteristic of an isolated oxygen. Now electron tunneling may occur from $2p_z$ to $3d_{z^2}$ for $O(A_{(1)})$-$Cu(P)$ during the "first" half of a vibrational period, as it may for $O(A_{(2)})$-$Cu(P)$ during the subsequent "second" part. We refer to the favorable molecular configuration at which $\varepsilon_d(Q) \simeq \varepsilon_p(Q)$ as the "crossover configuration" $Q = Q_c$.

Due to the 'odd parity' of the bond-stretching A_{2u} mode, there always is an apex oxygen $O(A_{(1,2)})$ approaching $Cu(P)$, as the one over the plane alternates with the one beneath it. Unlike A_{2u}, both apex oxygens $O(A_{(1)})$ and $O(A_{(2)})$ in the 'even parity' A_{1g} breathing mode either approach $Cu(P)$ during a vibrational half-period or go away from it altogether during the next half period. We see that the A_{1g} motion fails to secure any apex oxygen at tunneling distance from the in-plane copper during the "goaway" part, even though supplying two tunneling partners to the copper during the "goto" part. Inasmuch as there is no more than one electron tunneling from copper to oxygen at a time, the A_{2u} coupling seems more efficient than the A_{1g} coupling in promoting the leak electric current along the c-axis.

Once nearly degenerate, the $O(A)$ and $Cu(P)$ levels split by $2\delta t$ due to the tunneling interaction. In the axial configuration as above, $t_{dp} = \delta t$ is the hopping integral within the $Cu(P)$-$O(A)$ pair. The tunneling splitting is defined by $\delta t = <3d_{z^2}|V(\mathbf{r})|2p_z>$.

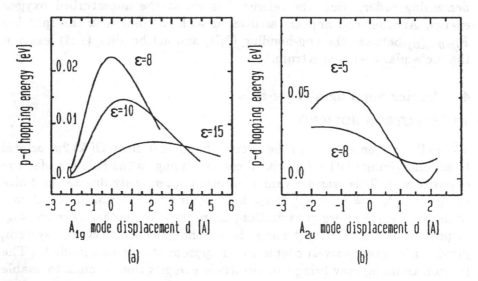

Fig.1 Dependence of the electron hopping energy t_{dp} on the apex oxygen displacement d within the $O(A)$-$Cu(P)$-$O(A)$ triatomic molecule. In (a), the hopping energy is modulated in the A_{1g} vibrational mode, whereas in (b) it is modulated in the A_{2u} vibrational mode [14].

The calculated result obtains through inserting the electron potential $V(\mathbf{r})$ and the wave functions into the integral of δt. We use two oxygen orbitals $|2p_{z'}>$ and $|2p_{z''}>$ and a copper orbital $|3d_{z^2}>$ in a medium with dielectric constant ε, whereas the coordinate notations for our specific geometry are: $r' = \sqrt{[x^2 + y^2 + (z-R)^2]}$, $z' = z-R$, $r'' = \sqrt{[x^2 + y^2 + (z+R)^2]}$, $z'' = z+R$, $r = \sqrt{(x^2 + y^2 + z^2)}$. We calculate the $O^{2-}\text{-}Cu^{3+} \Rightarrow O^{1-}\text{-}Cu^{2+}$ electron hopping energy in an electron-deficient $O^{2-}\text{-}Cu^{2+}\text{-}O^{2-}|$ triatomic molecule whose states are described by hydrogen-like wave functions. We place the origin at the copper ion of charge $Z_{Cu} = +3$ and two oxygens along the z-axis of charges $Z'_O = -1$ and $Z_O'' = -2$ at $\pm R$, respectively, to derive the off-diagonal Coulomb energy of an electron at $\mathbf{r}$:

$$\delta t = (-1/2)(e^2/\varepsilon) < 3d_{z^2}|[(Z'_O/r') + (Z_O''/r'') + (Z_{Cu}/r)](|2p_{z'}> + |2p_{z''}>)$$
(5)

Numerical results for the electron hopping integral $\delta t \equiv t_{pd}$ are presented in Fig.1 for two modes of modulating the O(A)-Cu(P) bond length R, as in the A_{1g} coupling (a) and in the A_{2u} coupling (b). The real ionic charges within an electron-deficient $O(A)^{2+}\text{-}Cu(P)^{3+}\text{-}O(A)^{1+}$ molecule were assumed. (a) and (b) are examples of a nonlocal electron-mode coupling.

4.2. DYNAMIC HOPPING BAND

The axial CT occuring across the O(A) gate, it involves both the Cu(P) $3d_{z^2}$ and the O(A) $2p_z$ orbitals. The nonvanishing CT gap ε_{dp}, hopping energy t_{pd}, and electron-mode coupling strength G make the PJT process a likely signature of the axial transfer rate. In the ionic picture, t_{pd} is vanishing and the CT gap is $\varepsilon_{pd}/2$. At finite t_{pd}, the hybridized gap is E_{NB-AB}. Due to the mode coupling, the CT hopping energy δt transforms into a tunneling splitting t_{CT}, viz. a CT band of magnitude:

$$t_{CT} = < \phi(Q + Q_o) \mid H \mid \phi(Q - Q_o) >$$
(6)

where $\phi(Q)$ are the renormalized ground state harmonic oscillator wavefunctions. Consequently the exact form of t_{CT} is through Gaussian integrands. If Gaussian exponentials are approximated for by delta-functions at $4E_{JT} \gg E_{(NB-AB)}$ ($\mu \equiv E_{(NB-AB)}/4E_{JT} \ll 1$), one derives an equation for the underdoped range where CT is carried by small polarons:

$$t_{CT} \sim (1/2)E_{(NB-AB)}exp(-2E_{JT}/\hbar\Omega)$$
(7)

At $\mu \leq 1$ ($4E_{JT} \geq E_{(NB-AB)}$), $t_{CT} \simeq E_{(AB-NB)}$: CT is by large polarons. The interwell energy barrier $E_B = E_{JT}(1-\mu)^2 = E_{JT}(1-E_{(NB-AB)}/4E_{JT})^2$ is low and in order to have a bound state the following condition must be met: $(1/2)\hbar\Omega_{ren} \leq E_B$. The phonon frequency renormalized by the electron-mode coupling is $\hbar\Omega_{ren} = \Omega\sqrt{[1 - \mu t_{dp}^2]}$, so that the single-bound-state condition reads

$$(1/2)\hbar\Omega\sqrt{(1 - \mu_{max}^2)} \leq E_{JT}(1 - \mu_{max})^2 \tag{8}$$

We get $\mu_{max} \sim 1 - 2^{1/6}(\hbar\Omega/2E_{JT})^{2/3}$.

5. Doping dependence

The chief structural change in the vicinity of a planar hole carrier is converting the relevant CuO_6 octahedron from tetragonal to octahedral symmetry by removing the $3d_{z^2} - 3d_{x^2-y^2}$ JT spliting. The restored 3d degeneracy stimulates a free hole flow to $3d_{z^2}$ orbitals and to its coupled in-plane a_1 ligand. This opens a way to further hole leaks along the c-axis through the dynamic $3d_{z^2} - 2p_z$ CT.

Due to the increasingly suppressed JT-distortion, i.e. the axial elongation of the CuO_6 octahedron, the doping x shrinks the observed Cu(P)-O(A) bond. With x standing for the number of free holes per Cu(P) atom, viz. the probability for finding a hole in a given CuO_6 octahedron, the average Cu(P) bond length is:

$$R(x) = (1 - x)R_T + xR_C = R_T - (R_T - R_C)x \tag{9}$$

where R_T and R_C are the bonds in tetragonal and cubic symmetries, respectively. The bond shrinking tends to level off the lattice parameters along all the three axes resulting in the hypothetical cubic phase. The $R(x)$ dependence results in a doping dependence of all bond dependent quantities, such as the electron hopping t_{pd}, the ionic energy gap ε_{dp}, and the hybridized energy gap $E_{(AB-NB)}$. With ε_{dp} vanishing as x increases, the AB-NB energy gap will be $E_{(AB-NB)}(R_{cross}) \simeq 2^{1/2}t_{dp}(R_{cross})$ near crossover where the electron tunneling occurs. With this assumption, the axial antibonding energy level splits from the nonbonding level by an amount of the order of the hopping energy. From the polaron point of view, as x increases so does the c-axis polaron size, evolving eventually from small polaron at $x \simeq x_{underdoped}$ to large polaron at $x \simeq x_{overdoped}$, provided $4E_{JT} > E_{(NB-AB)}(x)$.

The doping-suppressed JT distortion will also result in a drop of the electron-phonon mixing strength. We set

$$G \equiv G(x) = (1 - x)G_T \tag{10}$$

based on arguments similar to those leading to $R(x)$.

In the limit of an isolated center the vibrational frequency Ω of a vibronic oscillator is renormalized by the electron-mode coupling. The renormalized frequency reads:

$$\Omega_{ren} = \Omega\sqrt{[1 - (E_{(NB-AB)}/4E_{JT})^2]} \qquad (11)$$

This quantity is doping-dependent, as illustrated in Fig.2 based on the x-dependent quantities introduced above.

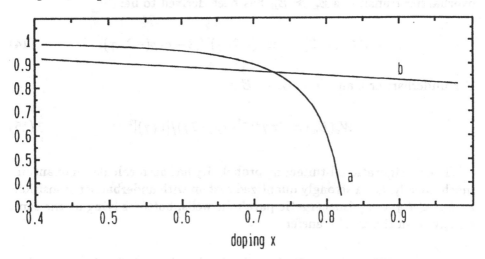

Fig.2 Doping dependence of the renormalized vibrational frequency Ω_{ren}/Ω (curve a), based on a x-dependent Cu(P)-O(A) bond length $R(x)/R_t$ (curve b), originating from a free-hole-suppressed Jahn-Teller distortion of CuO$_6$ octahedra. $R_t = 2.33\mathring{A}$, $E_{JT} = 0.39(1 - x)^2 eV$ [4].

6. Dynamic transfer rate

Our approach to the phonon-coupled transfer rate rests on an occurrence-probability formula accounting for both classical and quantal transitions. Introducing a renormalized-frequency in the harmonic approximation, the transition rate $k_{12}(T)$ reads [7] :

$$k_{12}(T) = (1/\pi)\Omega_{ren}sinh(\hbar\Omega_{ren}/2k_BT)\sum_n W_n(E_n)exp(-E_n/k_BT), \qquad (12)$$

the motion along the transfer path being quantized by n, E_n is the energy.

The transition probability factorizes as $W_n(E_n) = W_{Ln}(E_n)W_{en}(E_n)$ where W_L is the probability for configurational rearrangement and W_e is the probability for a change of the electronic state at the transition configuration $\{Q_C\}$ between the initial NB and final AB electronic states under the energy conservation condition.

Available calculations of the electron-transfer term W_e are based on Landau-Zener's approach. Introducing the parameter

$$\gamma(E_n) = (E_{12}^2/4\hbar\Omega_{ren})E_R^{-1/2}|E_n - E_C|^{-1/2} \tag{13}$$

where E_C is the crossover energy, the electron-transfer term for multiple overbarrier transitions $E_n \gg E_B$ has been derived to be:

$$W_e(E_n) = 2[1 - exp(-2\pi\gamma)]/[2 - exp(-2\pi\gamma)] \tag{14}$$

For underbarrier transitions $E_n \ll E_B$:

$$W_e(E_n) = 2\pi\gamma^{2\gamma-1}exp(-2\gamma)/[\Gamma(\gamma)]^2 \tag{15}$$

The configurational-tunneling probability has been calculated quantum-mechanically. For a strongly quantized system with underbarrier transitions at $E_n \ll E_B$ along isoenergetic parabolic wells, bottoms lying at the same energy as in the axial transfer:

$$W_L(E_n) = \pi[F_{nn}(q_o, q_c)/2^n n!]^2 exp(-E_R/\hbar\Omega_{ren}) \tag{16}$$

where $n = n_1 = n_2$ is the vibronic quantum number in the initial (final) electronic state, $q_o = q_2 - q_1$ is the interwell separation,

$$\begin{aligned} F_{nn}(q_o, q_c) &= q_o H_n(q_c)H_n(q_c - q_o) - 2nH_{n-1}(q_c)H_{n-1}(q_c - q_o) + \\ &\quad 2nH_n(q_c)H_{n-1}(q_c - q_o) \end{aligned} \tag{17}$$

where $H_n(q)$ are Hermite polynomials. For overbarrier transitions at $E_n \gg E_B$ one sets $W_L \leq 1$. $q = (M\Omega_{ren}/\hbar)^{1/2}Q$ is the scaled mode coordinate.

It is noteworthy that a finite zero-point rate of magnitude

$$k_{12}(0) = (\Omega_{ren}/2\pi)W_{eo}(E_o)W_{Lo}(E_o) \tag{18}$$

at $E_o = (1/2)\hbar\Omega_{ren}$ is predicted which is closely related to the small-polaron dynamic transfer band t_{CT}. From the relevant equations for W_{eo} and W_{lo}, we get:

$$t_{CT} = [\hbar\Omega_{ren}\hbar k_{12}(0)]^{1/2} \tag{19}$$

The temperature dependence of the dynamic transfer rate $k_{12}(T)$ is illustrated in Fig.3 for parameters taken from experimental axial resistivity data [8].

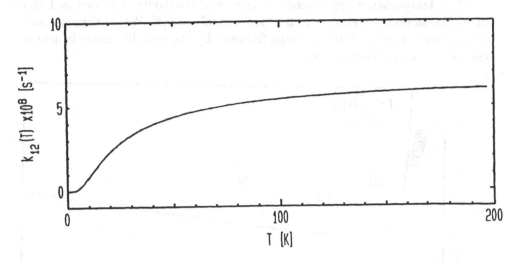

Fig.3 Temperature dependence of the dynamic charge transfer rate $k_{12}(T)$ at parameters implied by the experimental axial resistivity data: $E_B = 3.17 meV$, $\hbar\Omega_{ren} = 263 \mu eV$.

7. Axial resistivity

Given the c-axis interwell relaxation time $t_{dp}(x, T, F) = k_{dp}^{-1}$, where k_{dp} is the Cu(P)-O(A) dynamical transfer rate, the axial dc conductivity is:

$$\sigma = eN(r_A/t_{dp})/F = eN r_A \delta k_{dp}/F \tag{20}$$

so that the resistivity is proportional to $t_{dp}(x, T, F)/r_A(x)$. The temperature dependence of the resistivity is primarily of the semiconducting rather than the metallic type. The thermal activation energy $E_A = E_{JT}(1 - \mu)^2$ drops as x increases because of the increasing $\mu(x) = E_{(AB-NB)}(x)/4E_{JT}(x)$. The feature is consistent with experimental observations [2].

Due to transitions back and forth the resulting transfer rate along the applied electric field F is

$$\begin{aligned} \delta k_{dp} &= k_{dp} exp(+\mathbf{p}.\mathbf{F}/2k_BT) - k_{dp} exp(-\mathbf{p}.\mathbf{F}/2k_BT) \\ &= 2k_{dp} sinh(\mathbf{p}.\mathbf{F}/2k_BT) \end{aligned} \tag{21}$$

Here p is the dipole associated with the anharmonic oscillator. At small $(\mathbf{p}.\mathbf{F})/2k_BT \ll 1$,

$$\delta k_{dp} \sim k_{dp}(\mathbf{p}.\mathbf{F}/k_BT) \tag{22}$$

182

which gives for the single carrier axial mobility:

$$v_{axial} = r_A(x)k_{pd}(p|cos\Theta|/k_BT) = r_A(x)(p/k_BT)k_{pd} \qquad (23)$$

where $|cos\Theta| = 1$ if the external field F is along the c-axis .

The temperature dependence of the axial resistivity is shown in Fig.4 using parameters extracted from experimental data [8]. We see the semiconductor decreasing-resistivity range followed by the metallic range in which the resistivity increases $\propto$ T.

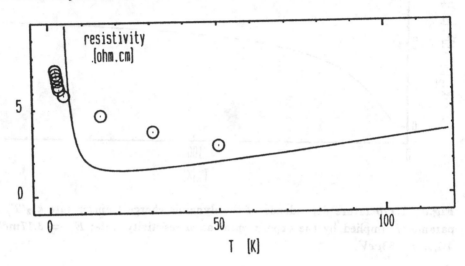

Fig.4 Temperature dependence of the axial resistivity (parameters as in Fig.3). The semiconductor range is seen to be followed by the metallic range as the temperature is increased.

8. Temperature and doping dependences

Equations above assumed temperature-independent vibrational frequencies. In order to derive a temperature dependence, we regard the manifold of off-centered O(A) ions in $La_{2-x}Sr_xCuO_4$ as a statistical ensemble of anharmonic oscillators coupled weakly through pair interactions. The free energy of the system is $\Phi = -k_BTlnZ$. We get using the vibronic eigenvalues $E_{\pm}(Q)$ to account for the effect of off-center displacements [9]:

$$\begin{aligned}\Phi = {}& -k_BTln\{2[1 + cosh([(E_{\alpha\beta}/2)^2 + (GQ)^2]^{1/2}/k_BT)]\} \\ & +(1/2)KQ^2 + U(R,Q)\end{aligned} \qquad (24)$$

where Q is the phonon coordinate, $K = M\Omega^2_{ren,bare}$, $U(R,Q)$ is the pair interaction at R. For a weak dipole-dipole coupling [10]:

$$U(R,Q) \equiv u(R)v(Q) = [R^2(\mathbf{p_1.p_2}) - 3(\mathbf{p_1.R})(\mathbf{p_2.R})]/kR^5 \times$$

$$4G_1Q_1G_2Q_2/\{[(2G_1Q_1)^2 + (E_{\alpha\beta}/2)^2][(2G_2Q_2)^2 + (E_{\alpha\beta}/2)^2]\} \quad (25)$$

G_i and Q_i are correspondingly the electron-mode mixing constants and vibrational coordinates associated with the i-th oscillator. We consider two cases, in-phase vibrations ($Q_1 = Q_2 = Q$) and out-of-phase vibrations ($Q_1 = -Q_2 = Q$), and setting $\kappa = Q_2/Q_1$ we have $\kappa = \pm 1$ and $\kappa = -1$, respectively. We focus onto the dipole-dipole attraction of two out-of-phase vibronic oscillators. With $\kappa = -1$ turning over the sign to u(R), we regard u(R) as positively definite. We seek the extrema of Φ in order to redefine the curvatures and thereby the renormalized vibrational frequency $\Omega_{renI}(T)$.

The renormalized frequency Ω_{ren} is calculated from

$$M\Omega_{ren}^2 = \partial^2\Phi/\partial Q^2|_{Q=<Q>} \quad (26)$$

where $< Q >$ is the average displacement coordinate. It gives

$$\Omega_{ren}(T, x) = \Omega_{bare}f(T, E_{(NB-AB)}(x)) \quad (27)$$

$\Omega_{renI}(T)$ is to compare with experimental phonon frequency data.

Details of the statistical model are presented elsewhere.[11] We now reproduce some of the basic conclusions. The free energy extrema require $\Phi' = 0$. A central maximum at $Q = 0$ yields

$$\Phi''(0) = K\{1 - (4E_{JT}/E_{\alpha\beta})[tanh(E_{\alpha\beta}/4k_BT) - 2u(R)/(E_{\alpha\beta}/2)]\} \quad (28)$$

which vanishes at $T = T_c$, the critical temperature for on-center to off-center conversion,

$$k_BT_c = (E_{\alpha\beta}/4)/tanh^{-1}\{(E_{\alpha\beta}/4E_{JT})[1 + 2u(R)/$$
$$(E_{\alpha\beta}/2) \times (4E_{JT}/E_{\alpha\beta})]\} \quad (29)$$

184

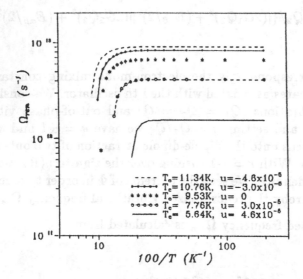

Fig.5 Temperature dependences of the renormalized vibrational frequency $\Omega_{ren}(T)$ in an ensemble of vibronic oscillators coupled by dipole-dipole attraction. $u = u(R)$ is the pair interaction function. The coupled vibrations in a pair are in-phase ($Q_1 = Q_2$) at $u < 0$ and out-of-phase ($Q_1 = -Q_2$) at $u > 0$. The parameters used are: $G = 60 meV/\text{Å}$, $E_{JT} = 1.8 meV$, $E_{\alpha\beta} = 7 meV$, $K = 1eV/\text{Å}^2$. $\mid u \mid = 10^{-5} eV$ corresponds to a pair separation $R \sim 10\text{Å}$ at $\mid p \mid = 0.3 e\text{Å}$.[11]

For the side minima at $Q = Q_0$:

$$1/2 E_{JT} = tanh(\Theta/2)/(\Theta k_B T) - 2u(R)(E_{\alpha\beta}/2)^2/(\Theta k_B T)^4 \quad (30)$$

$$\Phi''(Q_0) = K\{1 - (2E_{JT}/k_B T)[(1/\Theta)tanh(\Theta/2) + (GQ)^2/(\Theta k_B T)^2 \times$$
$$[1/2(cosh(\Theta/2))^2 - (1/\Theta)tanh(\Theta/2)]] - 4u(R)E_{JT}(E_{\alpha\beta}/2)^2 \times$$
$$[3(GQ)^4 + 2(GQ)^2(E_{\alpha\beta}/2)^2 - (E_{\alpha\beta}/2)^4]/(\Theta k_B T)^8\} \quad (31)$$

where for brevity, $\Theta = [((1/2)E_{\alpha\beta})^{1/2} + (GQ)^2]^{1/2}/k_B T$. The former equation obtains from $\mathbf{F'} = 0$ and is solved to give Q(T). Inserting into the latter yields $\Omega_{renI}(T) = \Omega_{bare}[...]^{1/2}$, the terms under the brackets yielding the renormalization factor. Numerical solutions to the foregoing equations for Q(T) and $\Omega_{renI}(T)$ at different pair potentials u(R) are shown in Fig.5.

9. Implication

We derived several essential theoretical dependences that should be easily deciphered and employed by the experimentalists. Among these are the doping dependences of the hybridized energy gap $E_{(AB-NB)}(x)$, the dynamic CT band $t_{CT}(x)$, and the renormalized vibrational frequency $\Omega_{ren}(x)$ as well as the ensemble temperature dependences of the dynamic CT rate $k_{12}(T)$, the axial resistivity $1/\sigma(T)$, and the renormalized vibrational frequency $\Omega_{ren}(T)$. All the above quantities have been subjected to intense experimental investigations over the years.

The CT gap seems to have been identified in NIR spectra and found to vary on doping, possibly as $E_{(AB-NB)}$ [12]. More complex is the problem posed by the vibrational frequencies $\Omega_{ren}(x,T)$ which might be degraded by renormalization down to the acoustic range and might not be easily identified. Yet, investigations over the accessible doping and temperature ranges have been performed on both $La_{2-x}Sr_xCuO_4$ and $YBa_2Cu_3O_{7-\delta}$ [5,13]. Vibrational frequencies have been deduced from IR and Raman spectra and found to soften or harden on doping. Our ensemble prediction is for a softening frequency as the temperature is raised at fixed doping or as doping is increased at fixed temperature. The dynamic CT rate can be derived from the temperature dependence of the axial resistivity. We believe these results demonstrate the capability of our methods.

References

1. J.G. Bednorz and K.A. Muller, Z. Phys. B - Condensed Matter 64 (2) 189 (1986): "Possible high T_c superconductivity in the Ba-La-Cu-O system".

2. P. Nyhus, M.A. Karlow, S.L. Cooper, B.W. Veal, and A.P. Paulikas, Phys. Rev. B 50 (18) 13898 (1994): "Dynamically assisted interlayer hopping in $YBa_2Cu_3O_{6+x}$".

3. M. Georgiev and M. Borissov, Phys. Rev. B 39, 11624 (1989): "Vibronic pairing models for high-T_c superconductors".

4. L. Mihailov, M. Ivanovich, and M. Georgiev, J. Phys. Soc. Japan 62 (7) 2431 (1993): "T_c-x phase diagrams and infrared spectra due to axial charge-transfer modes in $La_{2-x}Sr_xCuO_4$: A composite-boson model."

5. R. Henn, A. Wittlin, M. Cardona, and S. Uchida, Phys. Rev.B 56 (10) 6295 (1997): "Dynamics of the c-polarized infrared active modes in $La_{2-x}Sr_xCuO_4$".

6. D. Haskel, E.A. Stern, D.G. Hinks, A.W. Mitchell, and J.D. Jorgensen, Phys. Rev. B 56 (2) R521 (1997): "Altered Sr environment in $La_{2-x}Sr_xCuO_4$".

7. S.G. Christov, Phys. Rev. B 26 (12) 6918 (1982): "Adiabatic polaron theory of electron hopping in crystals: A reaction-rate approach".

8. Y. Ando, G.S. Boebinger, A. Passner, N.L. Wang, C. Geibel, and F. Steglich, Phys. Rev. Lett. 77 (10) 2065 (1996): "Metallic in-plane and divergent out-of-plane resistivity of a high-T_c cuprate in the zero-temperature limit".

9. I.B. Bersuker, *The Jahn-Teller Effect and Vibronic Interactions in Modern Chemistry* (Academic, New York, 1986). Russian translation: *Effekt Yana-Tellera i Vibronnie Vzaimodeystviya v Sovremennoi Khimii (Nauka, Moskva, 1988)*.

10. G. Baldacchini, U.M. Grassano, A. Scacco, F. Somma, M. Staikova, M. Georgiev, Nuovo Cim. 13D, 1399 (1991). "Revisited vibronic model for Li$^+$ ion and F_A(Li) centre in alkali halides".

11. P. Petrova, M. Ivanovitch, M. Georgiev, M. Mladenova, G Baldacchini, R.M. Montereali, U.M. Grassano, and A. Scacco, in: *Quantum Systems in Chemistry and Physics*, R.McWeeny et al. eds. (Kluwer, Dordrecht, 1997), pp. 373-395: "Revisiting the off-center impurity problem: Reorientational rates of hindered rotators".

12. C.J. Stevens, D. Smith, C. Chen, J.F. Ryan, B. Podobnik, D. Mihajlovic, G.A. Wagner, J.E. Evetts, Phys. Rev. Lett. 78 (11) 2212 (1997): "Evidence for two-component high- temperature superconductivity in the femtosecond optical response of YBa$_2$Cu$_3$O$_{7-\delta}$".

13. J. Schutzmann, S. Tajima, S. Miyamoto, Y. Sato and R. Hauff, Phys. Rev. B (1996): "Doping and temperature dependences of c-axis optical phonons in YBa$_2$Cu$_3$O$_y$ single crystals".

14. F. Mila, Phys. Rev. B 38 (16) 11364 (1988): "Parameters of a Hubbard Hamiltonian to describe superconducting Cu oxides".

COEXISTENCE OF SUPERCONDUCTIVITY WITH CHARGE- OR SPIN- DENSITY WAVES

A. M. GABOVICH AND A. I. VOITENKO

Crystal Physics Department, Institute of Physics, NASU
prospekt Nauki 46, 252650 Kiev, Ukraine

1. Introduction

Once the Bardeen-Cooper-Schrieffer (BCS) seminal theory had been developed and accepted by the superconducting community, it was successfully applied to metals and alloys. However, the validity of the BCS scheme was confirmed fully only for substances (but not all!) with low critical temperatures, T_c's. It is hardly surprising because the weak-coupling theory can be expressed through the reduced quantities in much the same way as the Van der Waals equation of state for real gases can be traced to the equation of corresponding states. Thus, all background physics contained in T_c, energy gap Δ, electronic specific heat, ultrasound attenuation coefficient, etc. is formally hidden in equations to be compared with experiment. The "only" (but extremely important) exception is the electrodynamics, where the microscopic quantities cannot be eliminated neither at the very beginning of the calculations nor from the final results. That is why the "BCS" relationships for reduced quantities do not exist here.

In reality, the so-called strong-coupling deviations from the BCS scheme become conspicuous even for simple metals such as Pb, so that the details of the electron-phonon interaction, originally included into the BCS theory, or any other possible attraction leading to the Cooper pairing (which may be incorporated into the original scheme on equal footing [1]) manifest themselves explicitly. The situation becomes even more entangled for compounds with magnetic ions [2], superconductors with low densities of current carriers [3, 4], layered and quasi-one-dimensional compounds [1] and so on. Such compounds were called "exotic" some time ago [5], and their list expanded enormously in recent years [6], now including the vast majority of superconductors. This expansion seems quite natural since, strictly speaking, all superconductors are "exotic". However, low T_c's in some of them result in a

M. Ausloos and S. Kruchinin (eds.), Symmetry and Pairing in Superconductors, 187–198.

strong reduction of the departure from the BCS simplicity [7]. Nevertheless, the BCS theory and the Cooper pairing concept remain the cornerstone of almost all current sophisticated theories of superconductivity [6, 7, 8, 9].

Among the exotic superconductors there is a very interesting large group involving substances with charge- or spin-density waves (CDW's and SDW's) [1, 4, 5, 9, 10, 11]. Here two kinds of pairings, namely, Cooper pairing and dielectric one, coexisting in the same microscopic cells of the sample, compete for the same Fermi surface (FS) and lead to the emergence of two interplaying order parameters Δ and Σ, respectively [4, 12]. The occurrence of two energy gaps Δ and $|\Sigma|$ makes the methods of tunnel (TS) and point-contact spectroscopies (PCS) or scanning tunnel microscopy (STM) the most suitable and powerful tools to investigate such a class of superconductors. Bearing in mind the Kramers-Kronig relationship between the quasiparticle current J and the nonstationary Josephson current I [13], one may expect the dielectric gap to reveal itself also in the $I(V)$ measurements, where V is voltage. From this point of view it is especially interesting and important to consider the most promising high-T_c oxides where the available tunnel data are somewhat confusing [14, 15, 16, 17, 18, 19]. At the same time, other various experimental methods clearly show that SDW's and CDW's manifest themselves for non-superconducting compositions [6, 8, 9, 10] and often exist as pseudogap phantoms above T_c in superconducting samples [9].

Below we present the theory of tunnel currents in junctions involving superconductors with CDW's and SDW's, founded on the BCS idea applied both to Cooper and to dielectric pairings, with the partial dielectrization (partial gapping) model of Bilbro and McMillan [20] used as the basis of the investigation. However, the original simplicity inherent to the TS of BCS superconductors or Peierls (excitonic) insulators is lost here due to the peculiar interplay between the underlying phenomena. Since the theories for SDW and CDW superconductors have much in common, we introduce also the notation DW to consider them simultaneously. The results describe well the data for high-T_c oxides, NbSe$_3$, heavy-fermion compound URu$_2$Si$_2$ and other compounds.

2. Theory

The model Hamiltonian of the DW superconductor has the form [14, 20, 21, 22]

$$\mathcal{H} = \mathcal{H}_0 + \mathcal{H}_{\text{BCS}} + \mathcal{H}_{\text{DW}}, \tag{1}$$

where $\mathcal{H}_0$ is the free-electron Hamiltonian, $\mathcal{H}_{\text{BCS}}$ is the original BCS Hamiltonian, and

$$\mathcal{H}_{\text{DW}} = -\Sigma \sum_{i=1}^{2} \sum_{\mathbf{p}\alpha} [1 + (2\alpha - 1)\Psi] \, a_{i\mathbf{p}\alpha}^{\dagger} \, a_{i,\mathbf{p}+\mathbf{Q},\alpha} + \text{H.c.} \qquad (2)$$

is the DW Hamiltonian describing the electron-hole pairing. The operator $a_{i\mathbf{p}\alpha}^{\dagger}$ $(a_{i\mathbf{p}\alpha})$ is the creation (annihilation) operator of a quasiparticle with a quasimomentum $\mathbf{p}$ and spin projection $\alpha = \pm\frac{1}{2}$ from the ith FS section. Namely, $i = 1$ and 2 for the nested sections where the electron spectrum is degenerate $\xi_1(\mathbf{p}) = -\xi_2(\mathbf{p} + \mathbf{Q})$, $\mathbf{Q}$ being the DW vector, while $i = 3$ for the rest of the FS where the dispersion relation for elementary excitations is described by the different function $\xi_3(\mathbf{p})$. Parameter $\Psi = 0$ (1) for CDW's (SDW's). The dielectric order parameter Σ emerges on the nested FS sections, so the summation in Eq. (2) is carried out over them only. On the other hand, the *single* superconducting order parameter Δ appears on the whole FS. The ratio

$$\nu = N_{nd}(0)/N_d(0), \qquad (3)$$

where $N_{d(nd)}(0)$ is the density of states for dielectrized (nondielectrized) part of the FS, characterizes the gapping degree of the metal.

The quantity Σ can be taken as a Δ independent phenomenological function of the temperature T [14, 23]. The order parameter Σ is real and can be of either sign [22, 24]. We performed most calculations using the BCS curve for $\Sigma(T)$ because the specific choice is not important from the conceptual point of view. The $\Delta(T)$ dependence for DW superconductors can be easily found from our theory [25, 26, 27, 28], using the function $\Sigma(T)$ and appropriate values of ν.

The normal $G_{ij}^{\alpha\beta}(\mathbf{p}; \omega_n)$ and anomalous $F_{ij}^{\alpha\beta}(\mathbf{p}; \omega_n)$ Matsubara Green's functions (GF) corresponding to the Hamiltonian (1) can be found from the Dyson-Gor'kov equations [11, 24, 26, 27, 28]. To calculate the tunnel currents we need temporal GF $F(\omega)$ and $G(\omega)$ rather than the temperature ones. The equivalence of the dielectrized FS sections 1 and 2, e.g., in the SDW superconductor reduces the number of GF for each electrode. Namely,

$$F_{11}(\omega) = F_{22}(\omega) \equiv F_d(\omega), G_{11}(\omega) = G_{22}(\omega) \equiv G_d(\omega),$$

$$F_{12}(\omega) = F_{21}(\omega) \equiv F_{is}(\omega), G_{12}(\omega) = G_{21}(\omega) \equiv G_{is}(\omega),$$

$$F_{33}(\omega) \equiv F_{nd}(\omega), G_{33}(\omega) \equiv G_{nd}(\omega), \qquad (4)$$

where the subscript nd corresponds to the nondielectrized part 3 of the FS, d to the nested (dielectrized) parts 1 and 2, and is to the intersection excitonic (electron-hole) pairing between a quasiparticle from the part 1 and a **Q**-shifted quasiparticle from the part 2. All other GF vanish. Functions $F_{d,nd,is}(\omega)$ and $G_{d,nd,is}(\omega)$ are generated in the conventional manner [29]. First, the thermal GF are integrated over the momentum **p** and the analytical continuation is made to the real axis of the variable $i\omega_n$. For SDW superconductors the result is

$$G_{nd}^R(\omega) = -\pi\omega s(\Delta), \tilde{F}_{nd}(-i\omega) = \pi\Delta s(\Delta),$$

$$G_d^R(\omega) = -\frac{\pi\omega}{2}\left[s(D_+) + s(D_-)\right], \tilde{F}_d(-i\omega) = \frac{\pi}{2}\left[D_+ s(D_+) + D_- s(D_-)\right],$$

$$G_{is}^R(\omega) = -\frac{\pi}{2}\left[D_+ s(D_+) - D_- s(D_-)\right], \tilde{F}_{is}(-i\omega) = -\frac{\pi\omega}{2}\left[s(D_+) - s(D_-)\right].$$

$$(5)$$

Here $D_\pm = \Delta \pm \Sigma$, $s(D) = 1/\sqrt{D^2 - (\omega + i0)^2}$, the superscript R reflects the retarded character of the GF. Second, the functions (5) are connected to the relevant temporal GF $F(\omega)$ and $G(\omega)$ by the dispersion relations[29].

For CDW superconductor the GF are much more simple [14]. The function $\tilde{F}_{is} \equiv 0$, and other functions involve a single Σ-dependent combined "gap" $D = \sqrt{\Delta^2 + \Sigma^2}$ rather than $D_\pm$.

To calculate the total tunnel current I through the junction we use the conventional tunnel Hamiltonian approach [29], according to which the Hamiltonian has the form:

$$\mathcal{H}_{\text{tun}} = \mathcal{H} + \mathcal{H}' + \mathcal{T}. \tag{6}$$

The left- and right-hand-side electrodes of the junction are described in Eq. (6) by the terms $\mathcal{H}$ and $\mathcal{H}'$, respectively, which coincide with the Hamiltonian (1) with an accuracy of notations. Hereafter primed entities including sub- and superscripts correspond to the right-hand-side of the junction. The tunnel term $\mathcal{T}$ is of the form

$$\mathcal{T} = \sum_{i,i'=1}^{3} \sum_{\mathbf{pq'}\alpha} T_{\mathbf{pq'}}^{ii'} a_{ip\alpha}^\dagger a_{i'q'\alpha} + \text{H.c.}, \tag{7}$$

where $T_{\mathbf{pq'}}^{ii'}$ are the tunnel matrix elements. The general expression for $I(T)$ obtained in the lowest order of the perturbation theory in $\mathcal{T}$ is a sum of functionals depending on temporal GF $F(\mathbf{p}, \tau)$ and $G(\mathbf{p}, \tau)$, where τ denotes time [29]. The GF $F(\mathbf{p}, \tau)$ and $G(\mathbf{p}, \tau)$, integrated over **p** variable, are connected to $F(\omega)$ and $G(\omega)$ by Fourier transformation. Making the assumptions [14] that (i) all matrix elements $T_{\mathbf{pq'}}^{ii'}$ are equal and not influenced by the existence of Δ and Σ, in the spirit of the standard Ambegaokar-Baratoff approach [30], and (ii) the current I is independent of the relative spatial orientation of the junction plane and the DW vector $\mathbf{Q}$, we introduce the universal tunnel resistance R:

$$R^{-1} = 4\pi e^2 N(0) N'(0) \left\langle |\mathsf{T}|^2 \right\rangle_{\mathrm{FS}}. \tag{8}$$

Here e is the elementary charge, the total density of states $N(0) = N_d(0) + N_{nd}(0)$, angular brackets $< \ldots >_{\mathrm{FS}}$ imply averaging over the FS. Then, in the adiabatic approximation $V^{-1}\frac{dV}{d\tau} \ll T$, for the ac bias voltage $V(\tau) \equiv V_{\mathrm{right}}(\tau) - V_{\mathrm{left}}(\tau)$ across the Josephson junction, we obtain the expression for the total current I through the junction made up of the DW superconductors(cf. with Ref. [29]):

$$I[V(\tau)] = \sum_{i=1}^{9} [I_i^1(V)\sin 2\phi + I_i^2(V)\cos 2\phi + J_i(V)], \tag{9}$$

where $\phi = \int^\tau eV(\tau)d\tau$, $I^1 = \sum_{i=1}^9 I_i^1$ is the Josephson current, $I^2 = \sum_{i=1}^9 I_i^2$ is the interference pair-quasiparticle current, and $J = \sum_{i=1}^9 J_i$ is the quasiparticle current. The explicit cumbersome expressions for $I_i^{1,2}$ and J_i are given in Ref. [14].

When obtaining Eq. (9), we assume the strong DW pinning. The phases of the superconducting order parameters are, as usual [30], considered free, with their difference obeying the given above Josephson relationship connecting it to the bias voltage.

3. Results

Below we shall confine ourselves to symmetrical junctions, i.e., when both electrodes are identical DW superconductors. At the same time, the junctions made up of *thermodynamically identical* superconductors ought to be further classified as genuinely symmetrical (*s*) or formally symmetrical with broken symmetry (*bs*). It is convenient to group nine components of each current amplitude in the following manner:

$$I_{(b)s1}^{1,2} = I_1^{1,2}, I_{(b)s2}^{1,2} = I_4^{1,2}, I_{(b)s3}^{1,2} = I_5^{1,2} + I_7^{1,2},$$

$$I_{(b)s4}^{1,2} = I_9^{1,2}, I_{(b)s5}^{1,2} = I_2^{1,2} + I_3^{1,2}, I_{(b)s6}^{1,2} = I_6^{1,2} + I_8^{1,2},$$

$$J_{(b)s1} = J_1, J_{(b)s2} = J_4, J_{(b)s3} = J_5 + J_7,$$

$$J_{(b)s4} = J_9, J_{(b)s5} = J_2 + J_3, J_{(b)s6} = J_6 + J_8. \tag{10}$$

For *s*-junctions $\nu = \nu', \Sigma = \Sigma', \Delta = \Delta'$, and (the notations $F_{ij}'' \equiv F_{i'j'}$ and $G_{ij}'' \equiv G_{i'j'}$ are used)

$$F_{11} = F_{11}' = F_d, G_{11} = G_{11}'' = G_d, F_{33} = F_{33}' = F_{nd}, G_{33} = G_{33}'' = G_{nd},$$
$$\tag{11}$$

$$F_{12} = F'_{12} = F_{ii}, G_{12} = G'_{12} = G_{ii}. \tag{12}$$

Then the initial current amplitude components $i = 2, 3$ and $i = 6, 8$ exactly compensate each other in pairs, and the total current (9) through the junction may be written as

$$
\begin{aligned}
I_s &= \sum_{i=1}^{4} [I^1_{si}(V) \sin 2\phi + I^2_{si}(V) \cos 2\phi + J_{si}(V)] \ for \ SDW \\
&= \sum_{i=1,3,4} [I^1_{si}(V) \sin 2\phi + I^2_{si}(V) \cos 2\phi] + \sum_{i=1}^{4} J_{si}(V) \ for \ CDW. \tag{13}
\end{aligned}
$$

For the current amplitudes $I^{1,2}_{si}$ and J_{si} the usual symmetry relations hold:

$$I^1_{si}(-V) = I^1_{si}(V), I^2_{si}(-V) = -I^2_{si}(V), J_{si}(-V) = -J_{si}(V), i = 1 \ldots 4. \tag{14}$$

Hence, these relations are also valid for total amplitudes. For s-junctions current-voltage characteristics (CVC's) of all three currents do not depend on the sign of Σ.

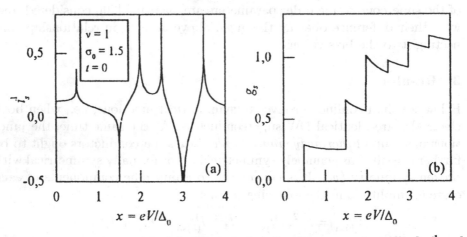

Figure 1. Dependences (a) of the dimensionless Josephson current amplitude i^1_s and (b) of the differential quasiparticle conductivity g_s on the dimensionless voltage x for symmetrical tunnel junction between SDW superconductors.

The CVC for the dimensionless Josephson current amplitude $i^1_s(x) \equiv I^1_s eR/\Delta_0$, where $x \equiv eV/\Delta_0$ and Δ_0 is the superconducting gap in the absence of the dielectrization, as well as the differential quasiparticle conductivity $g_s(x) = dj_s/dx$, where $j_s \equiv J_s eR/\Delta_0$ is the reduced quasiparticle current amplitude, are shown in Fig. 1 for the SDW case. The feature points are seen at voltages $eV = 2\Delta, 2|D_\pm|, |D_+| + |D_-|$, and $|D_\pm| + \Delta$. One can see that the structure of the curves is much more involved than for BCS superconductors. It is also remarkable that the *dielectric* gap influences strongly the CVC for coherent Josephson *supercurrent.*

The situation for junctions involving CDW superconductors is similar but less cumbersome. The respective $i_s^1(x)$ and $g_s(x)$ are depicted in Fig. 2. The feature points clearly seen in the figures correspond to the voltages $eV = 2\Delta, D + \Delta$, and $2D$.

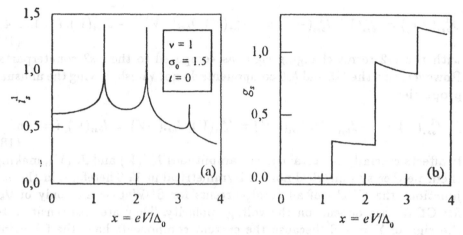

Figure 2. The same as in Fig. 1 but for CDW superconductors.

For formally symmetrical junctions involving identical SDW superconductors an alternative opportunity may be realized. Namely, the *symmetry breaking* can take place, i.e., the left-hand partially-gapped electrode possessing, say, a positive dielectric order parameter $\Sigma > 0$ and the right-hand one having a negative parameter $\Sigma' = -\Sigma < 0$, or vice versa. In both cases the junction is nonsymmetrical in reality, although $|\Sigma| = |\Sigma'|$ and all macroscopical properties of each separated electrode are *identical* due to the thermodynamical equivalence of SDW superconductors with equal Δ's and $|\Sigma|$'s [21, 26]. However, if the junction concerned is a part of the electric circuit, it will serve as a phase-sensitive indicator of the symmetry breaking between the electrodes [15, 31, 32]. Such a phenomenon comprises a new macroscopical manifestation of the symmetry breaking in many-body systems. The corresponding CVC's are substantially different from their genuinely symmetrical counterparts. Really, for this state of a junction $\nu = \nu', \Delta = \Delta', \Sigma' = -\Sigma$, Eqs. (11) hold but

$$F_{12} = -F'_{12} = F_{is}, G_{12} = -G'_{12} = G_{is}. \qquad (15)$$

Then the current amplitude components $i = 2, 3$ and $i = 6, 8$ do not compensate but enhance each other. So,

$$I_{bs} = \sum_{i=1}^{6} [I_{bsi}^1(V) \sin 2\phi + I_{bsi}^2(V) \cos 2\phi + J_{bsi}(V)] \; for \; SDW$$

$$= \sum_{i=1,3,4} [I_{bsi}^1(V) \sin 2\phi + I_{bsi}^2(V) \cos 2\phi] + \sum_{i=1}^{6} J_{bsi}(V) \; for \; CDW.$$

$$(16)$$

The symmetry relations for the bs-current components indexed by $bs\,1$ to $bs4$ are the same as for the s-junction [see Eq. (14)]

$$I_{bsi}^1(-V) = I_{bsi}^1(V), I_{bsi}^2(-V) = -I_{bsi}^2(V), J_{bsi}(-V) = -J_{bsi}(V), i = 1 \ldots 4,$$

$$(17)$$

with the $bs2$ terms changing signs as compared to their $s2$ counterparts. However, now the $bs5$ and $bs6$ components do not vanish, having the unusual properties

$$I_{bsi}^1(-V) = -I_{bsi}^1(V), I_{bsi}^2(-V) = I_{bsi}^2(V), J_{bsi}(-V) = J_{bsi}(V), i = 5, 6.$$

$$(18)$$

It affects crucially the total current amplitudes $I_{bs}^{1,2}(V)$ and $J_{bs}(V)$, making them neither symmetrical nor antisymmetrical in V. Therefore, in the bs-junctions the CVC's of all total currents for SDW case and only of J_{bs} for CDW case depend on the voltage polarity. They are also sensitive to the sign of $\Sigma = -\Sigma'$ because the current components have the following symmetry properties:

$$I_{bs1,2,3,4}^{1,2}(-\Sigma) = I_{bs1,2,3,4}^{1,2}(\Sigma), I_{bs5,6}^{1,2}(-\Sigma) = -I_{bs5,6}^{1,2}(\Sigma),$$

$$J_{bs1,2,3,4}(-\Sigma) = J_{bs1,2,3,4}(\Sigma), J_{bs5,6}(-\Sigma) = -J_{bs5,6}(\Sigma). \qquad (19)$$

With changing Σ sign, the different V-polarity branches are interchanged.

In Fig. 3, the dependences $i_{bs}^1(x) \equiv I_{bs}^1 eR/\Delta_0$ and $g_{bs}(x) = dj_{bs}/dx$ are displayed for SDW superconductors. Here l.h.s. and r.h.s. branches differ substantially due to peculiar compensations and amplifications of logarithmic singularities and jumps originating from various current components.

The plot $j_{bs}(x)$ in the CDW case resembles (with corresponding simplifications) that for the bs-junctions between SDW superconductors and was published elsewhere[15, 31].

The frustrated junction between DW superconductors can be treated as a discrete analog, with respect to the relative phase difference, of the Josephson junction. Nevertheless, it is radically different from the phase-coherent weak link between two Peierls insulators with sliding CDW's considered by Artemenko and Volkov [33]. Unlike these authors, we assume the pinning of the Σ and Σ' phases, therefore ruling out coherent effects.

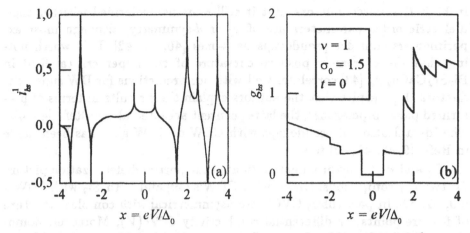

Figure 3. Dependences (a) of the dimensionless Josephson current amplitude i_{bs}^1 and (b) of the differential quasiparticle conductivity g_{bs} on the dimensionless voltage x for tunnel junction with broken symmetry between formally identical SDW superconductors.

Nevertheless, the junction concerned feels the difference or coincidence between the dielectric order parameter signs. Thus, the symmetry breaking in the symmetrical junction serves as a detector of the order-parameter phase multiplicity in electrodes. This is also common to nonsymmetrical junctions [23].

4. Discussion

Prior to comparing our results with experimental data, we should specify partially-dielectrized substances where Cooper and electron-hole pairings coexist. They are $2H\text{-}NbSe_2$, $NbSe_3$, $Tl_2Mo_6Se_6$, $R_5Ir_4Si_{10}$ (R labels various rare-earth elements), $Lu_5Rh_4Si_{10}$, $Li_{0.9}Mo_6O_{17}$, $BaPb_{1-x}Bi_xO_3$, Rb_xWO_3, $(PO_2)_4(WO_3)_{2m}$, Chevrel and Laves phases, A15 compounds for the CDW case (see references. in Refs. [4, 5, 6, 10, 11, 14]), $Cr_{1-x}Re_x$, URu_2Si_2, $(TMTSF)_2X$ ($X = AsF_6$, PF_6, ClO_4), RNi_2B_2C for the SDW case (see references in Refs. [2, 5, 10, 23]). As for high-T_c oxides, we compiled numerous observations of structural anomalies above T_c and indicated [10] as early as in 1992 that the partial dielectrization concept might be valid for them similar to their lower-T_c relatives [4].

The STM observations of CDW's in $YBa_2Cu_3O_{7-x}$ [34], and the pseudo-gap manifestations above T_c in neutron scattering for $La_{2-x}Sr_xCuO_4$ [35] and $YBa_2Cu_3O_{6.6}$ [36], Raman scattering for $Bi_2Sr_2(Ca_{0.62}Y_{0.38})Cu_2O_{8+\delta}$ and $YBa_2Cu_3O_{7-x}$ [37], NMR measurements for $HgBa_2Ca_2Cu_3O_{8+\delta}$ [38] and $YBa_2Cu_4O_8$ [39] seem to be the direct proof of the CDW existence in high-T_c superconductors. But it still remains unclarified whether their dielectric order parameters are of s- or d-symmetry, although most experimenters consider pseudogaps as d-ones [40, 41, 42]. It is worth noting that the extremely positive curvature of the upper critical field in $Bi_2Sr_2CaCu_2O_8$ [43] correlates well with our predictions for DW superconductors [11], despite that the authors explain their results in terms of preformed pairs (bipolarons), the latter concept accepted also in Refs. [44, 45]. The identification of pseudogaps with CDW or SDW gaps has been made in Refs. [9, 46, 47, 48] too.

TS and PCS measurements validated the partial dielectrization picture for two toy-compounds: $NbSe_3$ with CDW's [49] and URu_2Si_2 with SDW's [50, 51, 52]. In particular, CVC's are asymmetrical with complex structure of feature points for differential conductivity $G^{diff}(V)$. Moreover, homocontacts $URu_2Si_2-URu_2Si_2$ reveal the symmetry breaking predicted by us earlier for systems with DW's [32].

Again, the interpretation of TS and PCS results for high-T_c oxides is hampered by the uncertainties of the superconducting order parameter symmetry [9, 16, 17, 19]. Namely, s-type gaps Δ are seen for $HgBa_2CuO_4$ [53, 54] and $Nd_{1.85}Ce_{0.15}CuO_{4-\delta}$ [53, 55], whereas V-like $G^{diff}(V)$ are appropriate to $YBa_2Cu_3O_{7-x}$ [16, 53], $Bi_2Sr_2CaCu_2O_{8+\delta}$ [16, 17, 53, 56] and $Tl_2Ba_2CuO_6$ [19, 53]. On the other hand, our theory has been developed for s-like superconductivity. Thus, it is possible now to make only preliminary comparison with different oxides, although our treatment could be easily modified for the d-case.

Conductance asymmetry in $S-I-N$ structures was observed for $Bi_2Sr_2CaCu_2O_{8+\delta}$ [57, 58]. In $S-I-S$ junctions the peaks of $G^{diff}(V)$ were found [57] for $|eV| \approx 2\Delta$ and dips for $|eV| \approx 3\Delta$. If one considers, in accordance with Refs. [40, 41], $|\Sigma|$ and Δ to be of the same order of magnitude far below T_c, then these features should be attributed to our 2Δ and $\Delta + D$, with singularity at $|eV| \approx 2D$ being smeared. Two gap-like peculiarities were also seen for $HgBa_2Ca_2Cu_3O_{8-x}$ [59].

Applying simple BCS-based theories, to which ours positively belongs, to high-T_c superconductors one should bear in mind the feasibility of more complex "superspin" structure of the order parameters [60]. The relatively simple interplay between DW-driven gap $|\Sigma|$ and superconducting gap Δ assumed here might be replaced then by a $SO(n)$ picture including order parameter component mixing [60] and the emergence of current-density or spin-current-density waves [48]. So far, such a generalization lacks direct experimental evidence and lies beyond the scope of this article.

Acknowledgements

This work was supported, in part, by the Ukrainian State Foundation for Fundamental Researches (Grant No. 2.4/100).

References

1. *Problem of High-Temperature Superconductivity*, edited by V. L. Ginzburg and D. A. Kirzhnitz (Nauka, Moscow, 1977) (in Russian).
2. *Superconductivity in Ternary Compounds II, Superconductivity and Magnetism. Topic in Current Physics*, edited by M. B. Maple and Ø. Fischer (Springer-Verlag, Berlin, 1982).
3. N. M. Builova and V. B. Sandomirskii, Usp. Fiz. Nauk **97**, 119 (1969).
4. A. M. Gabovich and D. P. Moiseev, Usp. Fiz. Nauk **150**, 599 (1986) [Sov. Phys. Usp. **29**, 1135 (1986)].
5. *Superconductivity in Magnetic and Exotic Materials*, edited by T. Matsubara and A. Kotani (Springer, Berlin, 1984).
6. B. Brandow, Phys. Repts. **296**, 1 (1998).
7. A. M. Gabovich and A. I. Voitenko, Physica C **258**, 236 (1996).
8. V. M. Loktev, Fiz. Nizk. Temp. **22**, 3 (1996).
9. R. S. Markiewicz, J. Phys. Chem. Solids **58**, 1179 (1997).
10. A. M. Gabovich, in *High-T_c Superconductivity, Experiment and Theory*, edited by A. S. Davydov and V. M. Loktev (Springer, Berlin, 1992), p. 161.
11. A. M. Gabovich and A. S. Shpigel, Phys. Rev. B **38**, 297 (1988).
12. Yu. V. Kopaev, Tr. Fiz. Inst. Akad. Nauk SSSR **86**, 3 (1975).
13. A. B. Zorin, I. O. Kulik, K. K. Likharev and J. R. Schrieffer, Fiz. Nizk. Temp. **5**, 1139 (1979) [Sov. J. Low Temp. Phys. **5**, 537 (1979)].
14. A. M. Gabovich and A. I. Voitenko, Phys. Rev. B **55**, 1081 (1997).
15. A. M. Gabovich and A. I. Voitenko, J. Phys.: Condens. Matter **9**, 3901 (1997).
16. J. Lesueur, B. Leridon, M. Aprili and X. Grison, in *The Gap Symmetry and Fluctuations in High Temperature Superconductors*, edited by J. Bok and G. Deutscher (Plenum Press, New York, 1998).
17. H. Hancotte *et al.*, Phys. Rev. B **55**, 3410 (1997).
18. C-R. Hu, Phys. Rev. B **57**, 1266 (1998).
19. L. Ozyuzer *et al.*, Phys. Rev. B **57**, 3245 (1998).
20. G. Bilbro and W. L. McMillan, Phys. Rev. B **14**, 1887 (1976).
21. K. Machida, J. Phys. Soc. Jpn. **50**, 2195 (1981).
22. A. M. Gabovich and A. I. Voitenko, Phys. Rev. B **52**, 7437 (1995).
23. A. M. Gabovich and A. I. Voitenko, Phys. Rev. B (1998, to be published).
24. A. M. Gabovich, Fiz. Nizk. Temp. **19**, 641 (1993) [Low Temp. Phys. **19**, 457 (1993)].
25. A. I. Voitenko, A. M. Gabovich and A. S. Shpigel, Fiz. Nizk. Temp. **18**, 108 (1992) [Sov. J. Low Temp. **18**, 75 (1992)].
26. A. M. Gabovich and A. S. Shpigel, J. Phys. F **14**, 1031 (1984).
27. A. M. Gabovich, D. P Moiseev, A. S. Shpigel and A. I. Voitenko, Phys. Status Solidi B **161**, 293 (1990).
28. A. M. Gabovich, A. S. Gerber and A. S. Shpigel, Phys. Status Solidi B **141**, 575 (1987).
29. A. I. Larkin and Yu. N. Ovchinnikov, Zh. Eksp. Teor. Fiz. **51**, 1535 (1966) [Sov. Phys. JETP **24**, 1035 (1966)].
30. A. Barone and G. Paterno, *The Physics and Applications of the Josephson Effect* (Wiley, New York, 1982).
31. A. M. Gabovich and A. I. Voitenko, Europhys. Lett. **38**, 371 (1997).
32. A. M. Gabovich and A. I. Voitenko, Phys. Rev. B **56**, 7785 (1997).
33. S. N. Artemenko and A. F. Volkov, Zh. Eksp. Teor. Fiz. **87**, 691 (1984) [Sov. Phys. JETP **60**, 395 (1984)].
34. H. L. Edwards, A. L. Barr, J. T. Markert and A. L. de Lozanne, Phys. Rev. Lett. **73**, 1154 (1994); H. L. Edwards *et al.*, Phys. Rev. Lett. **75**, 1387 (1995).
35. K. Yamada et al., Phys. Rev. B **57**, 6165 (1998).
36. P. Dai, H. A. Mook and F. Doğan, Phys. Rev. Lett. **80**, 1738 (1998).
37. R. Nemetschek *et al.*, Phys. Rev. Lett. **78**, 4837 (1997).

198

38. M.-H. Julien *et al.*, Phys. Rev. Lett. **76**, 4238 (1996).
39. G. V. M. Williams *et al.*, Phys. Rev. Lett. **80**, 377 (1998).
40. H. Ding *et al.*, Nature **382**, 51 (1996).
41. J. M. Harris *et al.*, Phys. Rev. Lett. **79**, 143 (1997).
42. G. V. M. Williams *et al.*, Phys. Rev. Lett. **78**, 721 (1997).
43. A. S. Alexandrov, V. N. Zavaritsky, W. Y. Liang and P. L. Nevsky, Phys. Rev. Lett. **76**, 983 (1996).
44. J. Maly, K. Levin and D. Z. Liu, Phys. Rev. B **54**, 15657 (1996).
45. V. J. Emery, S. A. Kivelson and O. Zachar, Phys. Rev. B **56**, 6120 (1997).
46. R. A. Klemm, in *Ten Years after Discovery*, Proceedings of the International Workshop on High Temperature Superconductivity, Jaipur, India, edited by S. M. Bose and K. B. Garg (Narosa Publishing House, New Delhi, India, 1997) p.179.
47. T. Dahm, D. Manske and L. Tewordt, Phys. Rev. B **56**, 11419 (1997).
48. R. S. Markiewicz and M. T. Vaughan, Cond.-mat. #9709137; #9802078; R. S. Markiewicz and C. Kusko, Cond.-mat. #9802079.
49. J. P. Sorbier, H. Tortel, P. Monceau and F. Levy, Phys. Rev. Lett. **76**, 676 (1996).
50. A. Nowack *et al.*, Z. Phys. B **88**, 295 (1992).
51. Yu. G. Naidyuk *et al.*, Fiz. Nizk. Temp. **21**, 310 (1995).
52. Yu. G. Naidyuk, K. Gloos and A. A. Menovsky, J. Phys.: Condens. Matter **9**, 6279 (1997).
53. J. Zasadzinski *et al.*, in *Spectroscopic Sudies of High T_c Cuprates*, edited by I. Bozovic and D. van der Marel, SPIE Proceedings (SPIE, Bellingham, 1996), Vol. 2696, p. 338.
54. J. Chen *et al.*, Phys. Rev. B **49**, 3683 (1994).
55. T. Ekino, T. Doukan and H. Fujii, J. Low Temp. Phys. **105** , 563 (1996).
56. M. Itoh, S-I. Karimoto, K. Namekawa and M. Suzuki, Phys. Rev B **55**, 12001 (1997).
57. Y. DeWilde *et al.*, Phys. Rev. Lett. **80**, 153 (1998).
58. Ch. Renner, B. Revaz, J.-Y. Genoud and Ø. Fischer, J. Low Temp. Phys. **105**, 1083 (1996); Ch. Renner *et al.*, Phys. Rev. Lett. **80**, 149 (1998); Ch. Renner *et al.*, Phys. Rev. Lett. **80**, 3606 (1998).
59. G. T. Jeong *et al.*, Phys. Rev. B **49**, 15416 (1994).
60. S-C. Zhang, Science **275**, 1089 (1997).

INSTABILITY OF A 2-DIMENSIONAL LANDAU-FERMI LIQUID DUE TO UMKLAPP SCATTERING

N. FURUKAWA[1] AND T. M. RICE
Theoretische Physik, ETH-Hönggerberg, CH-8093 Zürich, Switzerland

Abstract A model for the breakdown of Landau theory for a 2-dimensional Fermi liquid is proposed drawing on the behavior of ladder systems, where a partial or complete truncation of the Fermi surface can be driven by opening spin and charge gaps in the absence of symmetry breaking. The latter is driven by a divergence of Umklapp scattering. The possibility of similar behavior in two dimensions is examined. For the cuprates, the spin and charge gaps would spread out from the saddle points and reduce the Fermi surface to four disconnected segments, a form which agrees with several recent phenomenological models and recent experiments.

1. Introduction

Although the symmetry of the superconducting state in the high-T_c cuprates is now generally accepted to be unconventional and d-wave and that this symmetry type points towards a magnetic mechanism, a comprehensive theory of these fascinating materials is still an open matter. Such a theory should describe how these materials evolve from a Mott insulating antiferromagnet when doped with holes with strong derivations from Landau theory and of course high temperature superconductivity, to a final nonsuperconducting state with normal metallic characteristics when the hole doping exceeds a critical value typically $\sim 20\%$.

This talk will concentrate on just one aspect of this generic phase diagram, namely the evolution from the usual metallic state as the hole doping is varied through the critical value to the unusual and anomalous metallic state that appears at temperatures $T > T_c$. This instability of the Landau-Fermi liquid would appear to be more than just an instability to superconductivity but rather one that causes a breakdown in the character of the metallic state in a fundamental way as has been emphasized by Anderson. Further this breakdown is triggered by a change in the hole (or electron) density through a critical value.

---

[1]on leave from I.S.S.P., Univ. of Tokyo, 7-22-1 Roppongi, Minato-ku, Tokyo 106, Japan

M. Ausloos and S. Kruchinin (eds.), Symmetry and Pairing in Superconductors, 199–209.

A number of proposals have appeared which ascribe this breakdown to the proximity to some critical point and a transition to a long range ordered state. But these have the difficulty that the only symmetry breaking observed is to the d-wave superconducting state or to long range antiferromagnetic (AF) order which only appears at a much smaller hole density. This suggests that the instability of the Landau-Fermi liquid is not associated with a symmetry breaking and that we should look elsewhere for its origin. In particular the instability could be a precursor to the Mott insulating state that occurs at half-filling. The charge gap of a Mott insulator signals the complete breakdown of the Fermi surface at the stoichiometric electron density. In the scenario where the magnetic interactions are dominant, the Fermi surface will be already partially truncated at finite hole density through the appearance of incommensurate AF order. The partial truncation becomes more complete until finally at half-filling there is a charge gap over the whole Fermi surface. But as we have said in the cuprates, experiments show a different behavior and point towards a partial truncation of the Fermi surface without any obvious symmetry breaking when the hole density is reduced below optimal doping.

Recently a lot of progress has been made on the understanding of ladder systems. These are intermediate between one and two dimensions and can be analyzed reliably in detail [1]. Their properties at low energies, or temperatures, are crucially dependent on their width. The single chain forms the well-known Tomonaga-Luttinger state which at half-filling develops a charge gap but not a spin gap. By contrast the 2-leg ladder forms a Luther-Emery liquid with holes bound in relative 'd-wave' pairs. At half-filling there is a charge gap but also a spin gap. In this case the spin-spin correlations are purely short range and one speaks of a spin liquid although one should be clear that this is not a disordered state but a unique quantum coherent groundstate. This behavior has been found in strong coupling, e.g. in numerical investigations of the t-J model, which is equivalent to the large U Hubbard model, but also in weak coupling qualitatively similar behavior is found [2]. This insulating spin liquid (ISL) can be regarded as a form of short range resonant valence bond (RVB) state. It is specially interesting to examine the weak coupling behavior. Away from half-filling the key criterion to obtain the Luther-Emery behavior is the presence of two bands (bonding and anti-bonding) at the Fermi energy. At half-filling additional Umklapp scattering processes enter the one-loop renormalization group (RG) equations and these scale to a strong coupling fixed point with a charge and spin gap [2]. Note that because of the purely short range order in the spin system, one cannot associate these gaps with symmetry breaking or long range order. Instead one has a Fermi system with a truncated Fermi surface which is not associated with a broken translational symmetry, — just the sort of behavior discussed earlier.

The 3-leg ladder is also very interesting. At half-filling the strong coupling limit is equivalent to the 3-leg Heisenberg $S = 1/2$ AF ladder. This has been well studied and reduces in the low energy sector to an effective single chain Heisenberg model with longer range but unfrustrated AF coupling [3]. The weak coupling limit shows a similar behavior (COS1 in the Balents-Fisher notation (zero charge and one spin gapless modes)). When holes are introduced one finds in strong coupling calculations that these enter the channel with odd parity w.r.t. reflection about the central leg and form a single channel Tomonaga-Luttinger liquid [4, 5]. The even parity channels remain at the stoichiometric filling and continue to form an ISL. Only after a critical hole density is reached, do the holes enter the even parity channels. The result is a finite region of hole doping where the original Fermi surface with 3 bands (or 6 Fermi points) is partially truncated to 2 Fermi points, but again without a broken translational symmetry. The key is again Umklapp scattering which has scaled to a strong coupling fixed point and which has introduced a charge gap in the even parity channels. This is then a clear example of a partially truncated Fermi surface through the formation of an ISL over part of the Fermi surface.

Recently we have examined the possibility of obtaining similar behavior in a two-dimensional system [6, 7]. The key to forming a charge gap lies in Umklapp scattering which clearly is responsible for the formation of a Mott insulator in one dimension. In two dimensions, the shape of the Fermi surface enters and there are two distinct possibilities that we will consider in turn. Either the Fermi surface bulges out towards the diagonals $(\pm 1, \pm 1)$ in the Brillouin zone or towards the saddle points. The former case is more straightforward to analyze but the latter case is the one which occurs in the cuprates and we will discuss them in turn.

2. 4–Patch Model

We start with a general 2-dim. dispersion relation, e.g. a form, $\varepsilon(\mathbf{k}) = -2t(\cos k_x + \cos k_y) - 4t' \cos k_x \cos k_y$ with $t(t')$ as (next) nearest neighbor hopping matrix elements. Taking $t > 0$ and $t' > 0$ and increasing the electron density, $n = 1 - \delta$, leads to a Fermi surface which touches the 4 points $(\pm \pi/2, \pm \pi/2)$. Following Haldane [8], we divide the Fermi surface into patches and examine the patches near the 4 points $(\pm \pi/2, \pm \pi/2)$. These 4 patches on the Fermi surface are connected through Umklapp processes which leads us to examine the renormalization group (RG) equations for the coupling constants [6]. The RG equations have similarities to those for a 2-leg ladder at half-filling which also has 4 Fermi surface points and which are known to scale to a strong coupling solution.

The 4 patches around $(\pm \pi/2, \pm \pi/2)$ are sketched in Fig. 1. The size of a patch is defined by a wave vector cutoff, k_c, and within each patch $\alpha(\alpha =$

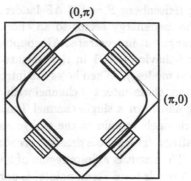

Fig. 1: Definitions of 4 patches, shown as hatched rectangular areas. The bold curve represents the 2-dimensional Fermi surface which touches the points $(\pm\pi/2, \pm\pi/2)$.

$1, \ldots, 4$) the electron energy relative to the chemical potential, μ is expanded as

$$\varepsilon_\alpha(\mathbf{q}) - \mu = vq_\alpha + uq_{\perp,\alpha}^2 \tag{1}$$

where $\mathbf{q}$ is the wavenumber measured from the center of α-th patch, and $q_\alpha(q_{\perp,\alpha})$ is the component of $\mathbf{q}$ normal (tangent) to the Fermi surface at the center of the patch. The Fermi velocity is given by v, and the energy cutoff is $E_0 = vk_c$.

The linear dispersion relation leads to logarithmic anomalies in the particle-hole (Peierls) and particle-particle (Cooper) channels as in one dimension but the transverse dispersion introduces an infrared cutoff, $E_T \approx uk_c^2$. The non-interacting susceptibility in the Peierls channel takes the form $\chi_p(\omega) = 1/2 \ln (\max(\omega, E_T)/E_0)$. In the parameter region $\omega > E_T$, the infrared cutoff from the transverse dispersion can be ignored and a set of RG equations can be derived as in one dimension.

In Fig. 2 we define the normal vertices g_1, g_2 and g_{1r} as well as Umklapp vertices g_3, g_{3p} and g_{3x}. Other interactions are not treated here since they are irrelevant within the framework of a one-loop approximation. Summing up all one-loop diagrams, we obtain the RG equations

$$\dot{g}_1 = g_1^2 + g_{1r}^2 + 2g_{3x}^2 - 2g_{3x}g_{3p} , \tag{2}$$

$$\dot{g}_2 = \frac{1}{2}\left(g_1^2 + 2g_{1r}^2 - g_3^2 - 2g_{3p}^2\right) , \tag{3}$$

$$\dot{g}_{1r} = (g_1 + g_2)\, g_{1r} , \tag{4}$$

$$\dot{g}_3 = (g_1 - 2g_2)\, g_3 + 2g_{3x}^2 - 2g_{3x}g_{3p} - g_{3p}^2 , \tag{5}$$

$$\dot{g}_{3x} = 2g_1g_{3x} - g_1g_{3p} - g_2g_{3x} + g_3g_{3x} - g_3g_{3p} , \tag{6}$$

$$\dot{g}_{3p} = -(g_2 + g_3)\, g_{3p} . \tag{7}$$

Here $\dot{g}_i \equiv x(dg_i)/(dx)$ and $x = \omega/E_0$.

These RG equations coincide with those obtained by Houghton and Marston in their study of a lightly doped flux state [9]. They differ from those studied by Zheleznyak et al. [10], who examined a 4-patch model with patches oriented along $(1,0)$ and $(0,1)$ directions. Umklapp scattering does not connect perpendicular patches in that case, so a close resemblance to single chain behavior follows.

We take repulsive Umklapp interactions, as $g_3 = g_{3x} = g_{3p} = U$, and treat g_1, g_2 and g_{1r} as parameters. The fixed points are obtained by numerically integrating the RG equations. In a wide region around $g_1 \sim g_2 \sim g_{1r} \sim U$, we find a strong coupling fixed point where both normal and Umklapp vertices diverge and a singularity appears at $\omega \sim \omega_c = E_0 \exp(-1/\Lambda)$ where $\Lambda \propto U$. In two dimensions, such an anomaly at finite ω is an artifact of the one-loop calculation and higher order terms will shift it to $\omega = 0$. Nevertheless, ω_c represents the energy scale where the system crosses over from weak coupling to strong coupling. We will explicitly assume that the interactions $\sim U$ are strong enough so that $\omega_c > E_T$ in which case the existence of a finite curvature

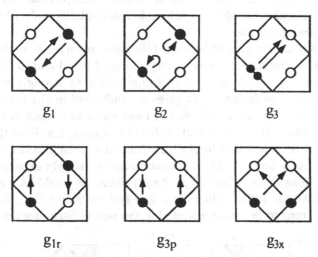

Fig. 2: The definitions of vertices for the 4-patch model.

becomes irrelevant at the strong coupling fixed point. In contrast, if the system had scaled to weak coupling, then E_T would always remain relevant. There is a limit with weak interactions and a dispersion relation with $t' \ll t$ where both conditions, $E_T \left(= 2t'k_c^2\right) \ll \omega_c$ and $U \ll E_0$, are satisfied and our approach based on one-loop RG equations is justified. We speculate that the qualitative nature of the anomaly obtained in this weak coupling region is also present in the strong coupling region $U \gtrsim t, t'$.

We examine the nature of the fixed point through the anomalies which these 4 patches contribute to the susceptibilities. At the fixed point with strong Umklapp coupling described above, the leading divergence is observed in the spin susceptibility $\chi_s(\mathbf{q})$ at $\mathbf{q} = (\pi, \pi)$ with the exponent $\alpha_s = -1.782$, while the exponents for charge and superconducting susceptibilities are positive so that these susceptibilities do not have a divergent contribution. The uniform spin $(\chi_s(0))$ and charge (κ) susceptibilities are also of interest. In the case of a 1d chain system, we have spin gap behavior when there is a divergence in g_1 and charge gap behavior from g_3. In the present case, both g_1 and g_3 flow to strong coupling which indicates a tendency to open up both spin and charge gaps.

We now compare the present results to those of a two-leg ladder at half-filling. In this case, as Balents and Fisher have shown [2], there are 9 vertices which are relevant within a one-loop calculation. Again the flow is too strong coupling in backward and Umklapp scattering channels. In this case the properties of the strong coupling fixed point are well established. The system is an insulating spin liquid (ISL) with both spin and charge gaps (C0S0 in the Balents-Fisher notation) and is an example of a short range RVB state, first proposed by Anderson for a $S = 1/2$ Heisenberg model [11]. The spin susceptibility $\chi_s(\pi, \pi)$ is strongly enhanced but remains finite.

In the present case we cannot be sure of the spin properties from the one-loop calculations especially since the spin susceptibility at (π, π) and $(0, 0)$ behave in a contradictory way. What is certain is the scaling to strong coupling with diverging Umklapp scattering. This gives us confidence in the result that the compressibility $\kappa = dk_{F,\alpha}/d\mu \rightarrow 0$ at the fixed point as it does in the two-leg ladder at half-filling. This has several profound consequences. First the condensate that forms is pinned and insulating. Secondly when additional electrons are added to the system, the Fermi surface does not simply expand along the $(\pm 1, \pm 1)$ directions beyond the $(\pm \pi/2, \pm \pi/2)$ points as would happen for non-interacting electrons. Instead the charge gap and vanishing $dk_{F\alpha}/d\mu$ force the additional electrons to be accommodated in the rest of Fermi surface.

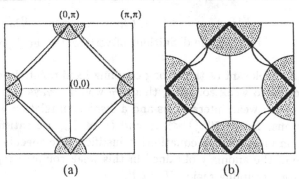

Fig. 3: Fermi surface (FS). (a) Two patches of the FS at the saddle points. (b) Truncated FS as electron density is increased.

3. 2–Patch (Saddle Point) Model

In the cuprates $t' < 0$ so that critical behavior can now be expected when the FS touches the saddle points at $(\pi, 0)$ and $(0, \pi)$. The leading singularity arises from electron states near the saddle points which leads one to consider a 2-patch model as illustrated in Fig. 3a [7]. The Cooper channel now has a log-square divergence and the Peierls channel at $\mathbf{q} = (\pi, \pi)$ has a similar form but crosses over to a single log form at a scale $\approx t'$. These singularities may be treated within a Wilson RG scheme [12]. There are four relevant interaction vertices $g_i (i = 1, 4)$ illustrated in Fig. 4. Note that normal and Umklapp processes are indistinguishable since the patches are at the zone edge. The one-loop RG eqns. in terms of the scaling variable $y = \ln^2(\omega/E_0)$ were obtained first by Lederer, Montambaux and Poilblanc [13] and take the form,

$$\dot{g}_1 = 2d_1(y)g_1(g_2 - g_1) , \tag{8}$$

$$\dot{g}_2 = d_1(y)(g_2^2 + g_3^2) , \tag{9}$$

$$\dot{g}_3 = -2g_3 g_4 + 2d_1(y)g_3(2g_2 - g_1) , \tag{10}$$

$$\dot{g}_4 = (g_3^2 + g_4^2) . \tag{11}$$

The function $d_1(y)$ describes the relative weight of the Cooper and Peierls channels with asymptotic forms, $d_1(y) \to 1$ at $y \approx 1$ and $d_1(y) \sim \ln(t/t')/\sqrt{y}$ as $y \to \infty$.

The case $d_1 \equiv 1$ arises in the limit $t' = 0$ and was studied by Schulz [14] and Dzyaloshinskii [15]. It was shown that $\chi_s(\pi, \pi)$ has the same exponent as d-wave pairing but is dominant to the next leading divergent terms. The fixed point is understood as a Mott insulator with long range AF order. The limit $d_1 = 0$ was treated by Dzyaloshinskii [16]. In this case (10) and (11) combine to give the simple flow equation

$$\dot{g}_- = -g_-^2 \tag{12}$$

with $g_- = g_4 - g_3$. Dzyaloshinskii considered the weak-coupling fixed point $g_- \to 0$ which arises from a starting value $g_- \geq 0$, and discussed the Tomonaga-Luttinger liquid behavior that results.

We concentrate on the RG equations with $0 < d_1(y) < 1$ which enables us to consider finite values of the ratios t'/t and U/t. We assume $t'/t \ll 1$ so that we are close to half-filling. The one-loop RG equations are solved numerically. Starting from a Hubbard-model initial value $g_i = U(i = 1 \sim 4)$, the vertices flow to a strong coupling fixed point with $g_2 \to +\infty$, $g_3 \to +\infty$ and $g_4 \to -\infty$, at a value $y_c \sim t/U_c$.

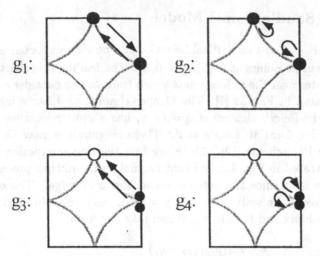

Fig. 4: The definitions of vertices for the 2-patch model.

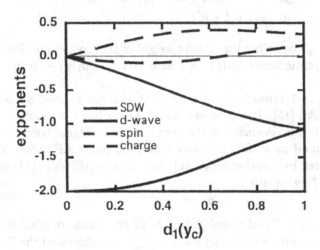

Fig. 5: Exponents for various susceptibilities. For uniform spin and charge susceptibilities, exponents are scaled by d_2/d_1

Although one cannot solve for the strong coupling fixed point using only one-loop RG equation, a qualitative description of the behavior comes from the examination of the susceptibilities. In Fig. 5 we show the exponents for d-wave pairing, $\chi_s(\pi, \pi)$, $\chi_s(0)$ and charge compressibility. Comparison of the values of the exponents shows us that the most divergent susceptibility is that of the d-wave pairing throughout the parameter region of $d_1(y_c)$, as found previously by Lederer et al. [13]. Note $\chi_s(0)$ has a positive exponent indicating spin gap behavior.

The exponent for the charge compressibility changes its sign at $d_1(y_c) \sim 0.6$. Namely there exists a critical interaction strength U_c such that for $U > U_c$ the charge compressibility is suppressed to zero. The critical point U_c is determined by t' in the form $U_c/t \propto \ln^{-2}(t/t')$. This implies a transition from a superconducting phase at $U < U_c$ with its origin in enhanced Cooper channel due to the van Hove singularity, to a phase with a charge gap at $U > U_c$ which can be regarded as a precursor of the Mott transition. The fixed point at $U > U_c$ resembles closely that of the half-filling two-leg ladder which has spin and charge gaps while the most divergent susceptibility is the d-wave pairing. The characteristics of this fixed point (C0S0 in the Balents-Fisher [2] notation) are well understood as an ISL of short range RVB form. The close similarity between the fixed points leads us to assign them to the same universality class.

Next we consider increasing the electron density. One possibility would be to follow the non-interacting FS which expands beyond the saddle points. But the flow to strong coupling and the opening of a charge and spin gap leads us to consider a second possibility, namely that the FS is pinned at the saddle point and does not expand beyond it. This proposal was put forward in ref. [7] after an examination of 8 FS-patches on the Umklapp surface (US) defined by lines joining the saddle points. Further support for this proposal comes from the lightly doped 3-leg ladder [4, 5] where in strong coupling a C1S1 phase occurs with an ISL with exactly half-filling in the even parity channels and an open FS only in the odd parity channel. This contrasts with the one-loop RG results which gives a C2S1 phase with holes immediately entering both odd and even parity channels [2]. Our proposal is sketched in Fig. 3(b) and it is based on a lateral spread of the spin and charge gap along the US leading to a set of 4 open FS segments consisting of arcs centered at the points $(\pm\pi/2, \pm\pi/2)$. Note the area enclosed by the surface defined by the US and these 4 arcs contain the full electron density, consistent with a generalized form of Luttinger's Theorem.

4. Discussion

Since the ISL is not characterized by any simple broken symmetry or order parameter, the resulting state cannot be described by a simple mean field or Hartree-Fock factorization. The proposal of a FS consisting of 4 disconnected arcs has strong parallels to the results of recent gauge theory calculation for the lightly doped strong coupling t-J model by Lee and Wen [17]. Signs of such behavior are also evident in a recent analysis of the momentum distribution using a high temperature series by Putikka et al. [18]. Note models which include only nearest-neighbor terms in the kinetic energy (i.e. $t' = 0$) are a special limit from the present point of view.

The proposal that an ISL truncates the FS along the US in the vicinity of the saddle points has some interesting consequences. There will be a coupling to the open segments in the Cooper channel through the scattering of electron pairs out of the ISL to the open FS segments. This process is reminiscent of the coupling of fermions to bosonic preformed pairs in the Geshkenbein-Ioffe-Larkin model [19]. They argued for an infinite mass for such pairs to suppress their contribution to transport properties. Such scattering process will be an efficient mechanism for d-wave pairing on the open FS segments.

In the normal state there is a close similarity to a phenomenological model proposed by Ioffe and Millis [20], to explain the anomalous transport properties. Here also the FS segments have usual quasi-particle properties (i.e. there is no spin-charge separation) but the scattering rate is assumed to vary strongly along the FS arcs. In our case we can expect the scattering rate will vary strongly since Umklapp processes will lead to the strong scattering at the end of the FS arcs where they meet the US. Ioffe and Millis justified their model by a comparison to the tunneling and ARPES experiments [21] which show a single particle gap opening in the vicinity of the saddle points similar to the form in Fig. 3b. Especially in underdoped cuprates there are clear signs in the ARPES experiments of single particle energy gaps at the saddle points at low temperature [21]. Lastly we refer the reader to the very recent preprint by Balents, Fisher and Nayak [22] which introduces the concept of a Nodal liquid with properties similar to the ISL discussed above.

Acknowledgment The work on the 2-path model was performed in collaboration with M. Salmhofer and we acknowledge useful conversations with S. Haas, R. Hlubina, D. Khveshchenko, M. Sigrist, E. Trubowitz and F.C. Zhang. N. F. is supported by a Monbusho Grant for overseas research.

References

1. For a review see, E. Dagotto and T.M. Rice, Science **271**, 618 (1996).

2. L. Balents and M.P.A. Fisher, Phys. Rev. B **53**, 12133 (1996); H.-H. Lin, L. Balents and M.P.A. Fisher, Phys. Rev. B **56**, 6569 (1997).

3. B. Frischmuth, S. Haas, G. Sierra and T.M. Rice, Phys. Rev. B **55**, R3340 (1997).

4. T.M. Rice, S. Haas, M. Sigrist and F.C. Zhang, Phys. Rev. B **56**, 14655 (1997).

5. S.R. White and D.J. Scalapino, Phys. Rev. B **57**, 3031 (1998).

6. N. Furukawa and T.M. Rice, J. Phys. Cond. Mat. **10**, L381 (1998).

7. N. Furukawa, T.M. Rice and M. Salmhofer, cond-mat/9806159.

8. F.D.M. Haldane, Proc. Int. School Phys. 'Enrico Fermi' Course 121 (1991) ed. J.R. Schrieffer and R.A. Broglia (New York, North Holland); also J. Fröhlich and R. Götschmann, Phys. Rev. B 55, 6788 (1997).

9. A. Houghton and J.B. Marston, Phys. Rev. B 48, 7790 (1993).

10. A.T. Zheleznyak, V.M. Yakorenko and I.E. Dzyaloshinskii, Phys. Rev. B 55, 1200 (1997).

11. P.W. Anderson, Science 235, 1196 (1987).

12. J. Feldman, M. Salmhofer and E. Trubowitz, J. Stat. Phys. 84, 1209 (1996).

13. P. Lederer, G. Montambaux and D. Poilblanc, J. Physique 48, 1613 (1987).

14. H.J. Schulz, Europhys. Lett. 4, 609 (1987).

15. I.E. Dzyaloshinskii, Sov. Phys. JETP 66, 848 (1987).

16. I.E. Dzyaloshinskii, J. Phys. I France 6, 119 (1996).

17. P.A. Lee and X.G. Wen, Phys. Rev. Lett. 78, 411 (1997).

18. W.O. Putikka, M.U. Luchini and R.R.P. Singh, preprint, cond-mat/9803141.

19. V.G. Geshkenbein, L.B. Ioffe and A.I. Larkin, Phys. Rev. B 55, 3173 (1997).

20. L.B. Ioffe and A.J. Millis, preprint, cond-mat/9801092.

21. M.R. Norman et al., Nature 392, 157 (1998), C. Kim et al., Phys. Rev, Lett. 80, 4245 (1998) and references therein.

22. L. Balents, M.P.A. Fisher and C. Nayak, preprint, cond-mat/9803086.

MOMENTUM BROADENING IN THE INTERLAYER PAIR-TUNNELING MODEL OF HIGH-T_C SUPERCONDUCTIVITY

J. O. FJÆRESTAD AND A. SUDBØ

Department of Physics, Norwegian University of
Science and Technology, N-7034 Trondheim, Norway

Abstract. We investigate how momentum broadening of the interlayer pairing kernel in the interlayer pair tunneling model of high-T_c superconductivity affects the critical temperature T_c and the zero-temperature gap function. We consider constant as well as k-dependent intralayer pairing kernels. For a sign-changing gap, the variation of T_c with broadening is slightly enhanced compared to a gap of constant sign. We identify the amount of broadening that will destroy the sharp k-space features of the gap obtained with a k-diagonal interlayer tunneling, explaining our results in terms of Fermi surface features of the gap. The parameter controlling this amount of k-space broadening is T_J/t, where T_J is the interlayer tunneling matrix element and t is the bandwidth parameter of the normal state intralayer motion.

1. Introduction

The interlayer pair tunneling (ILT) model of high-temperature superconductivity[1, 2, 3] has recently been the focus of much attention. In this model, the pairing of electrons in individual CuO_2-layers is considerably enhanced by the tunneling of Cooper pairs between neighbouring layers, causing a large increase of the critical temperature as compared to a purely two-dimensional (2D) theory.

A key aspect in justifying the model is the lowering of the c-axis kinetic energy upon entering the superconducting state. There now appears to be some experimental evidence suggestive of a frustrated c-axis kinetic energy in the normal state of the cuprates, at least in $La_{2-x}Sr_xCuO_4$ [4], and quite possibly also in $HgBa_2Cu_nO_{2n+2+\delta}$[5]. However, for Tl-2201 the situation is less settled [6, 7, 8, 9].

M. Ausloos and S. Kruchinin (eds.), Symmetry and Pairing in Superconductors, 211–218.

In Ref. [2] it was argued that in the tunneling process, the 2D momentum of the Cooper-pair electrons was conserved, due to the momentum conservation in the single-electron tunneling Hamiltonian in the absence of inelastic scattering. The high critical temperatures, as well as several of the unusual k-space features of the gap, are predicted within the ILT mechanism on the basis of this assumed momentum conservation, which, translated to real space, means that the e-e attraction associated with the interlayer tunneling has an *infinite* range. In real systems, one would expect the range to be finite, due to the inevitable presence of impurities etc.

In this article we investigate the effect of a finite range on the gap and the critical temperature by solving the ILT gap equation with a modified functional form of the pair tunneling term in which a phenomenological parameter $1/k_0$ is introduced as the characteristic range of the attraction. We expect that *qualitatively* correct conclusions may be drawn from our modelling of the momentum broadening.

2. Formulation of the problem

We consider compounds with two CuO_2-layers per unit cell. The generalization to an arbitrary number of CuO_2-layers per unit cell is straightforward [10]. Below the superconducting transition temperature, we will assume that the quasi-particle description is approximately valid. The total Hamiltonian is taken to be the sum of 2D BCS Hamiltonians for the individual layers, and an interlayer pair tunneling Hamiltonian, i.e. $H = H_{\text{layer}} + H_J$. Using the zero-momentum pairing assumption, the intralayer part is given by

$$H_{\text{layer}} = \sum_{k,\sigma,i=1,2} \varepsilon_k \, c_{k,\sigma}^{(i)\dagger} c_{k,\sigma}^{(i)} + \sum_{k,k',i=1,2} V_{k,k'} c_{k,\uparrow}^{(i)\dagger} c_{-k,\downarrow}^{(i)\dagger} c_{-k',\downarrow}^{(i)} c_{k',\uparrow}^{(i)}, \tag{1}$$

while the interlayer pair tunneling Hamiltonian is

$$H_J = -\sum_{k,k'} T_J(k,k') c_{k\uparrow}^{(1)\dagger} c_{-k,\downarrow}^{(1)\dagger} c_{-k'\downarrow}^{(2)} c_{k'\uparrow}^{(2)} + \text{h.c.} \tag{2}$$

Here $c_{k\sigma}^{(i)\dagger}$ is the creation operator of an electron in layer i ($i = 1, 2$) with intralayer wave vector k and spin projection σ, ε_k is the normal state dispersion measured relative to the Fermi level, and $V_{k,k'}$ is the intralayer pairing kernel. Writing the pair tunneling matrix element as $T_J(k,k') \equiv T_J f(k-k')$, Fourier-transforming (2) back to real space gives

$$H_J = -T_J \sum_{R_1,R_2,r} G(r) \, c_{R_1+r/2,\uparrow}^{(1)\dagger} c_{R_1-r/2,\downarrow}^{(1)\dagger} c_{R_2-r/2,\downarrow}^{(2)} c_{R_2+r/2,\uparrow}^{(2)} + \text{h.c.} \tag{3}$$

where r is the relative coordinate and R_i the center of mass coordinate in layer i of the two tunneling electrons (there are no restrictions on $|R_1 - R_2|$ due to the Schrieffer pairing condition $q = 0$, as in conventional superconductors). The characteristic decay length of the function $G(r) \propto \sum_k f(k)e^{ikr}$ represents the range of the effective interlayer tunneling attraction. In Ref. [2] $f(k) = \delta_{k,0}$, giving an infinite range of the attraction. We will relax this form here, allowing for a broadened distribution in k-space, which makes the interaction range finite.

The gap equation is [2]

$$\Delta_k = -\sum_{k'} V_{k,k'}\Delta_{k'}\chi_{k'} + \sum_{k'} T_J(k,k')\Delta_{k'}\chi_{k'}, \qquad (4)$$

where Δ_k is the gap function, and χ_k is the pair susceptibility, given by

$$\chi_k = \frac{\tanh(\beta E_k/2)}{2E_k}. \qquad (5)$$

Here $E_k = \sqrt{\varepsilon_k^2 + |\Delta_k|^2}$ and $\beta = 1/k_BT$, where k_B is Boltzmann's constant and T is the temperature.

For simplicity, we take the k's to be one-dimensional (1D). This should be considered merely as a mathematical simplification of the real problem. The main question asked in this paper is whether or not a k-space diagonal tunneling matrix element is a singular limit of a more realistic case where k-space broadening is included. In the diagonal case, the critical temperature and the maximum value of the gap is strictly independent of dimensionality, since they are determined by *local* k-space features of the pairing kernel. Moreover, the qualitative arguments we construct in Sec. 4 to explain our results, apply also to 2D systems.

Given this limited purpose of the calculation, it does make sense to consider the problem in one dimension. The calculation, however, implies nothing about superconductivity with true ODLRO in 1D systems, which is well-known not to exist for $T > 0$ [11], and prohibited by quantum fluctuations at $T = 0$.

The intralayer pairing kernel is taken to be of the form $V_{k,k'} = -V\, g_k\, g_{k'}$, where in the general D-dimensional case g_k belongs to the set of basis functions for irreducible representations of the point group of the underlying lattice, and $V > 0$ is an effective two-particle scattering matrix element. We consider two different functional forms for g_k. The first is the well-known BCS approximation, $g_k = \Theta(\omega_D - |\varepsilon_k|)$, where ω_D is an energy cutoff. The second case we consider is $g_k = \cos(ka)$. The gap obtained from (4) for the first case does not change sign in the Brillouin zone, while the gap for the second case in general does.

We assume a simple tight-binding dispersion form for ε_k,

$$\varepsilon_k = -2t \left[\cos(ka) - \cos(k_F a)\right], \tag{6}$$

where t is the single-electron tunneling matrix element, a is the lattice constant and k_F is the Fermi wave vector. Furthermore, we take $f(k)$ to be a 'lattice Lorentzian', given by

$$f(k) = \frac{k_0 a^2}{2L} \frac{1}{\sin^2\left(\frac{ka}{2}\right) + \left(\frac{k_0 a}{2}\right)^2}, \tag{7}$$

where L is the length of the system and $1/k_0$ is the typical range of the interlayer tunneling attraction in real space. The sine function ensures that the scattering is periodic in the reciprocal lattice. The prefactor in Eq. (7) is chosen to get the correct limit when $k_0 \to 0$, corresponding to a k-diagonal pair tunneling term in (4).

3. Results

We have calculated the dependence of the critical temperature T_c and the zero temperature gap on the parameter $k_0 a$ by solving Eq. (4) self-consistently in the thermodynamic limit $L \to \infty$. In Fig. 1 we show the results for T_c for $T_J = 30$ meV, $\omega_D = 20$ meV, $t = 25$ meV, $k_F a = \pi/4$ and $LV/2\pi a = 2.5$ meV. Note that T_c is slightly more sensitive to $k_0 a$ for $g_k = \cos(ka)$ than for $g_k = \Theta(\omega_D - |\varepsilon_k|)$.

In Fig. 2 we show the gap at $T = 0$ for four different values of T_J and fixed $k_0 = 0$, the other parameter values being the same as before. In this case, the gap is given implicitly by $\Delta_k = \Delta_0 g_k/(1 - T_J \chi_k)$, where $\Delta_0 \equiv V \sum_k g_k \Delta_k \chi_k$. The maximum of the gap occurs on the Fermi surface, where the enhancement factor $1/(1 - T_J \chi_k)$ is maximal. For $T_J = 0$, the k-space dependence of the gap is given entirely by g_k, while this is no longer the case for $T_J \neq 0$. However, as seen in Fig. 2, T_J does not affect the sign of the gap, which remains to be determined by g_k alone.

In Fig. 3 we show the gap at $T = 0$ for four different values of $k_0 a$ and fixed $T_J = 30$ meV. As $k_0 a$ increases, the k-space variation decreases.

4. Discussion

We finally discuss the criteria for when the momentum broadening will be effective in decreasing the maximum value of the gap, thereby smoothing out the sharp k-space structures obtained in the k-diagonal case.

Although the exact calculation is done self-consistently, it is instructive to consider how a slightly momentum-broadened interlayer term affects the

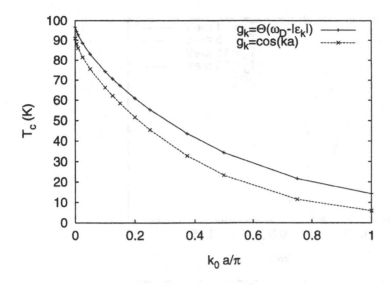

Figure 1. The critical temperature T_c as a function of $k_0 a/\pi$.

k-diagonal gap in the lowest approximation. For $k_0 = 0$, Δ_k is sharply enhanced on the Fermi surface, due to the large value of $1/(1 - T_J\chi_k)$ there. Since Δ_k varies more rapidly around the Fermi surface than χ_k, the variation of $\Delta_k\chi_k$ in this region is essentially determined by Δ_k. Moreover, the main contributions to $\sum_{k'} T_J(k_F - k')\Delta_{k'}\chi_{k'}$ roughly come from the region $|k_F - k'| < k_0$. Thus, as long as k_0 is much smaller than the characteristic width of the peak of the gap obtained for $k_0 = 0$, the broadened $T_J(k_F - k')$ essentially appears as a δ-function on this scale. Under such circumstances, the solution to the gap equation is little affected by the non-k-diagonality. A broadening of the order of the width of the peak of the gap for $k_0 = 0$ is therefore required for a substantial effect of the broadening to be felt. Note that these arguments would also be valid in the 2D case. Also note from Fig. 2 that the width of the peak of Δ_k increases with T_J. The detrimental effects of an increase of $k_0 a$ will therefore be reduced with an increase of T_J.

The Fermi surface features of the gap become sharper on increasing the bandwidth parameter t, since the factor $1/(1 - T_J\chi_k)$ drops more abruptly away from its peak value when the overall amplitude of the variation of ε_k is increased. Therefore, the gap will also drop more rapidly from its peak value when T_J is fixed and t increases. Note that one may scale the parameter t entirely out of Eq. (4) to obtain a gap equation in terms of the dimensionless quantities $\beta t, \varepsilon_k/t, T_J/t, \Delta_k/t, V/t$ (and ω_D/t, when $g_k =$

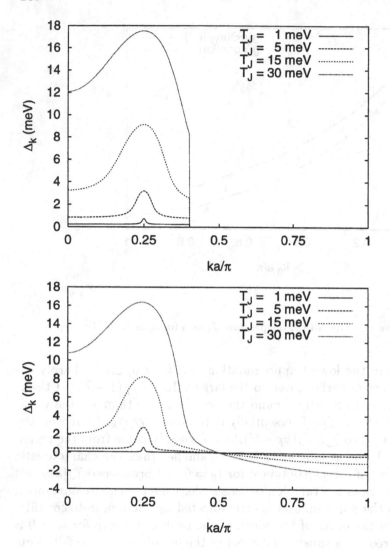

Figure 2. The gap function $\Delta_k = \Delta_{-k}$ at zero temperature plotted for four different values of T_J with $k_0 = 0$. The two cases $g_k = \Theta(\omega_D - |\varepsilon_k|)$ and $g_k = \cos(ka)$ are shown in the upper and lower panel, respectively. As T_J is increased, the maximum value of the gap, occuring on the Fermi surface, increases and the variation of the gap with k is enhanced. Note how the sign of the gap is always determined by g_k, the *intralayer* contribution to the pairing kernel.

$\Theta(\omega_D - |\varepsilon_k|)$). Enhancing t reduces the ratio T_J/t and therefore leads to a sharper reduction of T_c and the anomalous k-space features in the gap as a function of the k-space broadening. It should be mentioned that the value

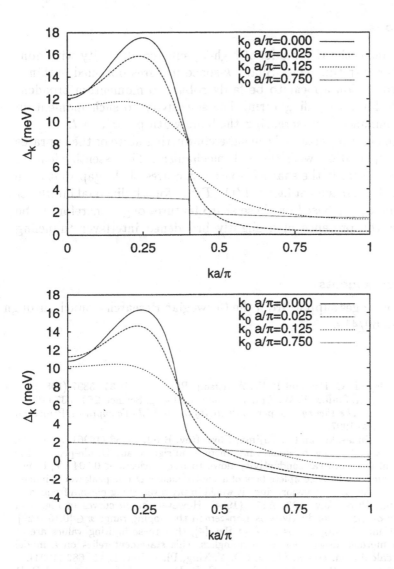

Figure 3. The gap function $\Delta_k = \Delta_{-k}$ at zero temperature plotted for four different values of $k_0 a/\pi$ with $T_J = 30$ meV. The two cases $g_k = \Theta(\omega_D - |\varepsilon_k|)$ and $g_k = \cos(ka)$ are shown in the upper and lower panel, respectively. As k_0 is increased, the maximum value of the gap and the variation of the gap with k decreases. For large enough values of $k_0 a/\pi$, this is reflected in the upper panel in a gap where the only k-space variation comes from the sharp discontinuity from the BCS approximation, while in the lower panel the gap eventually ceases to change sign.

we have used for the ratio T_J/t is quite uncertain at this stage. The effect of a variation of t with a non-Fermi liquid pair susceptibility also deserves attention.

218

5. Conclusions

We have considered the ILT model of high-T_c superconductivity with non-k-diagonal interlayer tunneling. Sharp k-space features obtained within a k-diagonal formulation appear to be fairly robust to momentum broadening of the interlayer tunneling term. The sensitivity to such momentum broadening is enhanced by increasing the bandwidth parameter t.

Several unusual properties of the superconducting state of the cuprates are given an explanation with the ILT mechanism. The essential feature of the ILT mechanism is the sharp k-space structures of the gap that arise from an unusual enhancement factor $1/(1-T_J\chi_k)$ for a k-diagonal interlayer tunneling. Conclusions based on these sharp features ought therefore to be reexamined in the presence of a slightly broadened interlayer tunneling term.

6. Acknowledgements

A. S. acknowledges support from the Norwegian Research Council through Grant No. 110569/410.

References

1. J. M. Wheatley, T. C. Hsu and P. W. Anderson, Phys. Rev. B **37**, 5897 (1988).
2. S. Chakravarty, A. Sudbø, P. W. Anderson and S. Strong, Science **261**, 337 (1993).
3. P. W. Anderson, *The theory of superconductivity in the high-Tc cuprates*, Princeton Series in Physics, 1997.
4. S. Uchida, K. Tamasaku, and S. Tajima, Phys. Rev. B **53**, 14558 (1996). It may be argued from Fig. 4 of this reference that there is no sign of any Drude-peak in the c-axis normal state optical conductivity down to frequencies $\omega < 0.004$ eV, in not overdoped compounds. A complete lack of a low-frequency Drude peak would indeed constitute non-metallic behavior, since it would imply a vanishing plasma frequency, see W. Kohn, Phys. Rev. **133**, A171 (1964). However, in *all* curves in Fig. 4 of Ref. [4], the c-axis dc-conductivity is *non-zero* in the doping range $x \in [0.15, 0.20]$. Although it may be argued, as done in Ref. [4], that these limiting values are so low that a metallic description is meaningless, this statement relies on a model dependent calculation, see M. Liu and D. Y. Xing, Phys. Rev. B **49**, 682 (1994).
5. C. Panagopoulos, J. R. Cooper, T. Xiang, G. B. Peacock, I. Gameson, and P. P. Edwards, Phys. Rev. Lett. **79**, 2320 (1997).
6. J. Schützmann *et al.*, Phys. Rev. B **55**, 11118 (1997).
7. K. A. Moler, J. R. Kirtley, D. G. Hinks, T. W. Li, and M. Xu, Science **279**, 1193 (1998).
8. P. W. Anderson, Science **279**, 1196 (1998). See also P. W. Anderson, Science **268**, 1154 (1995); A. J. Leggett, *ibid* **274**, 587 (1996).
9. S. Chakravarty, Europ. Phys. J. **B** (in press); cond-mat/9801025.
10. A. Sudbø, J. Low. Temp. Phys. **97**, 403 (1994).
11. P. C. Hohenberg, Phys. Rev. **158**, 383 (1967); N. D. Mermin and H. Wagner, Phys. Rev. Lett. **17**, 1133 (1967).

EFFECTS OF VAN HOVE SINGULARITIES ON THERMAL PROPERTIES OF HIGH-T_C SUPERCONDUCTORS

S. DORBOLO

S.U.P.R.A.S., Institute of Physics B5

AND

H. BOUGRINE

S.U.P.R.A.S., Institute of Electricity Montefiore B28
University of Liège,
B-4000 Liège, Belgium

Abstract. Specific heat, thermal conductivity and thermal diffusivity are calculated within a phenomenological model including Van Hove singularities. Two gap symmetries are considered namely *s-wave* and *d-wave*. Comparisons with experimental data are in good agreement with the *d-wave* scenario. A relation between the doping level and the oxygenation rate is found. Doping effects are then considered in calculation. Quite good agreement with experimental results is found.

1. Introduction

The existence of Van Hove singularities in ceramic superconducting cuprates is now an established fact [1]. Saddle points in the band structure are found in a large number of ARPES experiments [2, 3, 4] on several cuprates as Y-123 [5] and Bi-2212 [6]. Such singularities lead to divergences in the density of states, i.e. so called Van Hove singularities. This effect is enhanced by a quasi-2D band structure and a large effective mass of the charge carriers. Moreover, the Fermi level lies very close to the saddle point (100-200 K). The superconductivity is then enhanced by this particular type of density of states. The Van Hove singularities are used in each modern theory of superconducting including electron-phonon coupling [7], [8] ... In particular, Dagotto et al.[9] introduced a model based on antiferromagnetic fluctua-

219

M. Ausloos and S. Kruchinin (eds.), Symmetry and Pairing in Superconductors, 219–229.
© *1999 Kluwer Academic Publishers. Printed in the Netherlands.*

tions which also considers Van Hove singularities in the band structure. The $d_{x^2-y^2}$ gap symmetry is there found to be the most stable state. Moreover inter and intrablocks coupling models [10] which also consider Van Hove singularities can explain the pseudogap and find a *d-wave* gap symmetry. Theoretical implications are huge for transport properties.

In this work, we will emphasize the influence of those singularities on the thermal properties such as the the thermal conductivity κ, the thermal diffusivity α and the specific heat C. The consequences of the Van Hove scenario will be analyzed especially in view of different possible symmetries of the order parameter, namely *s-wave* or *d-wave* and the interaction between CuO_2 planes namely taking into account the Lawrence-Doniach model [11]. These physical ingredients will be included in the calculation through the quasi-particle spectrum. The point of view is phenomenological and does not make any hypothesis on the pairing mechanism.

Another success of the Van Hove scenario is to give an explanation of the variation of the critical temperature versus the doping level as soon as a relation between the doping level and the Fermi level is established. Such a dependence of the doping level on the critical temperature and the gap energy is considered here.

After some electronic band structure considerations, the paper is divided into two parts : the effect of Van Hove singularities on C, κ and α for an optimally doped compound (Sect.4) and the doping effect on C and κ (Sect.5). The theoretical calculations are compared in each section to the experimental results. Conclusions are drawn in Sect. 6.

2. Band Structure

If a strict 2D band structure is considered, there exist saddle points. That implies logarithmic divergences in the electronic band structure density of states. On the other hand, 1D Van Hove singularities, so called extended saddle points give square root divergences in the density of states. In this work, a simple 2D band structure is chosen as

$$\epsilon(\vec{k}) = \epsilon_F(1+\delta) + \frac{\hbar^2}{2m^*_{ab}} k_x k_y \qquad (1)$$

where ϵ_F is the Fermi energy, m^*_{ab} the effective mass in the plane and δ the relative energy distance between the Fermi level and the saddle point. This band structure is compatible with ARPES experiments and produces saddle points in (1,0) and (0,1) directions. This band structure can also be compared to the Hubbard band structure

$$\epsilon(\vec{k}) = -t(\cos(k_x) + \cos(k_y)) + t' \cos(k_x)\cos(k_y) \qquad (2)$$

where t and t' are the coupling energies between first and second neighbors respectively. In Fig.1, the Fermi surfaces of Eq.(1) and of Eq.(2) are presented. The saddle points are indicated by thick lines and circles.

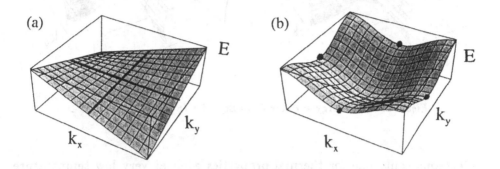

(a)

(b)

Figure 1. Comparison of 2D Fermi surfaces from (a) Eq.(1) and (b) Eq.(2).

Since cuprates are made of a stack of CuO_2 planes, holes can jump from one plane to another with a non zero probability. Those planes are coupled with an energy J given by

$$J = \frac{2\hbar}{s\gamma} \left(\frac{\epsilon_F}{m_{ab}^*}\right)^{1/2} \tag{3}$$

with $\gamma = (m_c^*/m_{ab}^*)$ the anisotropy parameter and s the interlayer distance. The energy J gives a relative idea about the dimensionality of the system; the smaller J, the more 2D the compound is. The band structure is then 3D and a new term can be considered, term which modulates the band structure in the c-direction as

$$\epsilon = \epsilon_F(1 + \delta) + \frac{\hbar^2}{2m_{ab}^*}k_x k_y + J\cos(k_z s) \tag{4}$$

In Fig.2 , energy surfaces of Eq.(4) are drawn with $\delta = 0$ and $J = 20 meV$ for $E < J$ and $E > J$. A topological change occurs for $E = J$. The Van Hove singularities are shifted by an energy equal to J as observed in the density of states (Fig.3a). Notice that the density of states does not diverge at $E = J$.

In this work, two symmetries of the order parameter are considered : the isotropic *s-wave* $(\Delta(\vec{k}) = 1)$ as the usual BCS theory and the *d-wave*, such that $\Delta(\vec{k})$ will have a $d_{x^2-y^2}$ symmetry, characterized by lines of nodes in the (1,1) and (-1,1) directions and a change of phases between the planes. The existence of line of nodes means that a non-zero density of states of normal

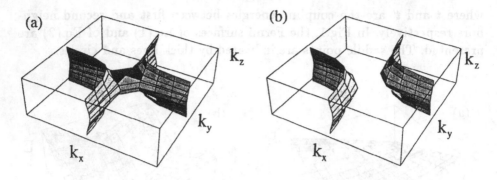

(a)

(b)

k_z

k_y

k_x

k_z

k_y

k_x

Figure 2. 3D energy surfaces from Eq.() for (a) $E < J$ and (b) $E > J$.

electrons contribue for thermal properties even at very low temperature. In Fig.3, the density of states of a quasiparticle spectrum with the band structure as in Eq.(4) and different gap parameters is represented. Two peaks in the density of states are pointed out : one for the superconduting state and one for the Van Hove singularity. Notice that states exist below the gap energy in the *d-wave* case.

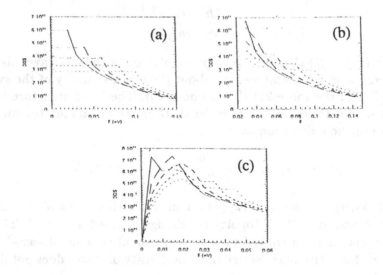

Figure 3. Density of states of a quasiparticle spectrum (Eq.(5)) with band structure of Eq.(4) (a) without gap, (b) *s-wave* gap symmetry and (c) *d-wave* gap symmetry .

3. Calculations

The electronic terms of each thermal properties have been calculated under the assumption that these terms give the most relevant contributions to the superconductor transition features. The quasiparticle spectrum is given by

$$E(\vec{k}, T) = E(\vec{k}) = \sqrt{(\epsilon(\vec{k}) - \epsilon_F)^2 + \Delta(\vec{k}, T)^2} \qquad (5)$$

where $\Delta(\vec{k}, T)$ is the function which describes the energy gap versus the temperature and the momentum. The gap energy dependence upon the temperature is taken as

$$\Delta(T) = \Delta(0) \tanh\left(\alpha\sqrt{\frac{T_c - T}{T}}\right) \qquad (6)$$

with $\alpha = 2.2$ [12]

Houssa et al. [14] have performed calculations of the electronic contribution to the thermal conductivity using the Köhler variational method [15]

$$\kappa_e^{\mu\mu} = \frac{1}{T}\left[\sum_{ij} U_i^{\mu\mu} Q_{ij}^{\mu\mu} U_j^{\mu\mu} - \frac{\left(\sum_{ij} J_i^{\mu\mu} Q_{ij}^{\mu\mu} U_j^{\mu\mu}\right)^2}{\sum_{ij} J_i^{\mu\mu} Q_{ij}^{\mu\mu} J_j^{\mu\mu}}\right] \qquad (7)$$

where $\mu = (x, y)$, $J_i^{\mu\mu}$ and $U_i^{\mu\mu}$ are the trial currents and $(i, j) = (1 \, or \, 2)$ and $Q_{ij}^{\mu\mu} = (P^{-1})_{ij}^{\mu\mu}$. The elements of the scattering matrix $P_{ij}^{\mu\mu}$ have to be calculated assuming dissipation effects to be the scattering of electrons by point defects and acoustic phonons as the most important heat carrier dissipation mechanisms. Three free physical parameters have to be taken into account, namely the interlayer coupling energy J, the impurity fraction N and the electron-phonon coupling λ.

The calculation of the electronic contribution to the specific heat, the calculation was done in a previous work [16] using the simple formula

$$C_e = \int E(\vec{k}, T) \frac{f(\vec{k}, T)}{dT} d\vec{k} \qquad (8)$$

with $f(\vec{k}, T)$ is the Fermi-Dirac distibution.

As far as the thermal diffusivity is concerned, it can be decomposed as follows [17] in terms of a phonon contribution α_{ph} and an electronic contribution α_e, both expressed in term of the electronic specific heat C_e (given by Eq.(8)), the electronic thermal conductivity κ_e, a phonon contribution to the thermal conductivity κ_{ph} and a phonon contribution to the specific heat C_{ph}, e.g. calculated within the Debye model, whence

$$\alpha_{tot} = \frac{\kappa_e + \kappa_{ph}}{(C_e + C_{ph})\rho} = \frac{\kappa_e}{\rho(C_{ph} + C_e)} + \frac{\kappa_{ph}}{\rho(C_{ph} + C_e)} = \alpha_e + \alpha_{ph} \qquad (9)$$

4. Theoretical results

The different thermal properties mentioned here above are calculated using the following physical parameters for illustration: $\Delta(0) = 20meV$, $T_c = 90K$, $m_{ab}^* = 8m_e$ (m_e is the mass of the electron), $\epsilon_F = 2000K$ and a Debye temperature $\Theta_D = 450K$.

4.1. THERMAL CONDUCTIVITY

In Fig.4, the theoretical electronic contribution to the thermal conductivity is plotted versus the temperature for different impurity fractions N. The normal state at high temperature does not depend upon N since the electron-phonon scattering dominates. An angular point is found at T_c. Then a bump occurs below T_c because of the increase of the mean-free path of normal electrons.

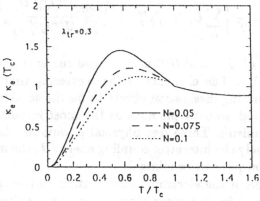

Figure 4. Theoretical thermal conductivity for different N values.

The electronic contribution to the thermal conductivity κ behaves at low temperature according an *exponential* law in the *s-wave* case (Fig.5a) and according to a *power* law with an exponent β in the *d-wave* case (Fig.5a)[14]. This exponent can be related to N, J and λ. The experimental behavior of the thermal conductivity at low temperature is represented in Fig.5c in the Y-123 case. A power law behavior is found in several high T_c superconductors, a feature which is compatible with the *d-wave* case plus Van Hove singularities [13].

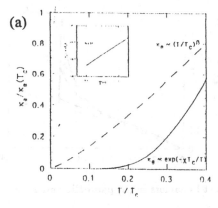
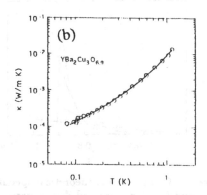

Figure 5. (a) Comparison of the behavior of κ at low temperature in the *s-wave* case (solid line) and in the *d-wave* case (broken line) (b) experimental measurements at low temperature on an Y-123 compound [16].

4.2. SPECIFIC HEAT

Several combinations of band structures (parabolic or with Van Hove singularity) and of order parameter symmetries (*s-* or *d-wave*) have been investigated. Experimentally, a linear term is found at low temperature [16] and a jump ΔC is found at the critical temperature. This normalized electronical jump defined as the ratio between the jump and the electronic specific heat in the normal state at T_c is about 2 [19]. In Fig.6, the normalized electronic specific heat is represented for two scenarios : (a) a parabolic band structure and an *s-wave* gap symmetry and (b) a band structure including saddle points and the *d-wave* case.

The only scenario in this model which is in agreement with the experimental data is the Van Hove singularities and the *d-wave* gap symmetry. Moreover the scenario of a parabolic band structure and a *d-wave* gap symmetry does not give any significant jump at T_c for realistic values of the energy gap.

4.3. THERMAL DIFFUSIVITY

Experimental results on the thermal diffusivity α show a change of slope in a log-log plot (Fig.8) for a $DyBa_2Cu_3O_{7-y}$ and of a $Y_{0.9}Ca_{0.1}Ba_2Cu_3O_{7-z}$ sample. That allows us to determine the critical temperature. Two different power laws are observed above and below the critical temperature. The power law exponent of the normal state behavior is the same in both compounds namely $-2/3$. On the other hand the exponent in the superconducting state is different in both cases and seems to be more sensitive

226

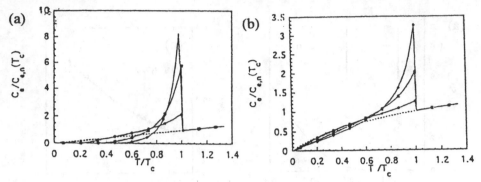

Figure 6. Comparison of theoretical specific heat behaviors in (a) parabolic and *s-wave* (BCS) and (b) Van Hove and *d-wave* case.

to impurities.

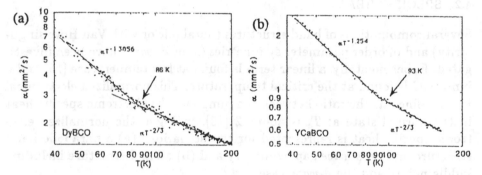

Figure 7. Total thermal diffusivity of (a) a $DyBa_2Cu_3O_{7-y}$ and of (b) a $Y_{0.9}Ca_{0.1}Ba_2Cu_3O_{7-x}$ sample [14]

Theoretical results are plotted in Fig.9 and allow to compare the *s-wave* case (broken line) to the *d-wave* case (solid line). A power law is found in the normal state and is impurity independent for the same reason as for the thermal conductivity in the latter case. A small jump is found at T_c due to the jump in the specific heat. Such a jump is also observed in experimental data [20, 21]. The thermal diffusivity behaves at temperatures below T_c with characteristic power laws which are consistent with experimental data. At very low temperature, α behaves as an *exponential* in the *s-wave* case and as a *power law* in the *d-wave* case. These different behaviors can be attributed to the remaining normal electrons on the Fermi surface at low temperature in the *d-wave* case.

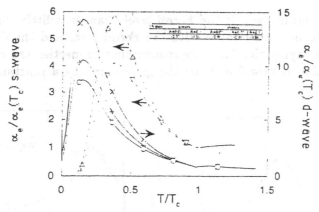

Figure 8. Theoretical comparison of the thermal diffusivity α in the *s-wave* case (broken line) and the *d-wave* case (solid line).

4.4. VAN HOVE AND *D-WAVE*

In order to explain low temperature behaviors, the first conclusion that we can drawn is that in this phenomenological model, the most probable scenario is a *d-wave* gap symmetry on which a band structure including Van Hove singularities is added in order to enhance the density of states in the direction where the gap is maximum, as sketched in Fig.9.

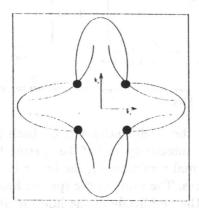

Figure 9. Sketch of the combination of *d-wave* and Van Hove singularities

5. Doping and Van Hove singularities

The doping level y in a $YBa_2Cu_3O_{7-y}$ compound can change the critical temperature as shown in Fig.4. Moreover the valence of the Cu(2) ion is also changed and the chemical potential is also modified. The similarity between both curves indicates that a change in the doping level corresponds to a change of the Fermi level. If the Fermi level changes a little, the saddle points do not coincide anymore with the Fermi level and the density of states becomes asymetric. The superconducting properties are degraded. This modification is also found for the gap energy as a function of δ in Eq.(1) from [22].

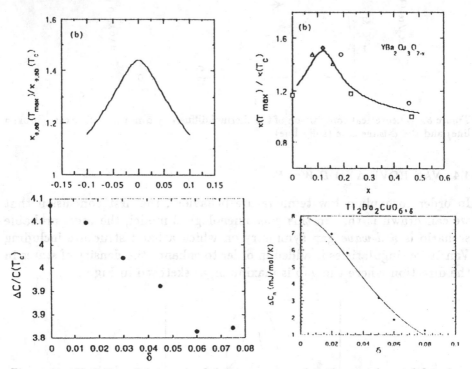

Figure 10. Variation of $\Delta\kappa_{norm}$ and ΔC_{norm} compared to the experimental data from [23,24,25] for κ and from [26] for C.

Taking into account the modification of δ on both T_c and $\Delta(0)$, calculations of the thermal conductivity and of the specific heat have been performed. As for the thermal conductivity, the bump is lowered and shifted toward high temperatures. The jump of the specific heat is shifted towards low temperatures and the amplitude of the jump is decreased. The theoretical normalized amplitude of the bump $\Delta\kappa_{norm}$ defined as $\kappa_{max}/\kappa(T_c)$ and the normalized jump of the specific heat $\Delta C_{norm} = \Delta C/C(T_c)$ are compared to experimental data in Fig.10. Good agreement is found.

6. Conclusions

The Van Hove scenario is compatible with band structure measurements and can explain the doping shift of the Fermi level and a change in the electronic density of states influence on the thermal transport properties and the specific heat. The demonstration of the effects of Van Hove singularities are useful in the controversy on the order parameter symmetry. Our calculations are in favour of a *d-wave* gap symmetry whenever no mixing has been considered, even in absence of a magnetic field. Universal laws are found in the low temperature behavior of the thermal conductivity characterized by an exponent β. Such a universal law is also found in the specific heat where $\beta = 1$.

Acknowledgements

Part of this work has been financially supported by the ARC 94- 99/174 contract of the Ministry of Higher Education and Scientific Research through the University of Liège Research Council. S. Dorbolo benefits from a FRIA research fellowship. Thanks to NATO for financially supporting this workshop and the organizers for providing such a fine setting for a meeting.

e-mail address :
(†) dorbolo@gw.unipc.ulg.ac.be
(‡) hassan@montefiore.ulg.ac.be

References

1. R.S. Markiewicz, *A survey of the Van Hove Scenario for High-T_c Superconductivity With Special Emphasis on Pseudogap and Striped Phases*, preprint.
2. D.S. Dessau, Z.X. Shen, D.M. King, D.S. Marshall, L.W. Lombardo, P.H. Dickinson, A.G. Loeser, J. DiCarlo, C.H. Park, A. Kapitulnik, and W.E. Spicer, Phys. Rev. Lett. 71, 2781 (1993).
3. K. Gofron, J.C. Campuzano, A.A. Abrikosov, M. Lindroos, A. Bansil, H. Ding, D. Koelling, and B. Dabrowski, Phys. Rev. Lett. 73, 3302 (1994).
4. J. Ma, C. Quitmann, R.J. Kelley, P. Alméras, H. Berger, G. Margaritondo, and M. Onellion, Phys. Rev. B 51, 3832 (1995).
5. J.C. Campuzano, G. Jennings, M. Fais, L. Beaulaigue, B.W. Veak, J.Z. Liu, A.P. Paulikas, K. Vanderwoort, H. Claus, R.S. List, A.J. Arko, and R.J. Bartlett, Phys. Rev. Lett. 64, 2308 (1990).
6. M.R. Norman, M. Randeria, H. Ding, and J.C. Campuzano, Phys. Rev. B52, 615 (1995).
7. E.A. Pashitskii, V.I. Pentegov, and A.V. Semenov, J. Low Temp. Phys 23, 101 (1997).
8. G.Litak, A.M. Martin, B.L. Györfyy, J.F. Annett, and K.I. Wysokiński, *Van Hove Singularity and D-Wave Pairing in Disordered Superconductors*, preprint.
9. E. Dagotto, A. Nazarenko, and A. Moreo, Phys. Rev. Lett. 74, 310 (1995).
10. J. Friedel, J. Physique 48, 1787 (1987).
11. W.E. Lawrence and S. Doniach, in *Proc. of the 12th Int. Conf. on Low Temp. Phys.*, Kyoto, 1971 edited by E. Kanda (Keigaku Publishing, Tokyo, 1972)p.361
12. R. Meservey and B.B.Schwartz, in *Superconductivity*, vol.1, R.D.Parks (M. Dekker, New-York, 1969) p.117.
13. S. Dorbolo, Physica C 276, 175 (1997).
14. M. Houssa and M. Ausloos, Europhys. Lett. 33, 695 (1996).
15. J.M. Ziman, *Electrons and Phonons* (Clarendon, Oxford, 1963).

MEAN FREE PATH EFFECTS ON THE ELECTRONIC
RAMAN SPECTRA OF d-WAVE SUPERCONDUCTORS

A. BILLE, C. T. RIECK AND K. SCHARNBERG

I. Institut für Theoretische Physik, Universität Hamburg
Jungiusstraße 9, D-20355 Hamburg, Germany

Abstract. Because of its polarization dependence, Raman scattering appears to be particularly well suited to study the anisotropy of quasiparticle interactions. It is likely that these interactions are responsible both for finite mean free paths and for superconductivity. Since d-wave superconductivity in a single band is possible only in the presence of a strongly anisotropic interaction, direct evidence for such anisotropy would lend support to the currently favored notion of a pair state with d-wave symmetry. Comparison of the low frequency B_{1g} and B_{2g} response in the normal state could provide an indication of the strength and anisotropy of the interaction. In the superconducting state a quasiparticle contribution to the Raman response, which would carry information on the anisotropy and temperature dependence of the interaction in the most easily identifiable way, should be observable in B_{2g} symmetry, if the experimental resolution at low frequencies is improved. Unfortunately, it is very unlikely that the corresponding data in B_{1g} symmetry will ever become available. In any case, high quality samples are required because disorder induced s-wave scattering would obscure effects due to the pairing interaction. The pair breaking peak also carries information on quasiparticle scattering, but it would take a careful quantitative analysis of the temperature dependence to find out whether pairing and scattering are described consistently by one interaction. The relative change in the Raman intensities at high frequencies should reflect the degree to which the quasiparticle interaction is separable with respect to frequency and momentum.

M. Ausloos and S. Kruchinin (eds.), Symmetry and Pairing in Superconductors, 231–244.

1. Introduction

In most conventional superconductors, finite quasiparticle mean free paths are the result of static disorder, while the pairing is due to exchange of virtual phonons. In the Cuprates, the transition temperatures T_c are so high that the boson exchange responsible for the pairing will also have a significant effect on all physical properties which reflect finite quasiparticle lifetimes. It follows from the temperature dependence of the normal state resistivity, which for optimally and overdoped samples extrapolates very closely to zero, that scattering is predominantly inelastic. If we make the plausible assumption that whatever interaction causes superconductivity in HTC-materials also determines the quasiparticle mean free paths above T_c, then the study of mean free path effects provides another perspective on the pairing interaction.

It has been concluded from the observation of a pronounced peak in the temperature dependence of the real part $\sigma_1(T, \omega)$ of the microwave conductivity, that the inelastic scattering is sharply reduced below T_c [1, 2]. This interpretation of the peak as a mean free path effect has been substantiated decisively by measurements on samples doped with Zn [3], which showed the expected reduction in peak height. The appearance of such a peak in $\sigma_1(T, \omega)$ is consistent with spin fluctuation exchange as pairing mechanism, but the agreement between this theory [4, 5] and experiment is not quantitative over the whole temperature range, possibly because the anisotropy that is responsible for d-wave pairing was not taken into account.

When some phenomenological pairing interaction, which captures both the momentum dependence and the frequency dependence of the spin fluctuation exchange, is used to calculate the electromagnetic response in the superconducting state, a peak in $\sigma_1(T, \omega)$ will be found even when the pairing interaction is assumed to be temperature independent [6]. This is due to the fact that quasiparticle mean free paths, which result from this anisotropic interaction, can vary substantially over the Fermi surface, with those states near the order parameter nodes having the longest lifetimes. Thus, as the temperature is lowered, only those quasiparticles will continue to contribute to $\sigma_1(T, \omega)$ which have the longest lifetimes. Together with the usual decrease of inelastic scattering, known from electron-phonon interaction which remains practically unaffected by the onset of superconductivity, this selection of long living quasiparticle states will more than compensate the reduction in the number of occupied states and thus will contribute to the observed peak.

Since the conductivity of the tetragonal CuO_2-planes is a scalar, it is not possible to separate such anisotropy effects from effects resulting from a temperature dependent modification of the pairing interaction due to the onset of superconductivity. The electronic Raman response has the attractive feature that by using different combinations of polarization vectors of the incoming and the reflected light, different parts of the Fermi surface can be sampled. It should thus be possible to detect anisotropy in the inelastic scattering rate. Scattering is, in fact, crucial for the observation of this response in the normal state because it relaxes the requirements of momentum and energy conservation. In the superconducting state, the possibility of breaking Cooper pairs renders the electronic Raman response finite even in the clean limit. Theoretical results derived for clean $d_{x^2-y^2}$ superconductors already seem to account for the gross features of many of the available experimental data taken at very low temperatures [7, 8, 9]. However, inelastic scattering has to be invoked even at the lowest temperatures in order to explain the finite Raman intensity at large frequencies. As the temperature is lowered, elastic scattering begins to compete with inelastic scattering at low frequencies. For a complete picture of mean free path effects in the superconducting state it is necessary, therefore, to include elastic scattering. This is done here in the usual t-matrix approximation where only s-wave scattering is taken into account [10, 11].

2. Band Structure and the Raman Scattering Cross Section

The scattering cross section for light undergoing a frequency shift ω when interacting inelastically with band electrons in a metal is, apart from some trivial factors, given by the Raman response function $\chi_{\gamma\gamma}(\mathbf{q}, \omega)$. [12, 13, 14, 9] The relevant wave vector $\mathbf{q}$ is that of the radiation inside the metal, i.e. the inverse of the penetration depth. Since in HTC-materials the penetration depth is very large, we need to consider only the limit $\mathbf{q} \to 0$. For a single band, $\chi_{\gamma\gamma}(0, \omega)$ is obtained by analytic continuation of

$$\chi_{\gamma\gamma}(0, i\nu_m) = -T \sum_{\mathbf{k}, i\omega_n} Tr\left[\hat{G}(\mathbf{k}, i\omega_n + i\nu_m)\tau_3\hat{G}(\mathbf{k}, i\omega_n)\tau_3\right]\gamma_{\mathbf{k}}^2$$

$$+ \text{ vertex corrections}. \tag{1}$$

The single particle Green's function

$$\hat{G}(\mathbf{k}, i\omega_n) = \frac{i\omega_n Z(\mathbf{k}, i\omega_n)\tau_0 + (\epsilon_{\mathbf{k}} + \chi(\mathbf{k}, i\omega_n))\tau_3 + \phi(\mathbf{k}, i\omega_n)\tau_1}{[i\omega_n Z(\mathbf{k}, i\omega_n)]^2 - [\epsilon_{\mathbf{k}} + \chi(\mathbf{k}, i\omega_n)]^2 - [\phi(\mathbf{k}, i\omega_n)]^2} \tag{2}$$

depends on some bare band energy $\epsilon_{\mathbf{k}}$, which also determines the dimensionless Raman vertex [15]

$$\gamma_{\mathbf{k}} = ma^2 \sum_{\alpha,\beta} e_{\alpha}^S \frac{\partial^2 \epsilon_{\mathbf{k}}}{\partial k_{\alpha} \partial k_{\beta}} e_{\beta}^I. \tag{3}$$

234

Here, e^I and e^S are the polarization vectors of the incoming and the scattered light, respectively, and a is the lattice constant. Momenta are given in units of $1/a$. For ϵ_k we shall adopt the widely used model [16]

$$\epsilon_k = -2t \left[\cos(k_x) + \cos(k_y) - 2B\cos(k_x)\cos(k_y) + \mu/2\right] \qquad (4)$$

which can be justified as a reasonable approximation to the results of band structure calculations and which can reproduce the Fermi lines observed in photoemission experiments.

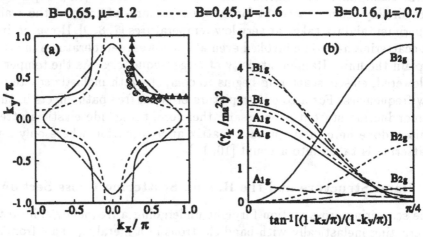

——— B=0.65, μ=-1.2   ····· B=0.45, μ=-1.6   ——— B=0.16, μ=-0.7

Figure 1. Panel (a): Fermi lines for the dispersion relation (4) for three different sets of parameters B and μ. Circles are ARPES data from Campuzano *et al.* [17], triangles are data from Rong Liu *et al.* [18, 19]. Panel (b): Variation of the squares of the Raman vertices (5) along the Fermi lines shown in panel (a).

Fig. 1a shows the Fermi lines for three different values of the next nearest neighbor hopping matrix element B and different chemical potentials μt, together with ARPES data for $YBa_2Cu_3O_{7-\delta}$. [17, 18, 19] The parameter set $B = 0.65$, $\mu = -1.2$ was used by Devereaux and Einzel [8], $B = 0.16$, $\mu \approx -0.7$, which may be more appropriate for $La_{2-x}Sr_xCuO_4$ [20], was used by Branch and Carbotte [21]. From Eqs. (3) and (4) one obtains the Raman vertices for the three polarization geometries used in experiments

$$\gamma_k^{A_{1g}} / ma^2t = \cos(k_x) + \cos(k_y) - 4B\cos(k_x)\cos(k_y) ,$$

$$\gamma_k^{B_{1g}} / ma^2t = \cos(k_x) - \cos(k_y) ,$$

$$\gamma_k^{B_{2g}} / ma^2t = 4B\sin(k_x)\sin(k_y) . \qquad (5)$$

In Fig. 1b we show the variation of the squares of these normalized Raman vertices along the Fermi lines from Fig. 1a. The position on the Fermi line is parametrized by the angle with respect to the $(0,1)$-axis, measured from the M-point. The variation of $\gamma_{\mathbf{k}}$ with angle reflects the expected symmetry [11] but the actual values differ considerably. Hence, even the predictions for the *relative* Raman intensities can vary dramatically. Since these Raman vertices represent the *curvature* of the dispersion relations, they are extremely sensitive to details of the band structure calculations. A microscopic derivation, therefore, appears to be hardly more reliable than the models used here, especially when strong correlation effects are present.

$\gamma_{\mathbf{k}}^{A_{1g}}$ and $\gamma_{\mathbf{k}}^{B_{1g}}$ are very similar for the examples with $B = 0.65$ and $B = 0.45$ (solid and short dashed lines). Not surprisingly, then, the Raman intensities calculated from Eqs. (1)-(4) without vertex corrections are also found to be very similar at all temperatures. In the case of the A_{1g} symmetry it is essential, though, to take vertex corrections into account in order to comply with the requirements of charge conservation. [11] Again depending on the band structure, these vertex corrections reduce the A_{1g} Raman intensity by one to three order of magnitude. [21] Because of these enormous uncertainties, which indicate that the single band Raman vertex Eq. (3) is probably an oversimplification, we shall not discuss the A_{1g} response any further in this paper. In the B_{1g} and the B_{2g} geometries vertex corrections are also nonvanishing when the interactions are momentum dependent, [22] but due to space limitations, further discussion of vertex corrections has to be deferred.

3. Interactions and Scattering Rates

Within a weak coupling approach, we choose a separable pairing interaction that gives the d-wave pair state $\phi(\mathbf{k}, i\omega_n) = \frac{1}{2}\Delta(T)[\cos(k_x) - \cos(k_y)]$. $\Delta(T = 0)$ is a parameter, and the temperature dependence of $\Delta(T)$ is essentially the same as in BCS theory. The selfenergy $\chi(\mathbf{k}, i\omega_n)$ in Eq. (2) vanishes. $Z(\mathbf{k}, i\omega_n)$ is calculated taking into account elastic scattering [10, 11, 23] as well as inelastic scattering in the nested Fermi liquid (NFL) model [24, 25]. This model not only describes the qualitative features of the resistivity and the Raman response in the normal state correctly, it can also be generalized in a simple and obvious way to the superconducting state. Since the NFL scattering rate is evaluated only at the Fermi surface, we shall calculate the Raman response assuming an infinite band width.

We also perform strong coupling calculations in which all three self-energy parts are determined selfconsistently from anisotropic Eliashberg equations. In addition to $\epsilon_{\mathbf{k}}$ Eq. (4) one requires as input an expression for the Eliashberg function. This we assume to result from spin fluctuation exchange, but we do not attempt to calculate it selfconsistently [26]. Instead we use a model introduced by Monthoux and Pines[27, 28]

$$\alpha^2 F_{sf}(\mathbf{q}, \Omega) = \frac{g_{sf}^2}{\pi} \frac{\chi_Q \frac{\Omega}{\omega_{sf}}}{\left[1 + \xi^2 (\mathbf{q} - \mathbf{Q})^2\right]^2 + \frac{\Omega^2}{\omega_{sf}^2}} \qquad \begin{array}{c} \mathbf{Q} = (\pi, \pi) \\ 0 \leq q_{x,y} \leq \pi \end{array} \qquad (6)$$

with temperature independent coefficients $\omega_{sf} = 7.7\,\text{meV}$, $\xi = 2.5$ and $\chi_Q = 51\,\text{states/eV}$. In the limited temperature range above T_c in which we are interested here, additional explicit temperature dependencies of these parameters [28] are expected to produce only minor quantitative changes [22]. The coupling constant g_{sf} is fixed in such a way that for the band structure under consideration a superconducting transition temperature T_c of around $90\,\text{K}$ is obtained. Possible modifications of Eq. (6) resulting from the onset of superconductivity are discussed below.

The quasiparticle scattering rates are given by the position of the poles in the complex ω-plane of the Green function Eq. (2). Obtaining the scattering rates would thus require knowledge of the selfenergy parts for arbitrary complex ω, which would involve an unwarrantable numerical expenditure. We evaluate the selfenergy parts on the imaginary frequency axis, continue to the real axis using Padé approximants, and then use the results as starting points for an iterative calculation of the imaginary parts on the real frequency axis. From these, the real parts are obtained through a Kramers-Kronig transformation. All three selfenergy parts contribute to the scattering rates [29], but since our numerical results are only meant to provide a better understanding of the dependence of the Raman response on polarization, frequency and temperature, we shall content ourselves with $\omega \operatorname{Im} Z(\mathbf{k}, \omega)$ as a representation of lifetime effects, even though this quantity can be a sizeable fraction of the band width.

In Fig. 2a we show $\omega \operatorname{Im} Z(\mathbf{k}, \omega)$ at $T = 100\,\text{K}$ and $T = 10\,\text{K}$ as function of frequency. At high frequencies we observe the expected decrease with decreasing temperature. In the superconducting state, $\omega \operatorname{Im} Z(\mathbf{k}, \omega)$ drops more precipitously over a frequency range determined by the amplitude of the order parameter, tending to zero for $\omega = 0$ in the clean limit. We have not adjusted the pairing interaction Eq. (6) for the expected loss of spectral weight at low frequencies. This effect is incorporated in the NFL model, which shows a greater reduction, persisting to higher frequencies, in the superconducting state (see inset).

The anisotropy in $\omega \operatorname{Im} Z(\mathbf{k}, \omega)$ evident from Fig. 2a is analyzed in more detail in Fig. 2b. The anisotropy is greatest for $B = 0.45$, $\mu = -1.6$ because in this case, $\mathbf{Q}$ connects points on the Fermi line close to the Brillouin zone boundary where the density of states is highest. Similar curves are obtained from FLEX-calculations [30]. Nesting between these "hot spots" is believed to be essential in high temperature superconductors [31]. For $B = 0.65$, $\mu = -1.2$, optimum nesting is obtained at some intermediate point on the Fermi line, so that the anisotropy is reduced. The overall size

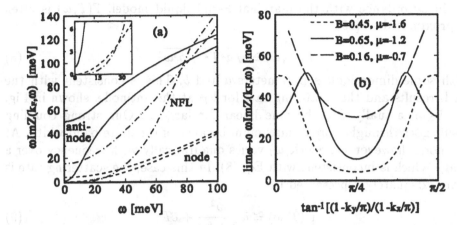

Figure 2. Panel (a): Scattering rates as function of frequency for the band structure (4) with $B = 0.45$ and $\mu = -1.6$ for two points on the Fermi line, corresponding to the node and the antinode of the superconducting order parameter. Also shown is the isotropic NFL result. The set of three curves starting at $\omega = 0$ is calculated in the superconducting state at $T = 10\,\text{K}$. The remaining three curves represent scattering rates in the normal state at $T = 100\,\text{K}$. Panel (b): Variation of the scattering rates at $T = 100\,\text{K}$ and $\omega = 0$ along the Fermi lines shown in Fig. 1a.

of the scattering rate is increased because as a result of poorer nesting the coupling constant g_{sf}^2 had to be increased from $0.9\,[\text{eV}]^2$ to $1.6\,[\text{eV}]^2$ in order to keep $T_c \approx 90\,\text{K}$. For the third example, we require $g_{\text{sf}}^2 = 1.4\,[\text{eV}]^2$. The quality of nesting is better than in the previous example but it does not vary much along the Fermi line so that there is very little anisotropy. It is evident from Eq. (6) that the anisotropy becomes smaller at higher frequencies. This would not be the case for separable interactions [32].

4. Approximate Results and Limiting Cases

Even in the normal state, $\text{Im}\,\chi_{\gamma\gamma}(\omega)$ involves integration with respect to momentum and frequency. Replacing the scattering rate in the integrand by some suitable temperature and polarization dependent average $\Gamma(T, \omega)$ [23], one arrives at the Drude-like expression

$$- \,\text{Im}\,\chi_{\gamma\gamma}(\omega) \cong \frac{\omega/\Gamma(T,\omega)}{1 + [\omega/\Gamma(T,\omega)]^2}. \tag{7}$$

Since the Raman vertices with their inherent uncertainties are buried in the definition of $\Gamma(T,\omega)$, this expression should not be used to determine the scattering rates.

In accordance with the marginal Fermi liquid model, $\Gamma(T,\omega)$ is often approximated as [25]

$$\Gamma(T,\omega) \cong \sqrt{a^2 T^2 + b^2 \omega^2} \tag{8}$$

with some dimensionless parameters a and b. This is consistent with the NFL results and the scattering rate for k_F at the antinode, shown in Fig. 2a. Γ_{NFL} actually does become linear for large ω. Our strong coupling result goes through a broad maximum because of the finite band width. At the node, however, $\omega \operatorname{Im} Z(k,\omega)$ varies quadratically with frequency over a range which is inconsistent with Eq. (8). In this case, the scattering rate is more adequately represented by

$$\Gamma(T,\omega) \cong \bar{a}\,\frac{\omega^2}{1+\bar{b}\omega} + \bar{c}T . \tag{9}$$

While the $q = 0$ electronic Raman response in the normal state is finite only if quasiparticles have a finite lifetime, breaking of Cooper pairs provides a mechanism for Raman scattering in the clean limit [8, 9]. In a Fermi surface restricted approach one finds

$$-\operatorname{Im}\chi_{\gamma\gamma}(\omega) = \frac{4\pi N(0)}{\omega}\tanh\left(\frac{\omega}{4T}\right)$$

$$\int\limits_{0}^{\pi/2} d\varphi\,\rho(\varphi)\,\gamma^2(\varphi)\,[2\Delta(\varphi,T)]^2\,\frac{\theta\,[\omega - |2\Delta(\varphi,T)|]}{\sqrt{\omega^2 - [2\Delta(\varphi,T)]^2}} , \tag{10}$$

where the angle φ parametrizes the position along the Fermi line. $\rho(\varphi) = p_F(\varphi)/4\pi^2 N(0)v_F(\varphi)$ is the normalized angle dependent normal state density of states. From this expression power laws for the low frequency Raman response are easily derived [8]. It is also obvious from this expression, that the B_{1g} response diverges at $\omega = 2\Delta(T)$.

5. Results and Discussion

In Fig. 3 we present normal state results for the interaction Eq. (6). The B_{1g} response shows the expected behavior, consistent with Eqs. (7) and (8). The shallow peak in the B_{2g} response is also consistent with Eq. (8), provided $b^2 < 1$. The sharp peak, followed by a minimum, that develops as the temperature is lowered, can be explained if the scattering rate is of the form given in Eq. (9). Fig. 3b shows that this behavior does not change qualitatively as long as the anisotropy of the scattering rate, which is related to the shape of the Fermi surface (cf. Fig. 2b) and the width of the interaction, persists. The structure becomes less pronounced as the strength g_{sf}^2 of the interaction increases. Observation of this structure in the B_{2g} response and

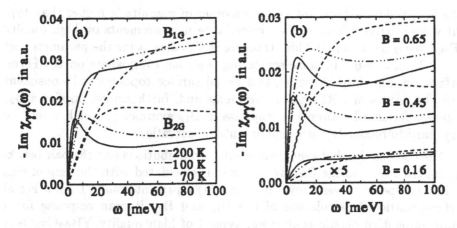

Figure 3. Panel (a): B_{1g} and B_{2g} Raman response in the normal state at three different temperatures for the band structure (4) with $B = 0.45$ and $\mu = -1.6$. Panel (b): Effect of band structure on the B_{2g} response at the same three temperatures as in Panel (a). Note, that results for $B = 0.16$ had to be multiplied by a factor of 5 to be visible in this display.

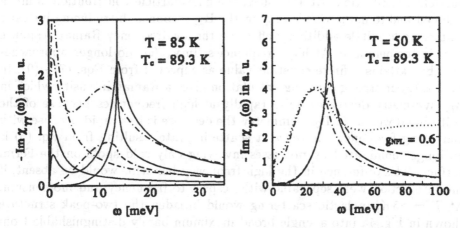

Figure 4. Panel (a): B_{1g} (solid lines) and B_{2g} (dot dashed lines) Raman response in the presence of elastic scattering $\Gamma_{imp} = 0.4\,\text{meV}$, scattering phase shift $\delta_N = 0.4\pi$. Band structure parameters are $B = 0.65$ and $\mu = -1.2$. Panel (b): Solid line and dot-dashed line represent the same quantities as in Panel (a), but at a lower temperature. The long dashed line and the dotted line show the effect of inelastic scattering within the NFL model. The coupling constant g_{NFL} is chosen such that a typical values of the resistivity above T_c are obtained.

its absence in the B_{1g} response would lend support to the current models of the Fermi surface topology and the pairing interactions. Experimental data available at present do not show this structure. Agreement between such data and this theory can be achieved by introducing an elastic scattering rate $\Gamma_{imp} \cong 4\,\mathrm{meV}$ [22], but this is an order of magnitude higher than typical values derived from surface impedance measurements on high quality YBa$_2$Cu$_3$O$_{7-\delta}$-films [23] This structure is also absent for the parameter set $B = 0.16$, $\mu = -0.7$ because scattering is strong everywhere on the Fermi surface (cf. Fig. 2b). However, this Fermi surface topology is inconsistent with the results of ARPES measurements and, furthermore, implies a B$_{2g}$ response far smaller than the response in B$_{1g}$ symmetry, which would be very hard to reconcile with experimental observations.

The competition between quasiparticle contributions and the pair breaking contributions to the Raman response is elucidated with the help of Fig. 4. The two-peak curves shown in Panel (a) represent results from a Fermi surface restricted calculation of the B$_{1g}$ and B$_{2g}$ Raman response for a small amount of elastic scattering, typical of high quality YBa$_2$Cu$_3$O$_{7-\delta}$ samples [23]. Curves with a single peak at low frequencies represent normal state results, described by Eq. (7) with constant Γ. Curves vanishing at $\omega = 0$ are clean limit results obtained from Eq. (10). Even that close to T_c, the contribution from Cooper pair breaking is comparable with the quasiparticle contribution. At $T = 50\,\mathrm{K}$, the quasiparticle contribution is much reduced and remains visible only in the B$_{2g}$ response. Introducing inelastic scattering has little additional effect on the low frequency Raman response at this temperature. At high frequencies, $\mathrm{Im}\chi_{\gamma\gamma}(\omega)$ no longer approaches zero but attains a finite constant value as expected from Eqs. (7) - (9). In contrast, our strong coupling calculation gives a Raman intensity which in B$_{1g}$ symmetry decreases quite rapidly at high frequencies because of the finite band width. In B$_{2g}$ symmetry the decrease is less rapid because with increasing frequency the relevant scattering rate resulting from Eq. (6) is no longer dominated by processes involving only cold spots on the Fermi surface. This difference in the high frequency behavior would be absent, if the interaction were separable with respect to frequency and momentum. At $T = 85\,\mathrm{K}$, inelastic scattering would broaden the two-peak structure shown in Fig. 4a into a single broad maximum barely distinguishable from the $T = 100\,\mathrm{K}$ result shown in Fig. 3b, unless the scattering rate were to collapse very fast below T$_c$.

Fig. 5 shows results from a strong coupling calculation involving the whole Brillouin zone. We have included in Fig. 5a results in the normal state at $T = 50\,\mathrm{K}$ and $T = 20\,\mathrm{K}$. The peak already noticed in connexion with Fig. 3 sharpens up and moves to even lower frequencies as T is lowered. The blow-up in Panel (b) shows that this peak survives in the superconducting state as a low-frequency shoulder, plateau, or even a satellite peak to the pair breaking peak. The exact behavior depends on the strength of the interaction and the nesting properties of the Fermi surface. The same features can be seen in Fig. 4b, where the isotropic NFL selfenergy most likely

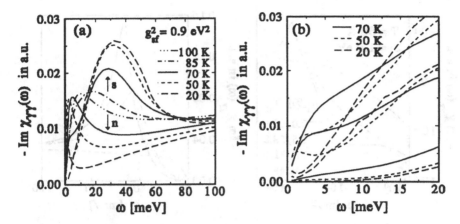

Figure 5. Panel (a): B_{2g} response for the band structure (4) with $B = 0.45$ and $\mu = -1.6$ at various temperatures. Normal state results at low temperatures are obtained by putting the order parameter equal to zero. Panel (b): Comparison of low the frequency B_{2g} response for two band structures. The set of three lines increasing most strongly towards large ω belongs to $B = 0.65$ and $\mu = -1.2$. The lowest set of three lines represent the B_{1g} response. The remaining lines have been taken from Panel (a).

overestimates the amount of scattering on the portion of the Fermi surface relevant for the B_{2g} response. Even so, this inelastic scattering causes only a small increase in the quasiparticle contribution over that caused by elastic scattering.

There is no obvious way how to generalize the pairing interaction Eq. (6) to the superconducting state. It is generally agreed that the spin fluctuations should reflect the reduction of the electron density of states at low frequency that results from the onset of superconductivity. If one multiplies $\alpha^2 F_{sf}(\mathbf{q}, \Omega)$ by ω^2 for $\omega \le \Delta(T)$, the order parameter amplitude calculated selfconsistently is found to be far too small. This is probably due to the choice of a rather small value $\omega_{sf} = 7.7$ meV in Eq. (6). Using a single d-wave DOS factor, we find the expected development of a low frequency quasiparticle peak in the B_{2g} response, with only minor changes to the pair breaking peak.

Lastly we examine the effect of strong disorder [10, 11] on the B_{1g} response in order to see whether this can change the low frequency ω^3-behavior to a linear behavior as observed by Sacuto *et al.* [33], without destroying superconductivity. Panel (a) of Fig. 6 is a double logarithmic plot of strong coupling results calculated for $T = 50$ K and a range of impurity concentrations n_i. The strength U_i of the individual scatterer is 1.0 eV which is already very close to the unitarity limit $\delta_N = \pi/2$ since increasing U_i to 5.0 eV has no visible effect on these results. Reducing U_i to $= .25$ eV nearly

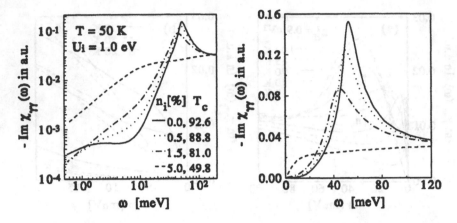

Figure 6. Effect of strong *elastic* scattering on the B_{1g} response. Band structure parameters are $B = 0.45$, $\mu = -1.6$. Panel (a): double logarithmic plot, Panel (b): linear plot of the same numerical data.

halves the normal state scattering rate and brings Im $\chi_{\gamma\gamma}$ closer to the clean limit, as expected, but this change can be compensated by increasing n_i. Thus, the scattering phase shift δ_N appears to be less important here than in the case of the low temperature microwave conductivity [23].

In the clean limit there is a plateau below $\omega = 10\,\mathrm{meV}$, which represents the quasiparticle contribution already discussed in connexion with Fig. 5b. This contribution is, however, two orders of magnitude lower than the Raman intensity at high frequencies so that there is little hope for quantitative experimental verification. Increasing the scattering does lead to the expected low frequency behavior [10, 11]. At the same time, however, T_c is reduced as listed in Fig. 6a. At the highest impurity concentration, T_c has dropped below the temperature used in the calculation. In a linear plot of the same data, shown in Fig. 6b, the change in slope is hardly recognizable while there is a substantial change in peak height and width. As in Fig. 4, the shift in peak position is due to the change in the order parameter amplitude with T/T_c. Since the samples investigated had T_c's around 126 K, disorder alone cannot account for the experimental observations of Sacuto *et al.* [33]. Changing the node structure to $\cos(4\phi) + s$ [33] or $\cos(2\phi) + r$ [34], does not appear to provide a simple and convincing explanation either.

6. Conclusions

Temperature and polarization dependencies of the Raman response in the normal state at low frequencies can provide information about the strength and the anisotropy of the inelastic scattering and hence on the pairing interaction. A low frequency peak structure in the B_{2g} response, which is absent in the B_{1g} response, is evidence for the presence of cold and hot spots on the Fermi surface. Whether or not a strongly anisotropic pairing interaction is separable with respect to frequency and momentum could be gleaned from a comparison of the high frequency behavior at low temperatures.

In B_{2g} symmetry this theory predicts quite a clear quasiparticle contribution at intermediate temperatures in the form of a low frequency shoulder or a distinct peak, which reflects the properties of the pairing interaction. Its experimental verification requires improved resolution at low frequencies, but this may well be within the scope of future experimental developments. Because of the rapid drop in the number of quasiparticles occupying states near the order parameter maximum, the low-frequency quasiparticle contribution to the B_{1g} response decreases below T_c so rapidly with temperature that there is little hope to find experimental evidence for it. An investigation of the pairing interaction through quasiparticle lifetimes is, therefore, only possible for the portion of the Fermi line near the node.

However, scattering also affects the properties of the condensate and here the pair breaking peak in B_{1g} symmetry provides a sensitive probe. An investigation of the temperature dependence of this peak would give complementary information on the decrease of the scattering rate in the superconducting state, albeit in a different frequency regime.

Disorder induced elastic scattering changes the low frequency power law observed in B_{1g} symmetry from cubic to linear. However, even for the largest scattering rates compatible with d-wave superconductivity, the predicted changes in Raman intensity are rather small. This can be attributed to the fact that for scattering rates small compared to the order parameter amplitude, far fewer states are created at the antinode, sampled in B_{1g} symmetry, than at the node of the d-wave gap. This aspect of the theory should be tested by measurements on a series of samples with controlled impurity content.

Acknowledgements

The authors are grateful to A. Bock, D. Manske. M. R. Norman, and M. A. Rübhausen for useful discussions. This work has been supported by the Deutsche Forschungsgemeinschaft through the Graduiertenkolleg "Physik nanostrukturierter Festkörper."

References

1. M. C. Nuss, P. M. Mankiewich, M. L. O'Malley, E. H. Westerwick, and P. B. Littlewood, Phys. Rev. Lett. **66**, 3305 (1991).
2. D. A. Bonn, R. Liang, T. M. Riseman, D. J. Baar, D. C. Morgan, K. Zhang, P. Dosanjh, T. L. Duty, A. MacFarlane, G. D. Morris, J. H. Brewer, and W. N. Hardy, Phys. Rev. B **47**, 11 314 (1993).

244

3. D. A. Bonn, S. Kamal, K. Zhang, R. Liang, D. J. Baar, E. Klein, and W. N. Hardy, Phys. Rev. B **50**, 4051 (1994).
4. P. J. Hirschfeld, W. O. Putikka, and D. J. Scalapino, Phys. Rev. B **50**, 10250 (1994).
5. C. T. Rieck, W. A. Little, J. Ruvalds, and A. Virosztek, Phys. Rev. B **51**, 3772 (1995).
6. A. Bille and K. Scharnberg, J. Phys. Chem. Solids (1998).
7. T. P. Devereaux, D. Einzel, B. Stadlober, R. Hackl, D. H. Leach, and J. J. Neumeier, Phys. Rev. Lett. **72**, 396 (1994).
8. T. P. Devereaux and D. Einzel, Phys. Rev. B **51**, 16336 (1995).
9. D. Einzel and R. Hackl, J. Raman Spec. **27**, 307 (1996).
10. T. P. Devereaux, Phys. Rev. Lett. **74**, 4313 (1995).
11. T. P. Devereaux and A. P. Kampf, Int. J. Mod. Phys. **11**, 2093 (1997).
12. A. A. Abrikosov and L. A. Fal'kovskii, Sov. Phys. JETP **13**, 179 (1961).
13. S. S. Jha, Nuovo Cimento B **63**, 331 (1969).
14. R. F. Wallis and M. Balkanski, *Many-Body Aspects of Solid State Spectroscopy* (North-Holland, Amsterdam, 1986).
15. A. A. Abrikosov and V. M. Genkin, Sov. Phys. JETP **38**, 417 (1974).
16. T. Schneider, H. de Raedt, and M. Frick, Z. Phys. B **76**, 3 (1989).
17. J. C. Campuzano, G. Jennings, M. Faiz, L. Beaulaigue, B. W. Veal, J. Z. Liu, A. P. Paulikas, K. Vandervoort, H. Claus, R. S. List, A. J. Arko, and R. J. Bartlett, Phys. Rev. Lett. **64**, 2308 (1990).
18. R. Liu, B. W. Veal, A. P. Paulikas, J. W. Downey, H. Shi, C. G. Olson, C. Gu, A. J. Arko, and J. J. Joyce, Phys. Rev. B **45**, 5614 (1992).
19. R. Liu, B. W. Veal, A. P. Paulikas, J. W. Downey, P. J. Kostic, S. Fleshler, U. Welp, C. G. Olson, X. Wu, A. J. Arko, and J. J. Joyce, Phys. Rev. B **46**, 11056 (1992).
20. P. Bénard, L. Chen, and A.-M. S. Tremblay, Phys. Rev. B **47**, 15217 (1993).
21. D. Branch and J. P. Carbotte, Phys. Rev. B **54**, 13288 (1996).
22. T. P. Devereaux and A. P. Kampf, cond-mat/9711039 5 Nov 1997.
23. S. Hensen, G. Müller, C. T. Rieck, and K. Scharnberg, Phys. Rev. B **56**, 6237 (1997).
24. A. Virosztek and J. Ruvalds, Phys. Rev. B **42**, 4064 (1990).
25. J. Ruvalds, Supercond. Sci. Technol. **9**, 905 (1997).
26. K. H. Bennemann, this volume.
27. P. Monthoux and D. Pines, Phys. Rev. B **47**, 6069 (1993).
28. P. Monthoux and D. Pines, Phys. Rev. B **49**, 4261 (1994).
29. D. J. Scalapino, in *Superconductivity*, edited by R. D. Parks (Marcel Dekker, New York, 1969), p. 449.
30. J. Altmann, W. Brenig, and A. P. Kampf, cond-mat/9707267.
31. Z.-X. Shen and J. R. Schrieffer, Phys. Rev. Lett. **78**, 1771 (1997).
32. H.-B. Schüttler and M. R. Norman, Phys. Rev. B **54**, 13295 (1996).
33. A. Sacuto, R. Combescot, N. Bontemps, C. A. Müller, V. Viallet, and D. Colson, cond-mat/9804060 6 Apr 1998.
34. M. T. Beal-Monod, J. B. Bieri, and K. Maki, Europhys. Lett. **40**, 201 (1997).

s- AND d-WAVE PAIRING IN SHORT COHERENCE LENGTH SUPERCONDUCTORS

JAMES F. ANNETT AND J.P. WALLINGTON
University of Bristol,
H.H. Wills Physics Laboratory,
Royal Fort, Tyndall Avenue,
Bristol BS8 1TL, UK.

1. Introduction

Over the past three years or so the evidence has become overwhelming that the cuprate high T_c superconductors have a d-wave pairing state[1]. The measurements of Wollman et al.[2] and of Tsuei et al.[3] are especially convincing since they do not depend on the microscopic physics of the energy gap, but instead depend only on the order parameter phase. Other experiments, such as photoemission[4, 5] and the temperature dependence of penetration depth[6, 7, 8], also strongly support the d-wave picture.

On the other hand, there is continuing controversy over whether the pairing state is a pure d-wave or an $s-d$ mixture[9, 10]. There is indeed evidence for a significant s-wave component in $YBa_2Cu_3O_7$[11]. A subdominant s-wave component could be compatible with the Wollman et al. and the Tsuei et al. experiments provided that it was not too large. The photoemission and penetration depth measurements also cannot rule out a small s-wave component (either $s \pm d$ or $s \pm id$), although they can possibly put upper bounds on the magnitude of the s component.

Any observations of an s component have important implications for the various theories of the pairing mechanism. For example, antiferromagnetic spin fluctuations lead to attraction in the $d_{x^2-y^2}$ pairing channel, but are pair breaking in the s-wave channel[12]. Similarly the Hubbard model with a positive on-site interaction U may have a $d_{x^2-y^2}$ paired ground state, but would presumably not support s-wave Cooper pairs. In the case of $YBa_2Cu_3O_7$ the orthorhombic crystal symmetry makes some non-zero s-wave component inevitable, but a large s-wave component would be dif-

M. Ausloos and S. Kruchinin (eds.), Symmetry and Pairing in Superconductors, 245–258.
© *1999 Kluwer Academic Publishers. Printed in the Netherlands.*

ficult to reconcile with either of these pairing mechanisms. On the other hand, pairing mechanisms based on electron-phonon interactions, polarons, or other non-magnetic excitations (e.g. excitons, acoustic plasmons) could be compatible with either *s*-wave or *d*-wave pairing states[13]. In these models, whether *s*-wave, *d*-wave or mixed pairs are more strongly favoured would depend on details of the model parameters, and could even vary from compound to compound. Indeed there is some evidence that the n-type cuprate superconductors are *s*-wave[14] (or at least they have no zeros in the gap $|\Delta(\mathbf{k})|$ on the Fermi surface). This would imply that either the pairing mechanism is different for the n- and p-type materials, or that the mechanism allows both *s*-wave or *d*-wave ground states depending on the band filling.

In this paper we examine the attractive nearest neighbour Hubbard model, which is the simplest model that allows *s*-wave, *d*-wave or mixed pairing states to occur[15]. We examine the overall phase diagram, paying particular attention to the regions near the phase transitions between the *s*, *d* and $s \pm id$ phases. We derive an appropriate Landau-Ginzburg-Wilson (LGW) effective action for the model. In the mean-field approximation this LGW functional reproduces the usual Hartree-Fock-Gor'kov equations. Beyond mean field theory this functional allows us to examine the large interaction limit, in which the superconducting phase transition becomes a Bose-Einstein condensation of preformed pairs. In this limit there will be a pseudogap in the normal state density of states. The strong coupling limit is especially interesting in this model because, unlike the case of on-site interactions, there are at least two 'species' of preformed pairs, $d_{x^2-y^2}$ and *s*. We show below that these become degenerate in the strong coupling Bose-Einstein limit.

2. The Nearest Neighbour Attractive Hubbard Model

Our starting point is the nearest-neighbour attractive Hubbard model in two dimensions,

$$\hat{H} = -t \sum_{\langle ij \rangle \sigma} \left(c_{i\sigma}^\dagger c_{j\sigma} + H.C. \right) - V \sum_{\langle ij \rangle} n_i n_j, \tag{1}$$

where $\sum_{\langle ij \rangle}$ denotes summation over all of the bonds between the nearest neighbour sites. For simplicity we ignore any on-site interaction terms.

The mean-field gap equations can be used straightforwardly to calculate T_c for this model (Section 4). The gap equation has stable solutions for (extended) *s*-wave, *p*-wave, and *d*-wave pairing states, depending on the model parameters. One can see in Fig. 1(a)-(c) that, for all V, the *d*-wave

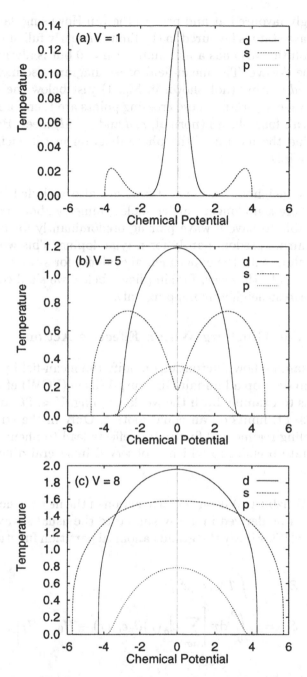

Figure 1. Mean-field transition temperature, T_c, versus chemical potential, μ, for s-, p- and d-wave superconductivity at different interaction strengths; (a) $V = 1t$, (b) $V = 5t$ and (c) $V = 8t$.

state is strongly favoured at and near to the van Hove singularity at $\mu = 0$, while the s-wave state is favoured in the limits of nearly full or empty bands. The p-wave solution also has a maximum at $\mu = 0$ but is always small compared with the d-wave. The mean-field phase diagram also includes a small region of $s \pm id$ pairing (not shown in Fig. 1) just below the points where the d- and s-wave T_c cross. These crossing points are therefore tetracritical points, where all four phases (normal, s, d and $s \pm id$) meet. Below we shall closely examine the nature of the phase diagram in the vicinity of these tetracritical points.

If next nearest neighbour hopping terms are also included in the Hamiltonian then there is no longer particle-hole symmetry about half-filling and and it is possible to have d-wave pairing predominantly for p-type doping and s-wave pairing predominantly for n-type doping. This would be qualitatively consistent with the experimental evidence for s-wave pairing in the n-type cuprates[14]. However, for simplicity, below we shall concentrate on the case of nearest-neighbour hopping only.

3. The Landau-Ginzburg-Wilson Effective Action

In order to examine how fluctuations modify the mean-field picture of Fig. 1 is is useful to develop a Landau-Ginzburg-Wilson (LGW) effective action. This allows us to examine both the weak coupling, $V \ll 8t$, and the strong coupling, $V \gg 8t$, limits on an equal footing. Only in the strong or intermediate coupling regime will fluctuation effects lead to phenomena such as the normal-state pseudogap which is observed in several different high T_c materials[16].

The Landau-Ginzburg-Wilson effective action of the nearest-neighbour Hubbard model can be derived in a way similar to the usual s-wave case of on-site interaction[17]. Firstly the grand canonical partition function is written:

$$\mathcal{Z} = \int \mathcal{D}[c, c^\dagger] e^{\mathcal{S}}, \tag{2}$$

$$\mathcal{S} = \int_0^\beta d\tau \left[\sum_{i\sigma} c_{i\sigma}^\dagger(\tau) \partial_\tau c_{i\sigma}(\tau) - \hat{H}_0 - \hat{H}_I \right], \tag{3}$$

where $\hat{H}_0$ is given by the hopping term in the Hamiltonian Eq. 1 (including a chemical potential term, $-\mu\hat{N}$) and $\hat{H}_I$ is the interaction term.

The interacting part may be neatly written in terms of pairing operators on nearest neighbour sites:

$$\hat{H}_I = -V \sum_{\langle ij \rangle} \text{tr} \left(F_{ij}^\dagger F_{ij} \right), \tag{4}$$

where F_{ij} is defined by:

$$F_{ij} = \begin{pmatrix} c_{j\uparrow} c_{i\uparrow} & c_{j\downarrow} c_{i\uparrow} \\ c_{j\downarrow} c_{i\uparrow} & c_{j\downarrow} c_{i\downarrow} \end{pmatrix}. \tag{5}$$

The physical nature of the interaction is best illustrated by introducing a new set of operators,

$$B_{ij}^{00} \equiv \frac{1}{\sqrt{2}} \left(c_{j\uparrow} c_{i\downarrow} - c_{j\downarrow} c_{i\uparrow} \right), \tag{6}$$

$$B_{ij}^{11} \equiv c_{j\uparrow} c_{i\uparrow}, \tag{7}$$

$$B_{ij}^{10} \equiv \frac{1}{\sqrt{2}} \left(c_{j\uparrow} c_{i\downarrow} + c_{j\downarrow} c_{i\uparrow} \right), \tag{8}$$

$$B_{ij}^{1\bar{1}} \equiv c_{j\downarrow} c_{i\downarrow}. \tag{9}$$

In terms of these, F_{ij} may be decomposed as

$$F_{ij} = \frac{i\sigma_y}{\sqrt{2}} \left(B_{ij} - \boldsymbol{\sigma} \cdot \mathbf{B}_{ij} \right), \tag{10}$$

where $\boldsymbol{\sigma}$ is the vector of Pauli matrices, and

$$B_{ij} \equiv B_{ij}^{00}, \tag{11}$$

$$\mathbf{B}_{ij} \equiv \left(\frac{B_{ij}^{1\bar{1}} - B_{ij}^{11}}{\sqrt{2}}, \frac{B_{ij}^{1\bar{1}} + B_{ij}^{11}}{-i\sqrt{2}}, B_{ij}^{10} \right). \tag{12}$$

The scalar, B_{ij}, is even under both time reversal and parity whilst the vector, $\mathbf{B}_{ij}$, is odd under both. We interpret B_{ij} and $\mathbf{B}_{ij}$ as the annihilation operators for singlet and triplet Cooper pairs on the bond $\langle ij \rangle$[18].

In terms of these operators the interaction Hamiltonian becomes:

$$\hat{H}_I = -V \sum_{<ij>} \text{tr} \left(F_{ij}^\dagger F_{ij} \right),$$
$$= -V \sum_{<ij>} \left(B_{ij}^\dagger B_{ij} + \mathbf{B}_{ij}^\dagger \cdot \mathbf{B}_{ij} \right), \tag{13}$$

where we have used $\text{tr}(\sigma_i) = 0$ and $\text{tr}(\sigma_i \sigma_j) = 2\delta_{ij}$.

Since this is a sum of squared bilinear Fermi operators, the Hubbard-Stratonovič transformation [17] can be employed. The Gaussian identity,

$$e^{+VA^\dagger A} = \frac{V}{2\pi i} \int d\phi d\phi^* e^{-V(|\phi|^2 + A\phi^* + A^\dagger \phi)}, \tag{14}$$

where A is a bilinear Fermi operator and ϕ is a c-number, allows us to decouple the quartic interaction term in terms of new bosonic fields. In the case of Eq. 13 this is accomplished by introducing a complex scalar field, $\psi_{ij}(\tau)$, and a complex vector field, $\mathbf{\Psi}_{ij}(\tau)$, for each bond in the lattice and every imaginary time. The partition function can now be expressed in terms of these Bose fields:

$$\mathcal{Z} \equiv \int \mathcal{D}[\psi, \psi^*; \mathbf{\Psi}, \mathbf{\Psi}^*] e^{S_b}. \tag{15}$$

The effective action is

$$S_b \equiv -V \int_0^\beta d\tau \sum_{\langle ij \rangle} \left(|\psi_{ij}(\tau)|^2 + |\mathbf{\Psi}_{ij}(\tau)|^2 \right) + \ln \int \mathcal{D}[c, c^\dagger] e^{S_f}, \tag{16}$$

with the remaining fermions contained in

$$\mathcal{S}_f \equiv \int_0^\beta d\tau \Big[\sum_{i\sigma} c_{i\sigma}^\dagger(\tau) \partial_\tau c_{i\sigma}(\tau) - \hat{H}_0 \tag{17}$$

$$-V \sum_{\langle ij \rangle} \left(\psi_{ij}^*(\tau) B_{ij}(\tau) + \mathbf{\Psi}_{ij}^*(\tau) \cdot \mathbf{B}_{ij}(\tau) + H.C. \right) \Big].$$

Equations 15-17 are a formally exact representation of the nearest neighbour attractive Hubbard model.

The bosonic fields introduced above are defined separately for each bond $\langle ij \rangle$ in the lattice. It is more convenient to form site-centred combinations with a definite symmetry. We define two singlet fields at site $\mathbf{r}_i$

$$\psi_s = \frac{1}{2}(\psi_x + \psi_{-x} + \psi_y + \psi_{-y}),$$

$$\psi_d = \frac{1}{2}(\psi_x + \psi_{-x} - \psi_y - \psi_{-y}), \tag{18}$$

where $\psi_a(\mathbf{r}_i, \tau) \equiv \psi_{ii+a}(\tau)$ for $a = \pm x, \pm y$ according to the direction of the bond $\langle ij \rangle$. Similarly we can define the symmetrised triplet fields

$$\mathbf{\Psi}_{p_x} = \mathbf{\Psi}_x - \mathbf{\Psi}_{-x},$$

$$\mathbf{\Psi}_{p_y} = \mathbf{\Psi}_y - \mathbf{\Psi}_{-y}. \tag{19}$$

In the limit of fields varying slowly in space and time these correspond to the Ginzburg-Landau order parameters for (extended) s-wave, $d_{x^2-y^2}$ and p-wave pairing, respectively. The s- and d- wave pairing fields are even under inversion symmetry about lattice site r_i, while the triplet fields are odd. In the notation of Ref. [19], these fields are the order parameters for superconductivity in, respectively, the A_{1g} and B_{1g} and E_u representations of the tetragonal point group D_{4h}.

4. The Saddle Point Solutions

The saddle points of the effective action generate Hartree-Fock-Gor'kov mean field theory, where the Bose fields become static and spatially uniform order parameters:

$$\psi_\alpha = -\frac{1}{\beta N} \int_0^\beta \sum_i \left\langle B_\alpha(r_i, \tau) \right\rangle_f, \qquad \alpha = s, d \tag{20}$$

$$\Psi_\alpha = -\frac{1}{\beta N} \int_0^\beta \sum_i \left\langle \mathbf{B}_\alpha(r_i, \tau) \right\rangle_f, \qquad \alpha = x, y \tag{21}$$

where N is the number of sites and $\langle \ldots \rangle_f$ denotes self-consistent averaging with respect to the fermionic part of the action, S_f.

The transition temperature for a given order parameter, in the absence of any of the others, is given by the solution of

$$1 = \frac{V}{2} \sum_\epsilon \frac{N^\alpha(\epsilon)}{\epsilon - \mu} \tanh\left(\frac{\epsilon - \mu}{2T_c^\alpha}\right), \tag{22}$$

in which the weighted density of states, $N^\alpha(\epsilon)$, is

$$N^\alpha = \frac{1}{N} \sum_k \zeta^\alpha(\mathbf{k})\zeta^\alpha(\mathbf{k})\delta(\epsilon - \epsilon_\mathbf{k}). \tag{23}$$

The form factors reflect the point group symmetries of the order parameters:

$$\zeta^\alpha(\mathbf{k}) = \begin{cases} \cos(k_x) + \cos(k_y) & \alpha = s, \\ \cos(k_x) - \cos(k_y) & \alpha = d, \\ \sin(k_x) & \alpha = \mathbf{p}_x, \\ \sin(k_y) & \alpha = \mathbf{p}_y. \end{cases} \tag{24}$$

The solutions of Eq. 22 are shown in Fig. 1 for three values of V. The s-wave solution dominates for small and large fillings with d-wave dominant near the van Hove peak at the centre of the band. The p-wave solutions are sub-dominant everywhere. It is expected that interaction with the large

d-wave order parameter will suppress the p-wave T_c even further. It is clear from this that p-wave pairing is irrelevant in the bulk superconductor, and henceforth we ignore it and concentrate on s- and d-wave pairs only.

5. Beyond the Saddle Point: s- and d-wave Mixing

Starting with the effective action, S_b, we Fourier transform in space and (imaginary) time and integrate out the fermions to give a purely bosonic action,

$$S_b = -\frac{V}{2} \sum_{q,\alpha} |\psi_\alpha(q)|^2 + \mathrm{Tr}\ln(1 - \mathbf{V}\mathbf{G}_0), \qquad (25)$$

where $q \equiv (\mathbf{q}, i\omega)$, and the trace is over both space-time and spinor indices. The fermions of the original theory live on in the form of the Nambu Green's function matrix,

$$\mathbf{G}_0(k, k') \equiv \begin{pmatrix} G_0(k) & 0 \\ 0 & -G_0^*(k) \end{pmatrix} \delta_{kk'}, \qquad (26)$$

in which $G_0(k) = (i\omega_n - \epsilon_\mathbf{k} + \mu)^{-1}$ is the Green's function for non-interacting fermions. The interaction of fermions and bosons occurs through the potential matrix,

$$\mathbf{V}(k, k') \equiv \frac{V}{\sqrt{2\beta N}} \begin{pmatrix} 0 & \psi_\alpha(k - k') \\ \psi_\alpha^*(-k + k') & 0 \end{pmatrix} \zeta^\alpha_{\mathbf{k},\mathbf{k}'}, \qquad (27)$$

where the Einstein summation convention has been used for repeated Greek indices and,

$$\zeta^\alpha_{\mathbf{k},\mathbf{k}'} \equiv \frac{1}{2}\Big(\cos(k_x) + \cos(k_x') \Big) \pm \frac{1}{2}\Big(\cos(k_y) + \cos(k_y') \Big), \quad \alpha = s, d. \quad (28)$$

When ψ_α is constant in real space (i.e. $\mathbf{k} = \mathbf{k}'$), this takes a particularly simple form,

$$\zeta^\alpha(\mathbf{k}) \equiv \zeta^\alpha_{\mathbf{k},\mathbf{k}} = \cos(k_x) \pm \cos(k_y), \quad \alpha = s, d \qquad (29)$$

as seen in the weighted densities of states.

Near T_c, where the ψ_α are small, we expand the logarithm as a power series up to fourth order;

$$\mathrm{Tr}\ln(1 - \mathbf{V}\mathbf{G}_0) = -\sum_{m=1}^{4} \frac{1}{m}\mathrm{Tr}\,(\mathbf{V}\mathbf{G}_0)^m + \mathcal{O}\left(\psi^5\right). \qquad (30)$$

The odd terms in the series are exactly zero but the even terms survive. The quadratic contribution is

$$\mathrm{Tr}\,(\mathbf{V}\mathbf{G}_0)^2 = V^2 \sum_q \psi_\alpha^*(q)\psi_\beta(q)\chi^{\alpha\beta}(q), \tag{31}$$

where the susceptibility is:

$$\chi^{\alpha\beta}(q) = -\frac{1}{\beta N}\sum_k G_0(k)G_0^*(k+q)\zeta_{\mathbf{k},\mathbf{k}+\mathbf{q}}^\alpha \zeta_{\mathbf{k}+\mathbf{q},\mathbf{k}}^\beta, \tag{32}$$

$$= \frac{1}{N}\sum_k \left\{ \frac{f(\epsilon_{\mathbf{k}+\mathbf{q}}-\mu)+f(\epsilon_{\mathbf{k}}-\mu)-1}{i\omega_\nu + \epsilon_{\mathbf{k}+\mathbf{q}}+\epsilon_{\mathbf{k}}-2\mu} \right\} \zeta_{\mathbf{k},\mathbf{k}+\mathbf{q}}^\alpha \zeta_{\mathbf{k}+\mathbf{q},\mathbf{k}}^\beta.$$

The quartic term is

$$\mathrm{Tr}\,(\mathbf{V}\mathbf{G}_0)^4 = \frac{V^4}{2\beta N}\sum_{\{q\}} \psi_\alpha^*(q)\psi_\beta(q')\psi_\gamma^*(q'')\psi_\delta(q-q'+q'')\chi^{\alpha\beta\gamma\delta}(q,q',q'')$$

$$\tag{33}$$

It is sufficient to evaluate the four body susceptibility at $q = q' = q'' = 0$ and then to treat it as a constant, $\chi \equiv \chi(0,0,0)$:

$$\chi^{\alpha\beta\gamma\delta} = \frac{1}{\beta N}\sum_k |G_0(k)|^4 \zeta^\alpha(\mathbf{k})\zeta^\beta(\mathbf{k})\zeta^\gamma(\mathbf{k})\zeta^\delta(\mathbf{k}), \tag{34}$$

$$= -\frac{1}{4N}\sum_k \frac{1}{\xi}\frac{d}{d\xi}\left(\frac{\tanh(\beta\xi/2)}{\xi}\right)\zeta^\alpha(\mathbf{k})\zeta^\beta(\mathbf{k})\zeta^\gamma(\mathbf{k})\zeta^\delta(\mathbf{k}),$$

where $\xi \equiv \epsilon_{\mathbf{k}} - \mu$.

Thus, after a trivial rescaling of ψ by $V^{\frac{1}{2}}$, the bosonic action to fourth order reads:

$$S_b \approx -\frac{1}{2}\sum_{q,\alpha\beta}\left(\delta^{\alpha\beta}+V\chi^{\alpha\beta}(q)\right)\psi_\alpha^*(q)\psi_\beta(q) \tag{35}$$

$$-\frac{V^2}{8\beta N}\chi^{\alpha\beta\gamma\delta}\sum_{\{q\}}\psi_\alpha^*(q)\psi_\beta(q')\psi_\gamma^*(q'')\psi_\delta(q-q'+q'').$$

In order to derive a Landau-Ginzburg-Wilson functional from S_b, we expand to lowest order in small $i\omega_\nu$ and $|\mathbf{q}|$:

$$\frac{1}{2}\left(\delta^{\alpha\beta}+V\chi^{\alpha\beta}(q)\right) \simeq a^{\alpha\beta} - id^{\alpha\beta}\omega_\nu + \sum_{\mu=x,y}\frac{q_\mu^2}{2m_{\alpha\beta}^\mu}, \tag{36}$$

where:

$$a^{\alpha\beta} = \frac{1}{2}\left(\delta^{\alpha\beta} - \frac{V}{2N}\sum_{\mathbf{k}}\left(\frac{\tanh(\beta\xi/2)}{\xi}\right)\zeta^{\alpha}(\mathbf{k})\zeta^{\beta}(\mathbf{k})\right), \tag{37}$$

$$d^{\alpha\beta} = -\frac{V}{8N}\sum_{\mathbf{k}}\left(\frac{\tanh(\beta\xi/2)}{\xi^2}\right)\zeta^{\alpha}(\mathbf{k})\zeta^{\beta}(\mathbf{k}), \tag{38}$$

$$\frac{1}{2m^{\mu}_{\alpha\beta}} = -\frac{V}{16N}\sum_{\mathbf{k}}\left\{\left(\frac{\partial^2\epsilon_{\mathbf{k}}}{\partial k_{\mu}^2}\right)\frac{d}{d\xi}\left(\frac{\tanh(\beta\xi/2)}{\xi}\right)\zeta^{\alpha}(\mathbf{k})\zeta^{\beta}(\mathbf{k})\right. \tag{39}$$

$$+\left(\frac{\partial\epsilon_{\mathbf{k}}}{\partial k_{\mu}}\right)^2\left(\frac{d^2}{d\xi^2}+\frac{1}{\xi}\frac{d}{d\xi}\right)\left(\frac{\tanh(\beta\xi/2)}{\xi}\right)\zeta^{\alpha}(\mathbf{k})\zeta^{\beta}(\mathbf{k})$$

$$+2\left(\frac{\partial\epsilon_{\mathbf{k}}}{\partial k_{\mu}}\right)\frac{d}{d\xi}\left(\frac{\tanh(\beta\xi/2)}{\xi}\right)\frac{\partial(\zeta^{\alpha}_{\mathbf{k},\mathbf{k}+\mathbf{q}}\zeta^{\beta}_{\mathbf{k}+\mathbf{q},\mathbf{k}})}{\partial q_{\mu}}\bigg|_{q=0}$$

$$\left.+2\left(\frac{\tanh(\beta\xi/2)}{\xi}\right)\frac{\partial^2(\zeta^{\alpha}_{\mathbf{k},\mathbf{k}+\mathbf{q}}\zeta^{\beta}_{\mathbf{k}+\mathbf{q},\mathbf{k}})}{\partial q_{\mu}^2}\bigg|_{q=0}\right\}$$

where $\xi \equiv \epsilon_{\mathbf{k}} - \mu$. The last two terms in Eq. 39 are due to the $\mathbf{q}$ dependence of ζ^{α}.

The resulting LGW functional is shown below expressed in real space:

$$S_{LGW} = -\int_0^{\beta}d\tau\sum_{\mathbf{r}}[\mathcal{K}+\mathcal{V}], \tag{40}$$

where $\mathcal{K}$ is the 'kinetic' part and $\mathcal{V}$ the 'potential' part of the action. They are given by,

$$\mathcal{K} = d^{ss}\psi_s^*\partial_{\tau}\psi_s + \frac{|\nabla\psi_s|^2}{2m_{ss}} + d^{dd}\psi_d^*\partial_{\tau}\psi_d + \frac{|\nabla\psi_d|^2}{2m_{dd}} \tag{41}$$

$$+\frac{1}{2m_{sd}}\left(\psi_s^*(\nabla_x^2-\nabla_y^2)\psi_d + C.C.\right),$$

$$\mathcal{V} = a^{ss}|\psi_s|^2 + a^{dd}|\psi_d|^2 + b_s|\psi_s|^4 + b_d|\psi_d|^4 \tag{42}$$

$$+\kappa|\psi_s|^2|\psi_d|^2 + \frac{\kappa}{4}\left(\psi_s\psi_s\psi_d^*\psi_d^* + C.C.\right),$$

where,

$$b_s = \frac{V^2}{8}\chi^{ssss}, \qquad b_d = \frac{V^2}{8}\chi^{dddd}, \qquad \kappa = \frac{V^2}{2}\chi^{ssdd}. \tag{43}$$

In the static and spatially uniform limit, $\mathcal{V}$ may be interpreted as the Landau free energy. We will examine the phases arising from this before considering further the effects of fluctuations. The free energy given in Eqs. 41 42 is of the form found earlier by Joynt[20] and by Feder and Kallin[21].

6. Landau Theory

The phase diagram generated by $\mathcal{V}$ is found by simultaneously solving,

$$\frac{\partial \mathcal{V}}{\partial \psi_s} = \frac{\partial \mathcal{V}}{\partial \psi_d} = 0. \tag{44}$$

When $\kappa = 0$, i.e. there is no coupling between the s- and d-wave order parameters, solving $a^{ss} = 0$ and $a^{dd} = 0$ as functions of μ gives the critical temperature for each. This is equivalent to solving the linearised gap equation, Eq. 22.

For non-zero κ it is possible to have mixed phases. To determine the stable phase, two parameters are needed. These are:

$$B_s = \frac{b_s}{b_d} \left| \frac{a^{dd}}{a^{ss}} \right|^2, \tag{45}$$

$$K_c = \frac{\kappa(1 + \frac{1}{2}\cos(2\Delta\theta))}{b_d} \left| \frac{a^{dd}}{a^{ss}} \right|, \tag{46}$$

where $\Delta\theta$ is the phase difference between ψ_s and ψ_d. Fig. 2 shows the stable superconducting phase as a function of these two dimensionless parameters. Note that in Fig. 2 the definition of K_c is different for $\kappa > 0$, where $(\Delta\theta = \pi/2)$, and $\kappa < 0$, where $(\Delta\theta = 0)$.

As Fig. 2 shows, depending on the parameter values s, d, $s \pm d$ or $s \pm id$ phases are possible. The transition from the pure s to the pure d phase can occur in a number of ways. For $K_c > 2$, a single first-order transition separates the two phases, analogous to a 'spin flop' transition. In this case, the $O(4)$ point (corresponding to $B_s = 1, K_c = 2$) represents the bicritical point where the line of first-order transitions $s \to d$ meets the two normal $\to$ superconducting second-order lines. When $K_c < 2$ the transition occurs via two second-order phase transitions with an intermediate mixed symmetry state, either $s \pm d$ or $s \pm id$. The $O(4)$ point here represents the meeting of the four second-order lines at a tetracritical point.

Using parameters derived from the nearest neighbour attractive Hubbard model, the dashed line in Fig. 2 shows the evolution of (B_s, K_c) for $V = 2$ as μ changes in the region near the crossing of T_c^s and T_c^d. As μ increases we move from the far right hand side of the figure, where T=$T_c^s < T_c^d$ and d-wave is dominant, to the far left, where T=$T_c^d < T_c^s$ and s-wave is dominant. In these extremes, the dominant d-wave (s-wave) order parameter suppresses the sub-dominant s-wave (d-wave) one even though T<$T_c^s(T_c^d)$.

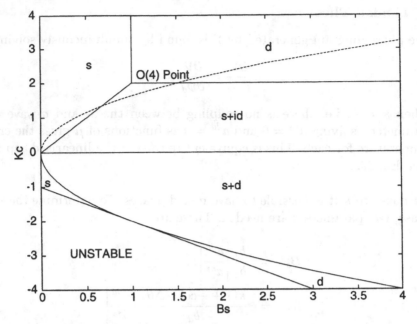

Figure 2. Landau stability plot for $T < T_c^s, T_c^d$. In the vicinity of the bicritical point, the dashed line shows the transition from d-wave to s-wave via a mixed $s \pm id$ state as μ increases (right to left).

Eventually, however, a mixed $s \pm id$ phase arises. This behaviour is seen for all V. In particular the mixed $s \pm d$ phase is never realised for any values of B_s and K_c derived from the nearest neighbour attractive Hubbard model unless an orthorhombic distortion is introduced which is beyond the scope of this paper.

7. The Large V Limit

In the large V limit it is necessary to retain the thermal and quantum fluctuations in the Bose fields derived in section 4. The superconducting phase transition then becomes the (Kosterlitz-Thouless in 2-dimensions) Bose condensation of these quantum fields. A detailed analysis is beyond the scope of the present paper, but it is possible to draw some qualitative conclusions based on the effective action derived in sections 3 and 5.

Fig. 1 shows the evolution of the T_c for s, d and p-wave pairing as a function of μ for various values of V/t. The small V limit, represented by Fig. 1(a), corresponds to the BCS weak coupling limit in which the pairing is a small perturbation on the filled Fermi sea. On the other hand as V is increased (Fig 1(b)-(c)) the behaviour changes markedly. One can see that the s-wave

T_c becomes significant even at $\mu = 0$, and that the s and d-wave curves become increasingly similar as V is increased. This has a simple physical interpretation: in the large V limit the pairs become nearly localised on bonds, and the most favourable state is a singlet pair (stable compared to triplet by an energy of order t^2/V). As V increases the effective hopping for pairs decreases (also as t^2/V) and so s- and d-wave pairs become nearly degenerate. This degeneracy implies that the Bose condensation of pairs in the large V nearest-neighbour Hubbard model is qualitatively different from that in the on-site attractive Hubbard model.

Figures 1(b) and 1(c) also shows that for large V superconductivity occurs even when the chemical potential is below the bottom of the electronic band at $-4t$. For these values of μ the fermions can only exist as bound pairs, and not as free fermions. This is analogous to the large V on-site s-wave case discussed by Randeria, Duan and Shieh [22], in which the criterion for Bose condensation to occur in the low density limit was shown to be the same as the criterion for formation of a two particle bound state. Figure 1(c) implies that in the low density, $\mu < -4t$, limit two-particle bound states are formed. An extended s-wave superconducting state then occurs when these preformed pairs Bose condense at temperatures below the mean-field T_c shown in Fig. 1(c).

8. Conclusions

We have derived an appropriate Landau-Ginzburg-Wilson effective action to describe the nearest neighbour attractive Hubbard model. This allows us to discuss both the small V BCS and the large V Bose-Einstein condensation limits on the same footing, as was done for the on-site attractive Hubbard model some years ago[23]. The nearest neighbour Hubbard model is especially interesting because the phase diagram includes regions of both (extended) s-wave superconductivity and d-wave pairing. We find that p-wave pairing is never stable. We have studied closely the region of the phase diagram near the cross-over from s- to d-wave pairing, and find that the two phases are always separated by two second-order phase transitions with an intermediate phase of $s \pm id$ superconductivity.

In future it will be interesting to examine more carefully the large V Bose-Einstein condensation limit of this model, since the extended s-wave and d-wave pairing states become nearly degenerate. It will also be interesting to see how closely features of this model, such as the existence of s, d and $s \pm id$ phases, correspond to the actual experimental features of the cuprates.

258

Acknowledgement

This work was supported by EPSRC Grant No. GR/L22454, and by the Office of Naval Research Grant No. N00014-95-1-0398.

References

1. J.F. Annett, N.D. Goldenfeld and A.J. Leggett in *"Physical Properties of High Temperature Superconductors V"*, D.M. Ginsberg (ED.), (World Scientific, Singapore 1996).
2. D.A. Wollman *et al.*, Phys. Rev. Lett. **71**, 2134 (1993).
3. C.C. Tsuei *et al.*, Phys. Rev. Lett. **73**, 593 (1994).
4. H. Ding *et al.*, Phys. Rev. B **54**, R9678 (1996).
5. Z.-X. Shen *et al.*, Science **267**, 343 (1995).
6. J. Mao *et al.*, Phys. Rev. Lett. **51**, 3316 (1995).
7. C. Panagopoulos *et al.*, Phys. Rev. B **53**, R2999 (1996).
8. D.M. Broun *et al.*, Phys. Rev. B **56**, 1146 (1997).
9. R.J. Kelley *et al.*, Science **271**, 1255 (1996).
10. R.A. Klemm, in this volume. R.A. Klemm, M. Ledvij and L. Shu, Chinese J. Phys. **34**, 201 (1996).
11. K.A. Kouznetsov *et al.*, Phys. Rev. Lett. **79**, 3050 (1997).
12. P. Monthoux and D. Pines, Phys. Rev. Lett. **67**, 3448 (1991).
13. G. Varelogiannis, Phys. Rev. B **57**, R723 (1998).
14. S.M. Anlage *et al.*, Phys. Rev. B **50**, 523 (1994).
15. R. Micnas, J. Ranninger and S. Robaszkiewicz, Rev. Mod. Phys. **62**, 113 (1990).
16. H. Ding *et al.*, Nature **382**, 51 (1996).
17. V.N. Popov, *"Functional Integerals and Collective Excitations"*, (Cambridge University Press, Cambridge UK, 1987).
18. G.E. Volovik and L.P. Gorkov, Zh. eksp. teor. Fiz. **88**, 1412 (1985) [Sov. Phys. JETP **61**, 843 (1985)].
19. J.F. Annett, Adv. Phys. **39**, 83 (1990).
20. R. Joynt, Phys. Rev. B **41**, 4271 (1990).
21. D.L. Feder and C. Kallin, Phys. Rev. B **55**, 559 (1997).
22. M. Randeria, J.M. Duan and L.Y. Shieh, Phys. Rev. B **41**, 327 (1990).
23. M. Randeria, C. DeMelo and J.R. Engelbrecht, Physica B **194**, 1409 (1994).

K-SPACE GAP ANISOTROPY WITHIN THE INTERLAYER PAIR-TUNNELING MECHANISM OF HIGH-T_c SUPERCONDUCTIVITY

GIUSEPPE G.N. ANGILELLA, RENATO PUCCI, FABIO SIRINGO
*Dipartimento di Fisica dell'Università di Catania, and
Istituto Nazionale per la Fisica della materia, Unità di Ricerca
di Catania, 57 Corso Italia, I-95129 Catania, Italy*

AND

ASLE SUDBØ
*Institutt for Fysikk, Norges Teknisk-Naturvitenskapelige
Universitet, Sem Saelandsvei 9, N-7034 Trondheim, Norway*

Abstract. We consider a mean-field approach to a $2D$ extended Hubbard model for a bilayer superconductor, in presence of coherent interlayer pair-tunneling and quenched coherent single particle tunneling. The functional form of the *intralayer* pairing potential is dictated by the symmetry character of the underlying crystal lattice. This gives rise to a competition between s- and d-wave symmetry, as the chemical potential is increased from the bottom to the top of a realistic band for most cuprates. It allows for mixed-symmetry paired state at temperatures below T_c, but never at T_c on a square lattice. The interlayer pair-tunneling mechanism contributes to the pairing potential as an effective k-diagonal term, which is responsible of a nonconventional k-dependence of the gap function. We study the evolution of such a gap structure with temperature and with band filling.

It is generally agreed that the dependence of the order parameter (OP) $\Delta_{\mathbf{k}}$ on the wave-vector $\mathbf{k}$ in reciprocal space carries an unambiguous image of the unconventional pairing mechanism in the high-T_c superconductors, which is still a debated issue among theorists. In that context, recent experimental findings have rekindled the earlier controversy over the symmetry of the paired state in the cuprates [1, 2]. Evidence for d-wave superconductivity in optimally doped Bi2212 single crystals is supported by angle-resolved photoemission spectroscopy (ARPES) [3, 4, 5]. Besides the symmetry character of the paired state, this technique also yields direct access to the

259

M. Ausloos and S. Kruchinin (eds.), Symmetry and Pairing in Superconductors, 259–267.
© *1999 Kluwer Academic Publishers. Printed in the Netherlands.*

modulus of the gap over the reciprocal space, and is suggestive of an unconventional temperature depending k-space gap anisotropy [6]. On the other hand, sparse results indicate that the detailed k-dependence of the OP could be a material specific property [7], as it the case for the critical temperature itself [8].

In this contribution, we shall consider a $2D$ extended Hubbard model in the case of a bilayer complex. The model is characterized by a realistic band dispersion, including nearest (N) and next-nearest (NN) neighbours hopping within the CuO planes, and a small-range in-plane potential, allowing for in-site, N and NN neighbours interaction, in the presence of coherent pair tunneling between adjacent layers. While coherent single-particle motion along the c-axis in the cuprates is blocked already in the normal state, due to the Anderson orthogonality catastrophe [9], Josephson-like coherent tunneling of pairs between adjacent layers may take place, as long as superconductivity is generated within each layer by an appropriate in-plane interaction.

We adopt the following model Hamiltonian to describe a collection of tightly bound interacting fermions in a bilayer complex:

$$H = \sum_{\mathbf{k}\sigma i} \xi_{\mathbf{k}}^i c_{\mathbf{k}\sigma}^{i\dagger} c_{\mathbf{k}\sigma}^i + \sum_{\mathbf{k}\mathbf{k}'ij} \tilde{V}_{\mathbf{k}\mathbf{k}'}^{ij} c_{\mathbf{k}\uparrow}^{i\dagger} c_{-\mathbf{k}\downarrow}^{i\dagger} c_{-\mathbf{k}'\downarrow}^j c_{\mathbf{k}'\uparrow}^j, \tag{1}$$

where $c_{\mathbf{k}\sigma}^{i\dagger}$ ($c_{\mathbf{k}\sigma}^i$) creates (destroys) a fermion on the layer i ($i = 1, 2$), with spin projection σ along a specified direction, wave-vector $\mathbf{k}$ belonging to the first Brillouin zone (1BZ) of a $2D$ square lattice, and band dispersion $\xi_{\mathbf{k}}^i = \varepsilon_{\mathbf{k}}^i - \mu$, measured relative to the chemical potential μ. Although associated with the gain in kinetic energy per tunneling pair, the interlayer pair-tunneling (ILPT) amplitude $T_J(\mathbf{k})$ is related to a second order effect, and therefore enters the spin-singlet pair interaction:

$$\tilde{V}_{\mathbf{k}\mathbf{k}'}^{ij} = \frac{1}{N} U_{\mathbf{k}\mathbf{k}'} \delta_{ij} - T_J(\mathbf{k}) \delta_{\mathbf{k}\mathbf{k}'} (1 - \delta_{ij}), \tag{2}$$

as an effective k-diagonal contribution [10]. Here, N is the number of sites in the square lattice and $U_{\mathbf{k}\mathbf{k}'}$ measures the coupling interaction within each plane.

Band structure calculations [11] as well as ARPES results [3] for various cuprates suggest the following tight-binding dispersion relation

$$\varepsilon_{\mathbf{k}} = -2t[\cos(k_x a) + \cos(k_y a)] + 4t' \cos(k_x a) \cos(k_y a), \tag{3}$$

a being the lattice step. In order to reproduce the most relevant properties common to the mainly $2D$ band structure of the majority of the cuprate compounds, nearest-neighbours ($t > 0$) as well as next-nearest neighbours

($t' > 0$) hoppings have at least to be retained with $t'/t \lesssim 0.5$. In the following, we shall assume $t = 0.25$ eV and $t'/t = 0.45$ [11, 5].

We shall not make any attempt at specifying the microscopic origin of the in-plane potential $U_{\mathbf{kk}'}$. However, it has to possess the symmetry of the underlying lattice, and may therefore be expanded as a bilinear combination of basis functions for the irreducible representations of the crystal point group, which is C_{4v} for the $2D$ square lattice [12]. Assuming a finite-ranged potential, a finite subset of all the basis functions (an infinite orthonormal set) will suffice. Retaining therefore only on-site, nearest and next-nearest neighbours in-plane interactions, and projecting out interaction terms in the spin triplet channel, one obtains the following expression for $U_{\mathbf{kk}'}$, which is *separable* in **k**-space:

$$U_{\mathbf{kk}'} = \sum_{\eta=0}^{4} \lambda_\eta g_\eta(\mathbf{k}) g_\eta(\mathbf{k}'). \tag{4}$$

Here, $g_0(\mathbf{k}) = 1$, $g_1(\mathbf{k}) = \frac{1}{2}[\cos(k_x a) + \cos(k_y a)]$, $g_2(\mathbf{k}) = \cos(k_x a)\cos(k_y a)$, $g_3(\mathbf{k}) = \frac{1}{2}[\cos(k_x a) - \cos(k_y a)]$, $g_4(\mathbf{k}) = \sin(k_x a)\sin(k_y a)$, and λ_η ($\eta = 0, 1, \ldots 4$) are phenomenological effective coupling constants. One immediately recognizes $g_0(\mathbf{k})$, $g_1(\mathbf{k})$, $g_2(\mathbf{k})$ to display (extended) s-wave symmetry, whereas $g_3(\mathbf{k})$ and $g_4(\mathbf{k})$ display d-wave symmetry. In the following, we shall assume repulsive on-site and attractive intersite coupling parameters ($\lambda_0 > 0$ and λ_1, $\lambda_3 < 0$), choosing their actual values in order to reproduce the correct order of magnitude for the critical temperature and gap maximum at $T = 0$ for the cuprates. Throughout this paper, we keep $\lambda_2 = \lambda_4 = 0$.

Finally, we assume the local dependence of the interlayer pair tunneling matrix element as $T_J(\mathbf{k}) = t_\perp^2(\mathbf{k})/t$, i.e. a second-order perturbation in the hopping matrix element $t_\perp(\mathbf{k})$ orthogonal to the CuO layers. Recent detailed band structure calculations [13] formally confirm the original choice of functional form made by Chakravarty et al. in Ref. [14], $t_\perp(\mathbf{k}) = t_\perp[\cos(k_x a) - \cos(k_y a)]^2/4$, as well as the range $t_\perp = 0.1$—0.15 eV.

A straightforward mean-field approximation allows to characterize the superconducting state through the complex gap $\Delta_\mathbf{k}$ within each layer. Standard diagonalization techniques then yield for $\Delta_\mathbf{k}$ the BCS-like gap equation:

$$\Delta_\mathbf{k} = F_\mathbf{k} \sum_\eta g_\eta(\mathbf{k}) \Delta_\eta, \tag{5}$$

where $F_\mathbf{k} = [1 - T_J(\mathbf{k})\chi_\mathbf{k}]^{-1}$, $\chi_\mathbf{k} = (2E_\mathbf{k})^{-1}\tanh(\beta E_\mathbf{k}/2)$ is the pair susceptibility at a finite temperature T, $\beta = (k_B T)^{-1}$, and $E_\mathbf{k} = \sqrt{\xi_\mathbf{k}^2 + |\Delta_\mathbf{k}|^2}$ is the upper band of the quasiparticle gapped spectrum. The auxiliary gap parameters

$$\Delta_\eta = -\lambda_\eta \frac{1}{N} \sum_{\mathbf{k}'} g_\eta(\mathbf{k}') \chi_{\mathbf{k}'} \Delta_{\mathbf{k}'} \qquad (6)$$

represent the projection of the gap function on $g_\eta(\mathbf{k})$, and yield therefore information about the overall symmetry character of the OP. However, they are not enough to characterize $\Delta_{\mathbf{k}}$ as a whole, its k-dependence being in fact ruled by the unusual pre-factor $F_{\mathbf{k}}$ in Eq. (5), which includes $|\Delta_{\mathbf{k}}|$ self-consistently via the pair susceptibility $\chi_{\mathbf{k}}$. We emphasize that Δ_η are *not* order parameters in themselves: the vanishing of only some of them simply signals the absence of the symmetry contribution which they represent to the full gap.

Equations (5) and (6) can be solved numerically at any temperature T and chemical potential μ within the band (see Ref. [15] for details). We find that the critical temperature T_c is characterized by the opening of a vanishing gap displaying *pure symmetry*. Therefore, we recover the known theorem, according to which no symmetry mixing is allowed at the critical point [16]. This result applies to purely nonlocal separable extended potentials acting on a square lattice only, and is here generalized to the case of local single-wave (*viz.*, s-wave) effective contributions to the potential, as in Eq. (2).

As temperature decreases, the auxiliary gap parameters Δ_η may soon depart from the limiting behaviour $\sim (T_c - T)^{1/2}$ at $T = T_c - 0$, which is expected in any mean-field theory for the OP. On the contrary, the maximum value Δ_M of $|\Delta_{\mathbf{k}}|$ over the the whole 1BZ displays a more conventional, BCS-like temperature dependence (Fig. 1).

Depending on the value of μ, other symmetry channels may begin contributing to the full order parameter $\Delta_{\mathbf{k}}$ as T decreases. Although the remaining auxiliary gap parameters Δ_η become different from zero, the temperature T_m at which this occurs does not characterize any new phase transition: the system is already in the superconducting state. Such an effect is made possible by the strong nonlinear character of the gap equations, which becomes more relevant as $T \to 0$. Our numerical analysis at $T = 0$ reveals that s-wave prevails at low band fillings, whereas d-wave wins out at larger, including optimal, band fillings (Fig. 2). The ILPT mechanism is seen to reduce considerably the μ–T region available to symmetry mixing, compared to the limit $T_J \to 0$. The occurrence of a mixed ground state below T_m may produce effects on the temperature dependence of observable quantities [15]. On the other hand, such effects should be hard to detect, since usually $T_m \ll T_c$, and samples of considerable purity would be required.

The most prominent superconducting property enhanced by the ILPT mechanism is the critical temperature itself [14]. This is best understood

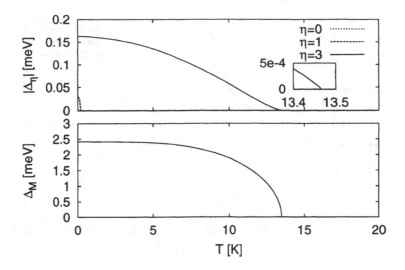

Figure 1. Temperature dependence of the gap parameters $|\Delta_\eta|$ (top), and of the gap maximum Δ_M (bottom), at $\mu = -0.4850$ eV, for $\{\lambda_0, \lambda_1, \lambda_2, \lambda_3, \lambda_4\} = \{0.01, -0.2125, 0.0, -0.2125, 0.0\}$ eV and $t_\perp = 0.08$ eV. Superconductivity sets in at $T_c \approx 13.4$ K, with a d-wave OP. The gap acquires an s-wave contribution at $T_m \approx 0.2$ K. The inset in the top figure shows that Δ_3 displays the expected critical behavior, with critical exponent $1/2$, only very close to T_c.

when considering the linearized version of the gap equations (6). At the critical temperature, these are easily seen to be equivalent to a homogeneous eigenvector problem for the auxiliary gap parameters Δ_η. The condition for such a problem to have a nontrivial solution can be cast in the form of a secular equation:

$$\det(\delta_{\eta\eta'} + \lambda_\eta M^0_{\eta\eta'}) = 0, \qquad (7)$$

to be treated as a nonlinear equation for T_c, where

$$M^0_{\eta\eta'} = \frac{1}{N} \sum_{\mathbf{k}} \tilde{\chi}^0_{\mathbf{k}} g_\eta(\mathbf{k}) g_{\eta'}(\mathbf{k}), \qquad (8)$$

and $\tilde{\chi}^0_{\mathbf{k}} = \lim_{\Delta_{\mathbf{k}} \to 0} F_{\mathbf{k}} \chi_{\mathbf{k}}$ is the linearized version of the renormalized pair susceptibility. Due to the definite symmetry character of $g_\eta(\mathbf{k})$, $M^0_{\eta\eta'}$ is block-diagonal. Therefore, Equation (7) factorizes into two equations, one for each single-wave symmetry block. It can be proved [15] that each equation admits *one* solution under the present work hypotheses on λ_η. The largest solution is to be identified as the critical temperature T_c. The spurious solution corresponds to the subdominant instability, which would take

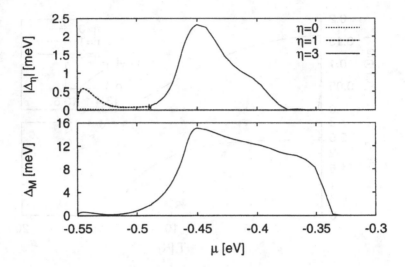

Figure 2. Gap parameters $|\Delta_\eta|$ (top), and gap maximum Δ_M (bottom), at $T = 0$, as a function of μ. Same values of the parameters as in Fig. 1. *s*-wave symmetry prevails at low filling, whereas *d*-wave symmetry wins out at larger, including optimal, filling. Mixing may occur at intermediate fillings, within a restricted range in μ. The maximum gap Δ_M is significantly different from zero only within a restricted region of the total band-width.

place in the absence of coupling in the dominant channel. Depending on the chemical potential, such an instability may evolve into a mixing state at a lower temperature T_m. However, symmetry mixing is thus excluded at exactly the critical point, except for accidental degeneracies.

Both solutions and, particularly, T_c are *bounded from below* by the temperature $T^\star = T^\star(\mu)$ at which $M^0_{\eta\eta'}$ first diverges. The analytical expression for $T^\star(\mu)$ shows that $T^\star(\mu) \lesssim T_J/4 = t_\perp^2/(4t)$, so that $k_B T_c \lesssim 0.01$ eV ($T_c \lesssim 110$ K) for the quoted values of $t_\perp$ [15].

The ILPT mechanism also provides the k-dependence of the OP with unusual features. A straightforward expansion of the gap equation in powers of $\beta^2|\Delta_{\mathbf{k}}|^2$ leads to the following expression for $|\Delta_{\mathbf{k}}|$ at $T = T_c - 0$:

$$|\Delta_{\mathbf{k}}| = \alpha_h \frac{T_c}{2}\left(1 - \frac{T}{T_c}\right)^{1/2} F^{0c}_{\mathbf{k}} \sum_{\eta \in h} W^{0c}_{\bar\eta\eta} g_\eta(\mathbf{k}), \qquad (9)$$

where α_h and $W^{0c}_{\bar\eta\eta}$ are closed functions of μ, T_c, and of the symmetry character ($h = s$- or d-wave) of the incipiently opening gap [15]. Equation (9) basically expresses $|\Delta_{\mathbf{k}}|$ as a linear combination of the basis functions $g_\eta(\mathbf{k})$ belonging to a prescribed symmetry ($\eta \in h$), times the unusual factor $F_{\mathbf{k}}^{0c}$, evaluated in the limit $\Delta_{\mathbf{k}} \to 0$, $T \to T_c$. The $\mathbf{k}$-dependence of the OP is therefore ruled by the strong variations of such a factor already at $T = T_c$. The linear combination of the basis functions provides an additional smearing over the 1BZ, and is responsible of the overall symmetry character (particularly, of nodes in the purely d-wave case), and possibly of symmetry mixing as T decreases.

Given the self-consistent definition Eq. (5), one observes that $|\Delta_{\mathbf{k}}|$ is maximum where $F_{\mathbf{k}}$ is maximum, $i.e.$ when $1 - T_J(\mathbf{k})\chi_{\mathbf{k}}$ is minimum. Therefore, the modulus of the OP presents a line of relative maxima along the $\xi_{\mathbf{k}} = 0$ locus, at all temperatures, and decays almost exponentially as one departs from the Fermi line, due to the presence of $\chi_{\mathbf{k}} \propto \tanh(\beta E_{\mathbf{k}}/2)$ in $F_{\mathbf{k}}$. Along the Fermi line, $|\Delta_{\mathbf{k}}|$ is seen to attain its absolute maximum value at the intersection between the $\xi_{\mathbf{k}} = 0$ locus and the Γ–X–M path in the 1BZ, its exact location depending on the topology of the Fermi line, and symmetry related points. The modulus of the OP decreases as one approaches $k_x = k_y$ along the Fermi line. Its value is here a relative minimum, which is zero in the case of pure d-wave symmetry.

In Fig. 3 we show our numerical results for $|\Delta_{\mathbf{k}}|$ evaluated along the Fermi line at $T = 0$, normalized with respect to its maximum value Δ_M over the 1BZ. Different curves correspond to different values of μ, and therefore to different topologies of the Fermi line. A wave-vector $\mathbf{k}$ on the Fermi line is parametrized by the angle ϕ formed by $\mathbf{k}$ with the Γ–X direction in the 1BZ.

The value which $|\Delta_{\mathbf{k}}|/\Delta_M$ assumes at $\phi = 45°$ is zero in the case of pure d-wave symmetry, and nonzero in presence of an s-wave contribution to the OP. In both cases, it is apparent that $|\Delta_{\mathbf{k}}|/\Delta_M$ decreases *more than linearly* as one departs from its maximum to $\phi = 45°$ along the Fermi line. We emphasize that such a decrease rate cannot be reproduced by a simply nonlocal, separable potential, and in fact has been claimed as a "smoking gun" for the ILPT mechanism by Anderson [17], with the support of ARPES data.

In conclusion, the ILPT mechanism provides the superconducting OP with an unusual dependence on the wave-vector $\mathbf{k}$ in the 1BZ. Both analytical results at T_c and numerical results at lower temperatures reveal a line of relative maxima for $|\Delta_{\mathbf{k}}|$ along the Fermi line, and *sharp* absolute maxima at the intersection with the Γ–X–M path, with an almost exponential decrease elsewhere in the 1BZ. Zero or nonzero minima are present along the $k_x = k_y$ direction, consistent with pure d-wave or (mixed) s-wave contributions to the OP, respectively. The resulting picture is that of a Fermi line which gets predominantly gapped in correspondence with the sharp maxima of $|\Delta_{\mathbf{k}}|$.

This is consistent with recent ARPES results [6], which describe the pro-

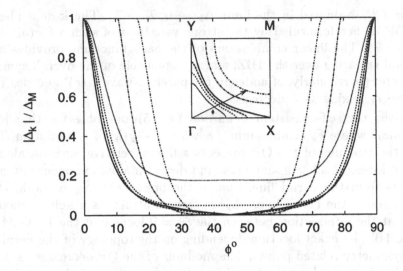

Figure 3. Normalized gap Δ_k/Δ_M at $T = 0$ along the Fermi lines corresponding to $\mu = -0.51$, -0.49, -0.47, $-0.45(= \mu_{VH})$, -0.298, -0.07 eV. Correspondingly, $\Delta_M = 0.38$, 1.71, 5.86, 15.12, 8.98, 5.22 meV. The angle ϕ measures the position of the wave-vector **k** with respect to the $\Gamma-X$ direction in the 1BZ. One easily recognizes the presence of an s-wave contribution at low band filling, signalled by a nonzero, relative maximum value in Δ_k at $\phi = 0°$. The gap is purely d-wave for larger band fillings, with a node at $\phi = 0°$. However, Δ_k is seen to vanish *more than linearly* as $\phi \to 0$, as a consequence of the ILPT mechanism.

gressive "erosion" of the Fermi line by the opening of a strongly anisotropic pseudogap in underdoped Bi2212 single crystals. The pseudogap first opens up near the X point in the 1BZ and removes larger sectors of the Fermi line, thus reducing it into disconnected, gapless arcs, which shrink with decreasing temperature. At $T = T_c$, the Fermi line is entirely but non-uniformly gapped (except at isolated points along $k_x = k_y$, in the pure d-wave case). The remarkable variations of the OP over the 1BZ closely follow the k-dependence of the pseudogap above T_c, and can be reproduced within the ILPT mechanism.

References

1. D. A. Wollman *et al.*, Phys. Rev. Lett. **71**, 2134 (1993); *ibid.* **73**, 1872 (1994); *ibid.* **74**, 797 (1995).
2. C. C. Tsuei *et al.*, Phys. Rev. Lett. **73**, 593 (1994); Science **271**, 329 (1996).
3. H. Ding *et al.*, Phys. Rev. Lett. **74**, 2784 (1995) [**75**, 1425 (1995)].
4. Z.-X. Shen *et al.*, Phys. Rev. Lett. **70**, 1553 (1993).
5. M. R. Norman *et al.*, Phys. Rev. B **52**, 615 (1995).
6. M. R. Norman *et al.*, Nature **392**, 157 (1998).

7. B. E. C. Koltenbah and R. Joynt, Rep. Prog. Phys. **60**, 23 (1997).
8. H. Zhang and H. Sato, Phys. Rev. Lett. **70**, 1697 (1993); H. Zhang, H. Sato, and G. L. Liedl, Physica C **234**, 185 (1994).
9. J. M. Wheatley, T. Hsu, and P. W. Anderson, Phys. Rev. B **37**, 5897 (1988).
10. A broadening in the k-dependence of $T_J(\mathbf{k})$ is expected, when inelastic scattering is taken into account. See J. O. Fjærestad and A. Sudbø, in this volume.
11. J. Yu and A.J. Freeman, Jour. Phys. Chem. of Solids **52**, 1351 (1991).
12. J. F. Annett, Adv. Phys. **39**, 83 (1990).
13. O. K. Andersen *et al.*, J. Low Temp. Phys. **105**, 285 (1996).
14. S. Chakravarty, A. Sudbø, P.W. Anderson, and S. Strong, Science **261**, 337 (1993).
15. G. G. N. Angilella, R. Pucci, F. Siringo, A. Sudbø, subm. to Phys. Rev. B.
16. F. Siringo, G. G. N. Angilella, and R. Pucci, Phys. Rev. B **53**, 2870 (1996).
17. P. W. Anderson, J. Phys. Chem. Solids **56**, 1593 (1995), proceedings of the Conference on Spectroscopies in Novel Superconductors. Stanford, CA, March 15-18, 1995.

PART II.

.... than experimental ....

POINT-CONTACT SPECTROSCOPY OF
SUPERCONDUCTING RNi$_2$B$_2$C (R=RARE EARTH, Y)

I.K. YANSON
*B. Verkin Institute for Low Temperature Physics and Engineering
National Academy of Sciences of Ukraine, 310164 Kharkiv,
Ukraine*

1. Introduction

Recently discovered nickel-borocarbides RNi$_2$B$_2$C where (R =rare earth, Y, Th), exhibit very interesting properties[1, 2]. They are superconducting at relatively high critical temperatures. In many of them the superconductivity coexists with the magnetic order (Table 1). Their crystal structure resem-

TABLE 1. The critical superconducting temperatures T_c, the Néel temperatures T_N, and the superconducting energy gaps Δ_0(meV) for RNi$_2$B$_2$C-compounds. T_c and Δ_0 are measured by point-contact Andreev-reflection spectroscopy.

Parameter	Y	Tm	Er	Ho	Dy	La
T_c [K]	15.6	10.9	10.8	8.6	6.05	-
T_N [K]	-	1.5	5.9	5.2	10.5	-
Δ_0 [meV]	2.4	1.3	1.7	1.0	0.99	-
$2\Delta_0/k_B T_c$	3.6	2.8	3.7	2.7	3.8	-

bles that of the high-T_c superconductors[3] although electronically they are rather three-dimensional isotropic metals [4, 5]. RNi$_2$B$_2$C-superconductors may be considered unconventional, since their heat capacity at low temperatures obeys power law [6] and the vortex lattice exhibits square symmetry instead of a triangular one[7, 8]. The Cooper pairing mechanism in this compound is unknown. One can find in the literature the indications that they are weak-to-moderately-strong superconductors[9], along with calculations

M. Ausloos and S. Kruchinin (eds.), Symmetry and Pairing in Superconductors, 271–285.
© *1999 Kluwer Academic Publishers. Printed in the Netherlands.*

that they are strong-coupling ones[5]. On one hand, there is a point of view that the mechanism of pairing has an ordinary electron-phonon origin and the high frequency boron vibrations play the main role[10]. However, on the other hand, the temperature dependence of resistivity does not obey the Bloch-Grüneisen theory[11, 12, 13, 14] and there is strong evidence that the low frequency phonon and magnon modes play an important role [15, 16, 17].

In the present brief review we add some more conventional and unconventional properties to these compounds by means of point-contact spectroscopy. The paper is divided in two parts. First, we extract the superconducting energy gaps and peculiarities of superconducting density of quasiparticle states by means of Andreev-reflection spectroscopy[18, 19]. In addition, we show how to measure the $H_{c2} - T$ phase diagram by means of the point contacts. Secondly, we apply the inelastic point-contact spectroscopy in the normal state to discover the strong interaction of conduction electrons with low-frequency modes[17, 20]. These modes are characteristic for the superconducting compounds and lead to an increase of quasiparticles relaxation at energy close to 2Δ which, in accord with Ref.[38], may be due to the appearance of a new phonon mode with the frequency close to the superconducting energy gap.

In experiments, as usual, the contact axis is directed along the ab-plane.

2. Part I. Andreev-reflection spectroscopy in the superconducting state

2.1. PRINCIPLES OF AR SPECTROSCOPY

The Andreev-reflection spectroscopy differs from the tunneling spectroscopy in that it uses junctions with zero or small potential barrier between electrodes[21, 22, 23]. The quasiparticle coming at the boundary from the normal metal has a certain probability, depending on the barrier transparency, to form a Cooper pair in the superconductor. An electron quasiparticle, being reflected into the normal metal as a coherent hole, moves in the opposite directions.

Just as the tunneling spectroscopy measures the local density of quasiparticle state at the surface of superconductor, the Andreev-reflection spectroscopy measures the transformation of quasiparticles to Cooper pairs which happens at least at a distance of coherence length ξ from the interface between a normal metal and superconductor. In the case of diffusive (but still conservative) motion of electrons, the energy position of the gap nonlinearity will be the same in the $I - V$ characteristics, only the intensity is decreased.

In the Blonder-Tinkham-Klapwijk (BTK) theory [23] which is based on the BCS theory of superconductivity, the contact is modeled as a one-dimensional ballistic junction with an arbitrary barrier strength Z. There is an important modification introduced in ref.[24] where a constant imaginary part Γ is added to the quasiparticle energy $E \Rightarrow (E - i\Gamma)$. This phenomenological correction enables one to account for various depairing interactions and greatly improves the fit of the BTK theory to an experiment. For fitting, one has three parameters: the energy gap at zero temperature Δ_0, the barrier strength Z, and the depairing factor Γ. For real superconductors with only a part of the contact region being in the superconducting state, and for diffusive motion of electron, one has to introduce a scaling factor SF. In the best case, this factor is of the order of unity.

2.2. SUPERCONDUCTING ENERGY GAP AND DENSITY OF STATES

In Fig.1 the examples of the BTK fits are shown. The fit is usually good for the energies not far above the energy gap $\Delta(T)$. For larger biases there are noticeable deviations from BTK curve (see Fig.1(a)). The typical deviation is a kink at biases $eV \approx 2\Delta_0$ which is common for all compounds measured. The point-contact spectroscopy in the normal state shows that there is a peak in the electron-phonon-interaction function at the close vicinity of this bias (see below). One can speculate that inelastic scattering of charge carriers at this energy leads to the stepwise increase of the parameter $\Gamma(eV)$. For the particular spectrum of Ho-compound (Fig.1 (b)) the kink is not visible due to the large enough parameter Γ. However, other contacts of Ho-compound, with smaller Γ, reveal this kink quite well (not shown here).

In Fig.2 the zero temperature energy gaps obtained by BTK fitting are shown as a function of critical temperature which we determine as an onset of changing the AR spectra near zero bias. Data points for Y-, Er-, and Dy-compounds well coincide with a straight line with a slope $2\Delta_0/k_B T_c = 3.63 \pm 0.05$ evidencing for moderately-strong coupling. Another two compounds, Ho- and Tm-, exhibit noticeable less slope $2\Delta_0/k_B T_c \simeq 2.8$. The tetragonal structure of these metals is such that the conducting Ni_2B ab-planes are intercalated with the R-C planes. The differences between these two groups lie in that the rare-earth magnetic moments for Ho- and Tm-compounds order ferromagnetically (FM) and exhibit the net FM component on the Ni_2B layers [25, 26, 27, 28]. On the contrary, in the first group the net ferromagnetic component on the NiB_2-plane is equal to zero. In Er-compound the magnetic moments order antiferromagnetically in the ab-layers [29, 30]. Y-compound is non-magnetic, and in the Dy-compound the commensurate antiferromagnetic (AF) ordering occurs at temperature much higher ($T_N = 10.5$ K) than the superconducting transition ($T_c = 6.05$

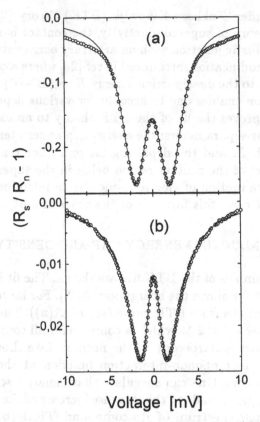

Figure 1. Andreev reflection spectra at $H=0$ for YNi_2B_2C-Cu (a) and $HoNi_2B_2C$-Ag (b) point contacts. R_s and R_n are the $dV/dI(V)$ characteristics in the superconducting and normal state , respectively. The contact parameters and measurement conditions are: (a) $R_0=16.6$ Ω, $T_c=15.4$ K, $T=4.2$ K; (b) $R_0 =0.77$ Ω, $T_c=8.5$ K, $T=1.6$ K. The solid curves stand for the BTK fits with the following parameters: (a) $\Delta_0=2.1$ meV, $\Gamma=0.6$ meV, $Z=0.5$, $SF=1.19$; (b) $\Delta_0 =1.05$ meV, $\Gamma=1.35$ meV, $Z =0.595$, $SF=0.58$.

K) although the magnetic moments also order FM in the ab-planes[31]. The situation for Ho-compound is more complicated. At about the superconducting transition temperature ($T_c = 8.6$ K) appears an incommensurate spiral magnetic order with wave vector along c-axis which creates a non-zero FM component on the Ni_2B-planes [27]. This order is detrimental to superconductivity which leads to the destruction of superconducting state with lowering temperature until the new AF transitions, incommensurate along a-axis ($T_N = 6.2$ K) and commensurate along c-axis ($T_N = 5.2$ K) occurs. The latter is the same as in the Dy-compounds. These AF transitions cancel the FM component at Ni_2 B-planes to zero.

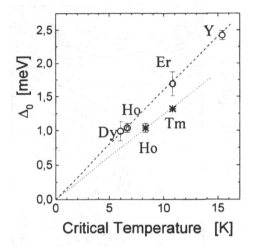

Figure 2. The superconducting energy gap as a function of the critical temperature as measured by AR spectroscopy.

The AR-spectroscopy shows a dramatic change in superconducting DOS at AF transitions (Fig.3) in $HoNi_2B_2C$. Above this temperature dV/dI-curves can't be fitted with any parameters of BTK theory and show very smeared gapless-like behavior. Below the temperature of about 6 K a new structure appears which indeed can be fitted by the BTK procedure. If we take the temperature where the BCS-like superconducting state appears in $HoNi_2B_2C$ as a "BCS-like critical-temperature" T_c^*, then the BTK-model fits well to the temperature dependence of the experimental curves [18] for $T < T_c^*$. Hence, we can place the second data point in Fig.2 for Ho-compound which matches well the dependency of the first group.

In Fig.4 the temperature series of AR spectra are shown for $ErNi_2B_2C$ compound. In spite of this material undergoes the AF transition at about 6 K, no changes in the overall shapes of dV/dI-curves occur at the T_N temperature. As we have already noted this is due to the AF ordering of Er-spins in the ab-planes. The same BCS-like behavior was observed for the temperature dependencies of non-magnetic YNi_2B_2C and AF-ordered $DyNi_2 B_2C$ compounds.

2.3. $H_{C2} - T$ PHASE DIAGRAM

Using the point contacts one can measure the $H_{c2} - T$ phase diagram. These measurements are in equilibrium, since an infinitesimal voltage might be

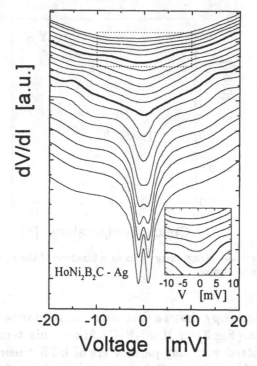

Figure 3. The temperature dependence of Andreev-reflection spectra for HoNi$_2$B$_2$C-Ag with R=0.58 Ω. Temperatures from top to bottom are: 12, 11, 10.6, 10, 9, 8.8, 8.6, 8.0, 7.6, 7.0, 6.6, 6.0, 5.4, 5.0, 4.2, 3.5, 3.0, 2.5, 2.0, and 1.6 K. The inset shows the portions of curves in the dotted rectangle enlarged. The curves are shifted vertically for convenience. The thicker curves mark the transitions to the superconducting and AF states.

applied near zero bias. An important advantage of this method is that the dimensions of a sample might be very small, since the size of a contact amounts only a few tens of nm.

In the normal state the magnetic field dependence of a point-contact resistance is negligible. When the S-electrode undergoes into the superconducting state the resistance drops down due to AR in the bulk. Hence, the point where a resistance saturates at the maximum value as a function of applied field may be taken as the critical value for H_{c2}. The series of such curves where the temperature serves as a parameter, is shown in the Fig.5 for HoNi$_2$B$_2$C. Clearly, it is seen the deep sink at the temperature of 5.0÷5.4 K which equals to the Neél temperature T_N for this compound.

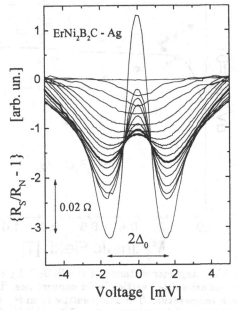

Figure 4. Andreev-reflection spectra for $ErNi_2B_2C$-Ag point contact with R=0.28 Ω at different temperatures below the superconducting transition. The normal state background is subtracted. The temperature from bottom to top are 1.4, 4.2, 4.6, 5.0, 5.6, 6.2, 6.8, 7.4, 8.0, 8.4, 9.0, 9.4, 9.8, 10.3, 10.6, 10.8, and 11.0 K, respectively. The thicker curve marks the transition to the AF state.

For $ErNi_2B_2C$ the sink is much less prominent due to the AF ordering of Er spins in the ab-plane mentioned above. The results for $H_{c2} - T$ phase diagram in two mutual perpendicular field orientations are almost isotropic, although there is an essential magnetic anisotropy in $HoNi_2B_2C$. Perhaps, the magnetic anisotropy compensates that of superconductivity connected with the NiB_2 layered structure. Note, that the extrapolated to zero temperature H_{c2} is about twice less than the results of Fig.5 in Ref.[18] for polycrystals. The H_{c2} -value for single crystals coincides rather well with the values obtained by other authors[32].

3. Part II. Electron-quasiparticle-interaction spectral function

3.1. PRINCIPLES OF POINT-CONTACT SPECTROSCOPY IN THE NORMAL STATE

The point-contact spectroscopy (PCS) involves studies of the nonlinearities of $I - V$ characteristics of metallic constrictions with characteristic size d

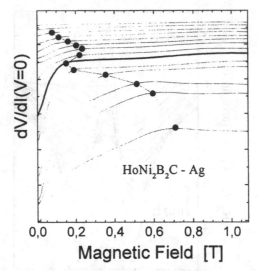

dV/dI(V=0)

HoNi$_2$B$_2$C - Ag

Magnetic Field [T]

Figure 5. The zero-bias magnetoresistance of HoNi$_2$B$_2$C-Ag contact versus different temperatures. The curves are shifted vertically for convenience. The solid dots mark the saturation field for each temperature. The temperature from top to bottom are: 9.0, 8.8, 8.6, 8.0, 7.6, 7.6, 6.6, 6.0, 5.4, 5.0, 4.2, 3.5, 3.0, and 1.6 K. the thicker curve marks the transition to the AF state.

smaller than the inelastic electron mean free path l_{in}[33]. In the case of a $S - c - N$ junction made of dissimilar electrodes, PCS has the advantage that the metal is probed into the depth of the current spreading region, which is of the order of the constriction size. On one side, this size may be large enough to make a negligible contribution of the spoiled interface layers inevitably created while producing a contact; on the other side, it may be not large enough with respect to the conditions of the spectroscopic regime of current flow:

$$d \leq \min\left(l_{in}, \sqrt{l_{in}, l_e}\right) \tag{1}$$

where l_e is the elastic mean free path. Thus, the movement of electrons should be conservative. In the constriction region the specific nonequilibrium distribution function is formed which has two pieces shifted in energy by applied bias voltage eV[34]. The relaxation of this state leads to the non-ohmic $I - V$ characteristics.

The voltage dependence of the differential resistance of a ballistic point contact reflects the energy-dependence of the scattering cross section of the conduction electrons. Hence, the point-contact spectra $[d^2V/dI^2(V)$ dependencies] are proportional to the electron-quasiparticle-interaction (EQI) spectral function. For scattering by phonons, this function is a transport version of electron-phonon-interaction (EPI) function which in many cases differs only little from the Eliashberg EPI function.

In the case of point contacts between different metals with strongly different Fermi velocities, only the spectrum of the metal with the smaller Fermi velocity v_F is seen[35]:

$$\frac{d\ln R}{dV}(V) = \frac{4}{3}\frac{ed}{\hbar v_F} g_{PC}(\omega)\mid_{\hbar\omega=eV}; \qquad (T \simeq 0) \qquad (2)$$

The function $g_{PC}(\omega) = \alpha_{PC}^2 F(\omega)$ is the product of the averaged EPI matrix element with kinematic restrictions imposed by the contact geometry, and the phonon density of states $F(\omega)$. The contact diameter d is determined by the normal-state resistance at zero bias R_0 via the Sharvin expression. In the case of copper and silver as a N-metal, $d \simeq 30/\sqrt{R_0[\Omega]}$ nm in a ballistic regime.

In the spectroscopic regime (1) no heating of the contact area occurs. However, if the contact size is large compared with the electron energy-relaxation length, then there is a local heating of the contact. The maximum temperature equals $eV = 3.63k_BT$ for the standard Lorentz number and the thermoconductivity obeying the Wiedeman-Franz law.

3.2. PCS SPECTRA OF RNi$_2$B$_2$C COMPOUNDS

In Fig.6 the PC spectra are shown for Y-, Ho-, Dy-, and LaNi$_2$B$_2$C compounds in the normal state (by applying the requisite magnetic field for superconductors). The contacts chosen for Fig.6 reveal the expected AR structure evidencing the presence of superconducting phase under the contact (see Part I). All the spectra of the superconducting compounds, both magnetic and non-magnetic, exhibit peaks at about 4–5 meV corresponding to the maxima of the soft phonon density of states observed by neutron in YNi$_2$B$_2$C [16] and LuNi$_2$B$_2$C[15]. There is also an essential spectral intensity at 9, 15, and 19 meV coinciding with a flattening of the phonon dispersion curves in LuNi$_2$B$_2$C[15].

The importance of the low-energy part of the EPI spectral function is strongly emphasized by the fact that the EPI spectrum of non- superconducting LaNi$_2$B$_2$C does not exhibit soft phonon modes. All the spectra show saturation at voltages of about $\sim$0.1 V corresponding to the boron vibration mode and marking the end of the phonon spectrum. Due to the scattering on the nonequilibrium phonons the background rises monotonically. We approximate it by the dotted line to obtain the zero values of spectral function $g_{PC}(eV)$ for energies greater than $\hbar\omega_{max}$ for phonons. In-

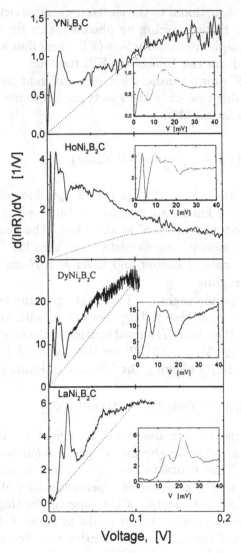

Figure 6. The second derivatives $d(\ln R)/dV$ of the $I - V$ curves $(R = dV/dI)$ for contacts of RNi_2B_2C (R = Y, Ho, Dy, La) with normal metal proportional to the electron-phonon-interaction spectral functions. The dotted lines stand for the supposed background. The contact parameters and measuring conditions are as follow. For YNi_2B_2C: R=16.6 Ω, $(H \perp c)$=7 T: for $HoNi_2B_2C$: R=2.3 Ω, $(H \perp c)$=0.5 T; for $DyNi_2B_2C$: R=27 Ω, H=0.65 T; for $LaNi_2B_2C$: R=0.91 Ω, $H = 0$. The insets show the same data for the low bias voltages.

tegrating of this function enables us to determine the λ-parameters [$\lambda =$

$2 \int_0^\infty g_{PC}(eV)d(eV)/(eV)]$ which are of the order of a few hundredths for the typical spectra, and reach several tenths for the most intensive spectrum. The discrepancy with the expected λ-value which is of the order of unity, may be partly due to the short electron mean free path l_e for elastic scattering which leads to the diffusive regime of the current flow in the contact region. In this case the contact diameter d in Eq.(2) should be replaced by l_e [35]. The additional suppression of the spectral intensity may be due to a barrier at the contact interface and to a possibly small fraction of the superconducting phase that contributes to the observed phonon structure. For Dy-compound we observed the most intensive spectrum with $\lambda = 1.7$ (using v_F=3.6x10^7 cm/s [5]). This scattering of λ-values shows that before to extract the absolute figures one has to establish the regime of current flow in a particular contact.

3.3. EVOLUTION OF SOFT MODE WITH TEMPERATURE AND MAGNETIC FIELD

From the dV/dI-spectra for single crystal HoNi$_2$B$_2$C- Ag point contact (Fig.3) we see that the nonlinearity of $I-V$ characteristic near zero bias first appears in the normal state while lowering the temperature down to 10–12 K. Judging from neutron data on powder samples [25, 27], at this particular temperature appears the first signs of magnetic ordering. With further decrease of temperature the structure at biases $\sim$4 mV (see Fig.6) preserves provided the superconductivity is destroyed by requisite magnetic field. Detailed temperature-dependent study of this structure by second derivative of $I-V$ curve confirms its abrupt appearance at the above mentioned temperature and its connection with magnetic order. At low temperatures, for fields greater than the saturation field $H \sim 2$ T in HoNi$_2$B$_2$C the intensity of this structure is greatly decreased and becomes almost not dependable on the magnetic field [20]. Since the Ho-magnetic moments are oriented along ab-plane (due to the crystal field splitting), the magnetic field which is parallel to c-axis does not influence both the magnetic structure and the 4-meV-peak intensity in PC spectra [20]. We interpret this structure as a soft phonon mode which appears due to the Kohn anomaly stimulated by the strong nesting of the Fermi surface [36], since it appears also in the non-magnetic LuNi$_2$ B$_2$C [15]. This low-frequency phonon mode interacts strongly with magnons which occur at about the same energy. One should recall that the Ho- and Dy-spins order ferromagnetically along ab-plane, and that for FM Ho-metal there exists a strong maximum in acoustical and optical branches of the magnon density of states at energies of about 4 meV[37] .

YNi$_2$B$_2$C is non-magnetic. Nevertheless, a maximum at about 4 meV is also seen in its spectra as shown in Fig.6, although not in a big range of magnetic field (in the normal state) [17]. Since H_{c2} for YNi_2B_2C is about $5 \div 6$ T ($T = 4.2$ K), we are forced to apply high enough requisite magnetic fields for this compound which possibly decreases the intensity of the low-frequency mode.

Recently Kee and Varma [38] proposed an explanation for the low- frequency (4 meV) phonon mode observed in superconducting Y- and Lu-compounds [15, 16]. They argue that the electronic polarizability of the superconducting state has a pole at energies near the superconducting gap 2Δ which leads to a sharp peak in the spectral function for the lattice vibrations near extremum vectors of the Fermi surface at ω just below 2Δ. The newly developed peak in the phonon DOS leads to the kink in the AR spectra like that shown in Fig.1 for YNi_2B_2C (see section 2.2). On one hand, the kink in the superconducting state follows well the $2\Delta_0$'s for Y-, Er- , Ho- and Dy-compounds. On another hand, it does not obey the temperature dependence of the gap, remaining almost constant up to the disappearance of the superconducting state. The peaks at about $eV \simeq 3$-4 meV are observed in *normal* compounds, including YNi_2B_2C at fields higher than H_{c2}, although its intensity is magnetic field dependent [17, 20]. For *normal* $HoNi_2B_2C$ the peak at $eV \simeq 2$ meV is also seldom observed in PC spectra (see Fig.4 in Ref.[20]. The situation is more like described in Ref.[40] where the low-frequecy peak has its origin from the lowering the energy of the soft phonon modes rather than appearing a new one. Further investigation is in progress to prove the existence of the mechanism proposed in Ref.[38].

3.4. MECHANISM OF COOPER PAIRING IN RNI_2B_2C

In PCS the high-frequency boron mode ($\hbar\omega = 106$ meV [10]) cannot be reached on the same grounds as the low-frequency modes, since the regime of current flow changes from spectroscopic (either ballistic or diffusive) to the thermal one. Hence, we can't compare the high- and low-frequency modes in relative intensities. But the observation of pronounced saturation with a shallow spectral peak at about 100 meV points to the considerable interaction of conduction electrons with boron phonons. Nonetheless, we cannot state that these phonons are of primary importance for the superconducting state. First, for nonsuperconducting $LaNi_2B_2C$ compound we observed the similar behavior of PC spectra at boron frequency (Fig.6). Second, for point contact oriented along the $c-$axis the observation of the superconductivity in $HoNi_2B_2C$ was very difficult, since we cannot make a fresh fracture of a crystal perpendicular to the $c-$ axis, and thus we have to use the dirty intact surface. The PC spectrum along c-axis of nonsuperconducting $HoNi_2B_2C$ surface is shown in Fig.7. The boron vibration feature is clearly seen, although there are no low-frequency modes and no traces of superconductivity which should be seen near zero bias at $H = 0$ due to AR. When we saw a superconductivity by means of AR spectra, the PC spectra in the normal state always revealed the low-frequency modes.

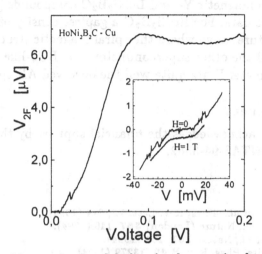

Figure 7. The second harmonic of modulation signal proportional to the second derivative of the $I-V$ curve for $HoNi_2B_2C$-Cu with the contact axis $\parallel c$. The contact parameters and measuring conditions are as follow. $R=1\ \Omega$; $H=0$ T; $V_{mod}=0.47$ mV; $T=4.2$ K. The inset shows the same data for low voltages and $(H \parallel ab)=0$, and 1 T.

Returning to the discussion of the Fig.6, we compare the low-frequency parts of spectra for superconducting Ho- and Y-compounds with a non-superconducting homolog $LaNi_2B_2C$. The low-frequency mode at about 4 meV exists only in the superconducting compounds, and its intensity is strongly magnetic field dependent, unlike the spectra of nonsuperconducting $LaNi_2B_2C$. We conclude that EPI with soft phonon modes is important for the superconducting compounds. The magnetic-field dependence of the point-contact spectra points to an enhancement of EPI by magnetism. This effect competes with the destruction of superconductivity by magnetism.

4. Conclusion

In conclusion, we have shown the effectiveness of point-contact spectroscopy in the study of the electron-phonon-interaction spectral function both in superconducting and normal, magnetic and nonmagnetic, rare-earth nickel borocarbides. To our knowledge up to now this is the only experimental technique able to yield energy-resolved information on the electron-phonon interaction in these compounds. The strong EPI with soft phonon modes is clearly demonstrated to be important for the superconducting compounds. In magnetic superconductors the low-frequency magnon mode interacts with soft phonon increasing the overall electron-phonon-magnon interaction.

AR spectroscopy yields superconducting energy gaps for both magnetic and nonmagnetic compounds. The best tunneling determination of this quantity for nonmagnetic Y- and $LuNi_2B_2C$ compounds [8, 39] are in good accord with AR data. For $HoNi_2B_2C$ a gapless density of states was found in the temperature range where the spiral magnetic structure exists ($T_N < T < T_c$). For all the other superconductors the BCS-like DOS modified by depairing parameter Γ fits quite well the observed AR spectra.

Acknowledgement

The author acknowledges the financial support by the European Community grant INTAS-94-3562.

References

1. R.Nagarajan et al.., Phys. Rev. Lett. 72, 274 (1994).
2. R.J. Cava et al., Nature (London) 367, 146 (1994).
3. T. Siegrist et al., Nature 367, 254 (1994).
4. L.F. Mattheiss, Phys. Rev. B 49, 13279 (1994).
5. W.E. Pickett and D.J. Singh, Phys. Rev. Lett. 72 , 3702 (1994).
6. C. Godart et al.., Phys. Rev. B 51, 489 (1995).
7. U. Yaron et al.., Nature 382, 236 (1996).
8. Y. DeWilde et al.., Phys. Rev. Lett. 78, 4273 (1997).
9. S.A. Carter et al., Phys. Rev. B 50, 4216 (1994).
10. L.F. Mattheiss et al., Solid State Commun. 91, 587 (1994).
11. I.R. Fisher et al., Phys. Rev. B 52, 15086 (1995).
12. I.R. Fisher et al., Phys. Rev. B 56, 10820 (1997).
13. K.D.D. Rathnayaka et al., Phys. Rev. B 55, 8506 (1997).
14. A.K. Bhatnagar et al., Phys. Rev. B 56 , 437 (1997).
15. P. Dervenagas it et al., Phys. Rev. B, 52, R9839 (1995).
16. H. Kawano it et al., Phys. Rev. Lett. 77, 4628 (1996).
17. I.K. Yanson et al., Phys. Rev. Lett., 78, 935 (1997).
18. L.F. Rybaltchenko et al., Europhys. Lett. 33, 483 (1996).
19. L.F. Rybaltchenko et al., Physica B 218, 189 (1996).
20. I.K. Yanson et al.., Fiz. Nizk. Temp. 23, 951 (1997) [Low Temp. Phys. 23, 712 (1997)].
21. A.F. Andreev, Zh. Eksp. Teor. Fiz. 46, 1823 (1964) [Sov. Phys.-JETP 19, 1228 (1994)].
22. S.N. Artemenko, A.F. Volkov, A.V. Zaitsev, Pis'ma Zh. Eksp. Fiz. 28, 637 (1978) [JETP Lett. 28, 589 (1978)], Zh. Eksp. Teor. Fiz. 76, 1816 (1979) [Sov. Phys.-JETP 49, 924 (1979)], Solid State Commun. 30, 771 (1979).
23. G.E. Blonder, M. Tinkham, and T.M. Klapwijk, Phys. Rev. B 25, 4515 (1982).
24. R.C. Dynes et al., Phys. Rev. Lett. 41, 1509 (1978).
25. T.E. Grigereit et al., Phys. Rev. Lett. 73, 2756 (1994).
26. A.I. Goldman et al., Phys. Rev. B 50, 9668 (1994).
27. J.W. Lynn et al., Phys. Rev. B 53, 802 (1996).
28. L.J. Chang et al., Phys. Rev. B 54, 9031 (1994).
29. S.K. Sinha et al., Phys. Rev. B 51, 681 (1995).

30. J. Zarestky *et al.*, Phys. Rev. B **51**, 678 (1995).
31. P. Dervenagas *et al.*, Physica C, **212**, 1 (1995).
32. K.D.D. Rathnayaka *et al.*, Phys. Rev. B **53**, 5688 (1996).
33. I.K. Yanson, Zh. Exp. Teor. Fiz. **66**, 1035 (1974) [Sov. Phys.-JETP **39**, 506 (1974)].
34. I.O. Kulik *et al.*, Fiz. Nizk. Temp. **3**, 1543 (1977) [Sov. J. Low Temp. Phys. **3**, 740 (1977)].
35. A.V. Khotkevich and I.K. Yanson, *Atlas of Point Contact Spectra of Electron-Phonon Interactions in Metals* (Kluwer Academic, New York, 1995).
36. J. Rhee *et al.*, Phys. Rev. B **51**, 15585 (1995).
37. M.W. Stringfellow *et al.*, J. Phys. C: Metal Phys. Suppl. No.2, **3**, S189 (1970).
38. H.-Y. Kee and C.M. Varma. Phys. Rev. Lett. **79**, 4250 (1997).
39. T. Ekino *et al.* Phys. Rev. B **53**, 5640 (1995).
40. M. Bullock *et al.* Phys. Rev. B **57**, 7916 (1998).

THEORY AND EXPERIMENTAL EVIDENCE FOR PAIR FQHE STATE CAUSED BY ZERO-POINT HARMONIC OSCILLATION IN BLOCK-LAYER POTENTIAL

MASANORI SUGAHARA, SYUHEI MITANI, HONG FEI LU, YASUHIKO KUMAGAI, HISAYOSHI KANEDA, NOBUO HANEJI, AND
NOBUYUKI YOSHIKAWA
Faculty of Engineering, Yokohama National University, Hodogaya, Yokohama, 240-8501, Japan

Abstract. The zero point oscillation state of hole carriers confined in block layer potential of $La_{2-x}Sr_xCuO_4$ is similar to the zero point state in a Landau gauge, when the effective "magnetic field" is greater than $10^3 T$. It is shown that a strong effective field and space charge form layered fractional quantum hall effect (FQHE) state for c-axis oriented $La_{2-x}Sr_xCuO_4$ films with localization character, for which there is an anomalous dielectric polarization with negative dielectric constant is expected to appear. Experimental study of the dielectric property of c-axis oriented $La_{2-x}Sr_xCuO_4$ film with oxygen deficiency is made by "ac method"" and "dc method". The negativeness of the electric field in $La_{2-x}Sr_xCuO_4$ film and of the film capacitance are observed with hysteretic polarization. *

(*) Note from the editors. Due to technical circumstances the contribution of Dr. Sugahara et al. cannot be printed at this place in the Proceedings. We regret the inconvenience.

M. Ausloos and S. Kruchinin (eds.), Symmetry and Pairing in Superconductors, 287.
© *1999 Kluwer Academic Publishers. Printed in the Netherlands.*

DIFFERENT ESTIMATES OF THE ANISOTROPY FROM RESISTIVE MEASUREMENTS IN HIGH T_c SUPERCONDUCTORS

Ö. RAPP, M. ANDERSSON, J. AXNÄS, YU. ELTSEV,
B. LUNDQVIST AND A. RYDH
Solid State Physics, Kungliga Tekniska Högskolan,
SE-100 44 STOCKHOLM, SWEDEN

1. Introduction

The anisotropy is one key parameter in understanding the physics of high temperature superconductors (HTSC). In addition to quantifying the variation of physical properties in different crystal directions, it contributes to strongly enhanced thermal fluctuations, and thereby to the rich variety of new phenomena observed in HTSC. Some examples are the existence of a mixed state solid-to-liquid phase transition (which depending on the amount of disorder can be a melting transition or a glass transition), the sometimes extreme broadening of the resistive transition in magnetic fields, and the strong decrease of the critical current, even far below T_c for increasing temperature or magnetic field (see e.g. Cohen and Jensen, 1997).

In this brief review, three different methods to estimate the anisotropy

$$\gamma = \sqrt{\frac{m_c}{m_{ab}}} = \frac{\xi_{ab}}{\xi_c} = \frac{\lambda_c}{\lambda_{ab}} \approx \sqrt{\frac{\rho_c}{\rho_{ab}}} \tag{1}$$

from the resistive transition curve will be described. Subscripts ab and c refer to in-plane and out-of-plane properties respectively. The first members define γ in terms of mass ratios and coherence lengths, and the two following instead employs directly measurable quantities, i.e. the penetration depths and resistivities.

The outline is as follows. In Section 2 we show how to determine γ from the glass transition line, i.e. the onset of (linear) in-plane resistivity, in Section 3 we use the vortex liquid, and in particular the onset of out-of-plane resistivity in parallel magnetic field, and in Section 4 the superconducting fluctuation region above T_c is used. In each section, examples from pure and

M. Ausloos and S. Kruchinin (eds.), *Symmetry and Pairing in Superconductors*, 289–300.

doped $YBa_2Cu_3O_{7-\delta}$ (YBCO) single crystals will be given. In Section 5 the results are briefly summarized.

2. The glass transition line

2.1. METHOD

The mixed state of HTSC is divided into a low temperature vortex solid with non-zero critical current and a high temperature dissipative vortex liquid. For increasing anisotropy, the solid phase is more suppressed, i.e. the phase boundary comes closer to the temperature axis in the $B - T$ phase diagram. Therefore, the position of this phase boundary can be used, in principle, as a method to determine the anisotropy. However, the determination of γ from this phase boundary is not straightforward, both since it depends on the type of disorder in the sample, and also since there is lack of consensus between different theories. For example, it has been suggested by Houghton *et al.* (1989) that the melting transition close to T_c should follow a relation

$$B_m = B_1 \left(\frac{1 - T/T_c}{T/T_c} \right)^2. \tag{2}$$

In a vortex glass, on the other hand, it has been theoretically proposed (Fisher *et al.*, 1991) and experimentally verified close to T_c that the glass line $B_g(T)$ should follow

$$B_g = B_2(1 - T/T_c)^n, \tag{3}$$

with $4/3 \leq n \leq 3/2$. Recently we found (Lundqvist *et al.*, 1998) that an improved description of the glass line, valid over orders of magnitude in B_g/B_0, could be obtained by

$$B_g = B_0 \left(\frac{1 - T/T_c}{T/T_c} \right)^\alpha, \tag{4}$$

with α as an adjustable parameter, assuming values close to a characteristic number for each particular class of compounds. The differences between these expressions, and the complexities and uncertainties of the form of the prefactors B_1, B_2, and B_0, make it unfeasible at the present stage to directly determine γ by fitting experimental data. However, the prefactors in all cases above can be taken to be of the form $B_\nu \propto \gamma^{-2}$, $\nu = 0, 1, 2$. This gives a method to determine the anisotropy variation by the following route: (i) Determine $T_g(B)$ experimentally for a series of samples. (ii) Fit an expression for the glass line for each sample. (iii) Choose a sample, where measurements of γ have been performed by another method, and calibrate the prefactor of the glass line at that point. (iv) Determine γ from the fitted

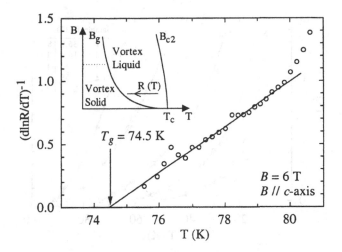

Figure 1. Extraction of the vortex glass temperature T_g at $B = 6$ T for a twinned and fully oxygenated YBCO single crystal using the Vogel-Fulcher relation $(dlnR/dT)^{-1} \propto (T - T_g)$. Inset: Schematic phase diagram of high-T_c superconductors. The arrow indicates how the measurement and the extraction of T_g are made from the vortex liquid side.

prefactors for the other samples, using this calibration. In resistive measurements $T_g(B)$ can be determined e.g. from current-voltage characteristics or from the disappearance of the (linear) in-plane resistance. The equivalence between these procedures has been demonstrated experimentally (Hou *et al.*, 1997; Lundqvist *et al.*, submitted).

2.2. EXAMPLE

The anisotropy in the YBCO system can easily be changed by varying the oxygen content in the CuO chains (Chien *et al.*, 1994). Five YBCO single crystals of varying anisotropy were prepared by annealing in air at different temperatures (Lundqvist *et al.*, 1998). Measurements of the in-plane resistance for constant magnetic fields $0 \leq B \leq 12$ T along the c-axis were made in order to study the vortex glass line from the vortex liquid side, as indicated by the arrow in the inset of Fig. 1. According to the vortex glass model, the resistance at low currents should vanish at the glass temperature T_g as $R \propto (T - T_g)^s$, where s is a scaling exponent (Fisher *et al.*, 1991). Consequently, $T_g(B)$ can be extracted from the linear relations obtained when the Vogel-Fulcher relation $(dlnR/dT)^{-1} \propto (T - T_g)$ is applied to the resistive tails, as shown for one sample in Fig. 1.

As shown in Fig. 2, B_g is shifted towards lower magnetic fields when

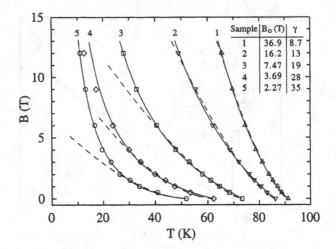

Figure 2. Vortex glass lines for $B \parallel c$-axis as determined from the disappearance of resistivity for five YBCO samples with different anisotropies. Solid lines are fits to Eq. (4) with $\alpha \approx 1$. Dashed lines are described by Eq. (3) with $n \approx 1.5$.

the anisotropy is increased. For temperatures below $T_c/2$, Eq. (3) fails to describe the data, as has also been observed previously (see e.g. Hou *et al.*, 1997). However, Eq. (4) better describes the temperature dependence of the vortex glass lines, including temperatures on both sides of $T_c/2$. The solid curves in Fig. 2 are fits to data using this relation.

Provided that the anisotropy γ of one sample is known, γ for the other ones can be determined by using the relation $B_0 \propto \gamma^{-2}$. The anisotropy of a sample with T_c in between the 60 K- and 90 K-plateaus is well defined, since there is a strict one-to-one correspondence between γ and T_c in this region. We therefore used torque measurements of the anisotropy by Chien *et al.* (1994) to obtain the anisotropy for a sample with $T_c \approx 73$ K. By calibrating the prefactor in this way, we obtained very good agreement with their data, see Fig. 7. However, we did not observe the strong increase of anisotropy below the 60 K-plateau, as seen by Chien *et al.* (1994).

3. The Vortex Liquid

3.1. METHOD

Other important changes in the $B - T$ phase diagram take place in the vortex liquid phase, above the glass line studied in Section 2. For instance, approaching T_c as a function of B or T, vortices will loose their coherence over extended lengths, and transform e.g. into independent pancake vortices

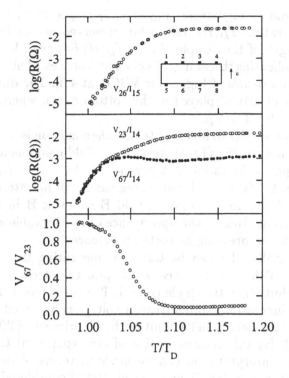

Figure 3. Top panel: Contact configuration and temperature dependence of voltage drop through contacts $2 - 6$ with current applied between contacts $1 - 5$. Middle panel: Temperature dependence of voltage drop through contacts $2 - 3$ and $6 - 7$ with current applied between contacts $1 - 4$. Bottom panel: Temperature dependence of the voltage ratio V_{23}/V_{67} with current applied between contacts $1 - 4$. The magnetic field $B = 4$ T is applied parallel to the c-axis of an YBCO single crystal, with $T_D(4\text{T}) = 83.8$ K.

or into a complicated entangled system.

A powerful experimental technique to study vortex properties in the liquid state is the use of a pseudo-flux transformer, with four contacts on each of two opposite ab-planes of a single crystal. In the first experiments on HTSC performed on $Bi_2Sr_2CaCu_2O_8$ (BSCCO) single crystals at magnetic fields $B > 0.1$ T (Busch *et al.*, 1992; Safar *et al.*, 1992), no coherent vortex state along the c-axis was observed. For YBCO however, with smaller anisotropy, it was found independently by Safar *et al.* (1994) and Eltsev *et al.* (1994) that a transition from correlated to uncorrelated vortex motion along the c-axis occurred at fields $B > 0.3$ T. Later, this transition from coherent to broken vortices was observed by Keener *et al.* (1997) in BSCCO at fields below 0.1 T. With reference to the contact arrangement shown in Fig. 3, such a transition can be described as follows. With a magnetic field along the c-axis, and with current fed through contacts 1 and 4, I_{14}, the top and bottom voltage drops, V_{23} and V_{67} respectively, are equal up to a certain temperature $T_D(B)$, indicating preserved vortex integrity through the c-axis length of the sample. Above $T_D(B)$ however, V_{67} is found to fall below V_{23}, indicating breaking of coherent vortices. This is illustrated in the two bottom panels of Fig. 3, for YBCO at 4 T. The difference between V_{23} and V_{67} is clearly displayed in the bottom panel, where V_{67}/V_{23} is seen to fall below 1 for $T > T_D$.

An alternative way to investigate this phenomenon is to study the c-axis resistance. When vortices break up at T_D, a field component parallel to the ab-planes appears. With current parallel to the c-axis, I_{15} in Fig. 3, the Lorentz force, $\mathbf{I} \times \mathbf{B}$, on such vortex segments will initiate a vortex motion $\perp \mathbf{B}$ in the plane, and an electric field $\mathbf{E} = -\mathbf{v} \times \mathbf{B}$ in the direction of the c-axis current. Hence, the appearance of a detectable c-axis resistance coincides with the breaking of vortex coherence.

Experimentally, this can be tested by measuring the voltage V_{26} with current I_{15}, which roughly gives ρ_c, despite the strongly inhomogeneous current distribution in the single crystal. The top panel in Fig. 3 illustrates $V_{26}(I_{15})$. For increasing temperature, a voltage is observed at $T = T_D$, corresponding to the detectability limit of the experiment of 20 nV (Eltsev and Rapp, 1995a). By this transformation of the experiment to c-axis resistive properties, an interpretation can be made in terms of theories for c-axis conduction. In particular, Daemen et al. (1993) considered the high-T_c superconductor as a stack of junctions along the c-axis coupled by Josephson weak links. By associating the loss of coherence along the c-axis with the appearance of c-axis resistance they derived for the magnetic field B_D the expression

$$B_D = \frac{\phi_0^3}{4\pi^2 \mu_0 sek_B T \lambda_{ab}^2(T)\gamma^2} \tag{5}$$

where ϕ_0 is the flux quantum, s the interplanar distance, $e = 2.718...$, and $\lambda_{ab}(T)$ is the in-plane penetration depth which e.g. for an Y-based system can be taken as $\lambda(T) = \lambda_0(T)[1 - (T/T_c)^2]^{1/2}$ with $\lambda_0 = 1400$ Å (Hardy et al., 1993; Lee et al., 1994). γ is therefore the only adjustable parameter. Thus, fitting Eq. (5) to the decoupling line $B_D(T)$, determined experimentally as the onset of c-axis dissipation in parallel magnetic field, opens a nice possibility to determine the anisotropy of layered superconductors.

3.2. EXAMPLE

This method has been used in studies of Fe- (Eltsev and Rapp, 1995b) and Zn-doped (Eltsev and Rapp, 1996) single crystals of YBCO. In all cases Eq. (5) describes the phase boundary quite well with a single adjustable parameter γ, as shown in Fig. 4. The anisotropy increases faster for Fe- than

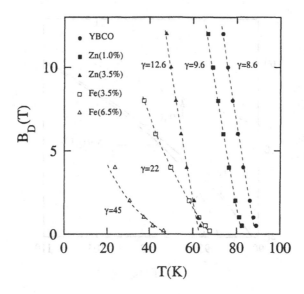

Figure 4. Data for the decoupling line determined from the onset of c-axis resistivity for YBCO single crystals doped by Fe and Zn. Dashed lines are fits to Eq. (5) with γ as a free parameter.

for Zn-doping. This can be associated with Fe mainly substituting for Cu in the chains, while Zn substitutes for Cu in the planes. Varying charge in the chains apparently affects the coupling between planes and the anisotropy. A similar strong effect is also observed from oxygen depletion in the chains, as discussed in Section 2. Zn-doping on the other hand would seem to mainly influence charge balance in the planes, resulting in a strong depression of T_c without appreciable effect on γ.

4. Superconducting Fluctuations

4.1. METHOD

Superconducting fluctuations, i.e. the presence of short-lived superconducting electron pairs above T_c, have a number of physical effects. The most apparent one is probably the increased conductivity just above T_c.

There are well established theories for superconducting fluctuations and their effects on the resistivity. In order to compare with experiments, it is necessary to know what the resistivity would be in the absence of fluctuations, which can only be estimated from an extrapolation of the linear resistivity at higher temperatures (where fluctuations are negligible). A more reliable way is to study the magnetoresistivity $\Delta\rho(B) = \rho(B) - \rho(0)$, which is independent of any extrapolation. The usual situation in e.g. YBCO close

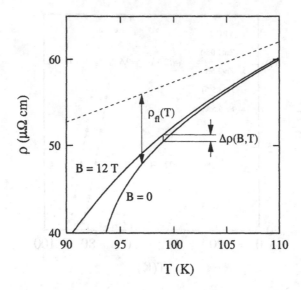

Figure 5. Resistivity vs temperature in two different applied magnetic fields (solid curves). The magnetoresistivity, $\Delta\rho(B,T)$, is independent of any extrapolation of the normal state (dashed line). ρ_{fl} is the resistivity decrease caused by the fluctuations.

to T_c is schematically shown in Fig. 5. The Aslamazov-Larkin term (AL) is the dominating contribution in this regime: when a magnetic field B is applied, the contribution to a supercurrent from short-lived pairs is destroyed, and the magnetoresistivity is positive.

A second contribution comes from the Maki-Thompson terms, which arise from indirect effects, such as the increased probability for normal quasiparticles to be scattered off superconducting electron pairs in the forward direction. This leads to a reduced resistivity, and the destruction of electron pairs by a magnetic field thus again implies $\Delta\rho(B,T) > 0$. A third contribution, the density of states effect (DOS), is a renormalization in the quasiparticle density of states, with a contribution to $\Delta\rho(B,T)$ of opposite sign; formation of short-lived superconducting pairs depletes the charge carriers of the normal state conduction channel, and increases the resistivity. A magnetic field opposes this effect, and $\Delta\rho(B,T) < 0$. The importance of the DOS term has been realized only in recent years, and presumably due to its mathematical complexities, it is not very often employed in the analysis of experimental data. Here we will therefore present results both with and without the DOS term included.

The magnetoconductivity $\Delta\sigma(B,T) = 1/\rho(B,T) - 1/\rho(0,T)$ is thus

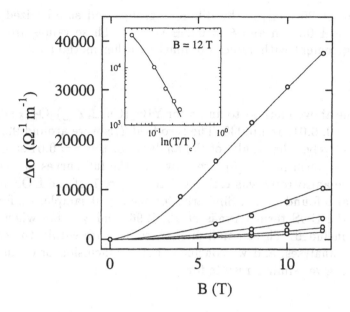

Figure 6. Magnetoconductivity $\Delta\sigma(B) = 1/\rho(B) - 1/\rho(0) \approx -\Delta\sigma(B)/\rho^2$ for the sample $YBa_2(Cu_{1-x}Zn_x)_3O_{7-\delta}$ with $x = 0.01$. Circles are experiments and solid curves are theoretical calculations using all four terms in Eq. (6), with $\xi_{ab} = 1.5$ nm and $\xi_c = 0.21$ nm, which gives $\gamma \approx 7$. The main panel shows the field dependence at five different temperatures, from top to bottom 90.0, 95.3, 100.2, 105.6, and 110.3 K. The inset displays the temperature dependence for $B = 12$ T. B is always parallel to the c-axis.

composed of several terms

$$\Delta\sigma(B,T) = \Delta\sigma^{AL} + \Delta\sigma^{MT(reg)} + \Delta\sigma^{MT(an)} + \Delta\sigma^{DOS}$$
$$= \Delta\sigma(B, T, T_c, \xi_{ab}(0), \xi_c(0), \tau_\phi) \tag{6}$$

where in the last member the variables and parameters have been explicitly listed, with the zero temperature coherence lengths, $\xi_{ab}(0)$ and $\xi_c(0)$, and the phase breaking time τ_ϕ. T_c is usually not fitted but taken close to the midpoint of the transition curve. For references to detailed expressions for the four terms in Eq. (6) and to original publications see e.g. Axnäs *et al.* (1997) or Axnäs *et al.* (submitted). The scheme to determine the anisotropy $\gamma = \xi_{ab}/\xi_c$, is thus to measure $\Delta\rho(B,T)$ over a wide range of B and T, to analyze data in terms of Eq. (6), and to determine the coherence lengths.

In view of the number of free parameters in Eq. (6), we briefly comment on the accuracy of this method to determine microscopic parameters. A rough estimate of the uncertainties could be the spread of the values obtained in existing studies of pure 90 K YBCO. Holm *et al.* (1995) compiled results from almost 20 analyses by different groups and summarized the data as $\xi_{ab} = 1.36 \pm 0.2$ nm and $\xi_c = 0.2 \pm 0.05$ nm. These values are also in reasonable agreement with values obtained by other methods.

4.2. EXAMPLE

We have used the above method to find γ for $YBa_2(Cu_{1-x}Zn_x)_3O_{7-\delta}$ single crystals with $x = 0, 0.01$, and 0.035. The zero field T_c:s were around 92, 86, and 70 K, respectively. The results of the analysis for the $x = 0.01$ sample, including the DOS term in Eq. (6), are shown by the full curves in Fig. 6. In particular, the anisotropy was estimated to be $\gamma \simeq 7$. If the DOS term is excluded we also found $\gamma \simeq 7$. Similarly, for the $x = 0$ sample, we found $\gamma \simeq 6$ without the DOS term (Axnäs et al., 1996) and $\gamma \simeq 6.5$ when the DOS term was included. These small differences in γ are within the error margin for these analyses, and we conclude that the inclusion or exclusion of the DOS term gives similar results for γ.

5. Summary

The results from all methods described here are collected in Fig. 7 in the form of γ vs depression of the transition temperature, $-\Delta T_c$. In addition, results from torque measurements on oxygen depleted samples by Chien et al. (1994) have been included. For the fluctuation method, results both with and without density of states term included in the analysis are shown.

It can be seen that γ vs $-\Delta T_c$ behaves in two distinctly different ways, with little scatter between the results within each group: the strong increase of γ for chain doping, in contrast to the weak increase for plane doping. This difference is further emphasized by the fact that Zn and Fe doping produce roughly the same depression of T_c per at%, illustrated in Fig. 4. Apparently T_c is mainly affected by charge variations in the plane for Zn-doping, while by chain doping, charge transfer, anisotropy, and T_c variations are all interrelated. For the purpose of the present study however, the agreement between different results in each subgroup, illustrates the consistency of the results from these three methods to determine γ.

A fourth method to determine γ could also be mentioned. From the last member of Eq. (1) one can measure the resistivities of a single crystal in different crystallographic directions. However, γ determined in this way would refer to normal state properties, and turns out to be strongly temperature dependent and therefore less informative. E.g., Eltsev and Rapp (1995b) compared the resistivity ratios for Fe-doped YBCO crystals evaluated at $1.1T_c$, with results of the method of Section 3. In spite of such an attempted uniform treatment of all data, the results were scattered. For increasing Fe doping, $\sqrt{\rho_c/\rho_{ab}}(1.1T_c)$ was 6.3, 27, 55, to be compared with

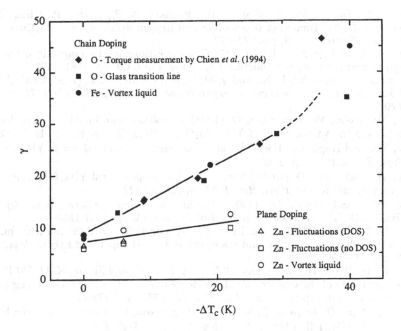

Figure 7. Anisotropy as a function of depression of T_c in doped YBCO single crystals.

the results for γ in Fig. 7 of 8, 22, 45. These differences are larger than the scatter illustrated in Fig. 7, summarizing the results of the three main methods.

6. Acknowledgement

Grants and support from the Swedish Natural Science Research Council, The Swedish Research Council for Engineering Sciences, The Swedish Superconductivity Consortium, The Göran Gustafsson Foundation, and NATO Advanced Research Workshop are gratefully acknowledged.

References

Axnäs, J., Holm, W., Eltsev, Yu., and Rapp, Ö (1996) Increased phase-breaking scattering rate in Zn-doped $YBa_2Cu_3O_{7-\delta}$, *Phys. Rev. B* **53**, pp. R3003–R3006

Axnäs, J., Eltsev, Yu., Holm, W., Lundqvist, B., and Rapp, Ö (1997) Density of states fluctuations in the magnetoresistivity of $YBa_2Cu_3O_{7-\delta}$, *Physica C* **282–287**, pp. 1535–1536

Axnäs, J., Lundqvist, B., and Rapp, Ö, submitted to *Phys. Rev. B*

Busch, R., Ries, G., Werther, H., Kreiselmeyer, G., and Saemman-Ischenko, G. (1992) New aspects of the mixed state from six-terminal measurements on $Bi_2Sr_2CaCu_2O_x$ single crystals, *Phys. Rev. Lett.* **69**, pp. 522–525

300

Chien, T. R., Datars, W. R., Veal, B. W., Paulikas, A. P., Kostic, P., Chun Gu, and Jiang, Y. (1994) Dimensional crossover and oxygen deficiency in $YBa_2Cu_3O_x$ single crystals, *Physica C* **229**, pp. 273–279

Cohen, L. F., and Jensen, H. J. (1997) Open questions in the magnetic behaviour of high-temperature superconductors *Rep. Prog. Phys.* **60**, pp. 1581–1672

Daemen, L. L., Bulaevskii, L. N., Maley, M. P., and Coulter, J. Y. (1993) Critical current of Josephson-coupled systems in perpendicular fields, *Phys. Rev. Lett.* **70**, pp. 1167–1170

Eltsev, Yu., Holm, W., and Rapp, Ö. (1994) Transition from intact to short decoupled vortices in the vortex liquid of $YBa_2Cu_3O_{7-\delta}$, *Phys. Rev. B* **49**, pp. 12333–12336

Eltsev, Yu., and Rapp, Ö. (1995a) Local versus nonlocal conductivity in $YBa_2Cu_3O_{7-\delta}$, *Phys. Rev. Lett.* **75**, p. 2446

Eltsev, Yu., and Rapp, Ö. (1995b) Vortex liquid in single-crystal $YBa_2(Cu_{1-x}Fe_x)_3O_{7-\delta}$ of varying anisotropy, *Phys. Rev. B* **51**, pp. 9419–9422

Eltsev, Yu., and Rapp, Ö. (1996) Weakly increasing anisotropy in Zn doped $YBa_2Cu_3O_{7-\delta}$, *Czechoslovak Journal of Physics* **46**, pp. 1557–1558

Fisher, D. S., Fisher, M. P. A., and Huse, D. A. (1991) Thermal fluctuations, quenched disorder, phase transitions, and transport in type-II superconductors, *Phys. Rev. B* **43**, pp. 130–159

Hardy, W. N., Bohn, D. A., Morgan, D. C., Liang, R.-X., and Zhang, K. (1993) Precision measurements of the temperature dependence of λ in $YBa_2Cu_3O_{6.95}$: Strong evidence for nodes in the gap function, *Phys. Rev. Lett.* **70**, pp. 3999–4002

Holm, W., Rapp, Ö, Johnson, C. N. L., and Helmersson, U. (1995) Magnetoconductivity in $YBa_2Cu_3O_{7-\delta}$ thin films, *Phys. Rev. B* **52**, pp. 3748–3755

Hou, L., Deak, J., Metcalf, P., McElfresh, M., and Preosti, G. (1997) Dependence of the vortex-solid phase transition of $YBa_2Cu_3O_{7-\delta}$ thin films on anisotropy: Evidence for a universal phase boundary, *Phys. Rev. B* **55**, pp. 11806–11815

Houghton, A., Pelcovits, R. A., and Sudbø, A. (1989) Flux lattice melting in high-T_c superconductors, *Phys. Rev. B* **40**, pp. 6763–6770

Keener, C. D., Trawick, M. L., Ammirata, S. M., Hebboul, S. E., and Garland, J. C. (1997) Nonlocal conductivity in $Bi_2Sr_2CaCu_2O_{8+\delta}$ crystals, *Phys. Rev. B* **55**, pp. R708–R711

Lee, J. Y., Paget, K. M., Lemberger, T. R., Foltyn, S. R., and Wu, X. (1994) Crossover in temperature dependence of penetration depth $\lambda(T)$ in superconducting $YBa_2Cu_3O_{7-\delta}$ films, *Phys. Rev. B* **50**, pp. 3337–3341

Lundqvist, B., Rydh, A., Eltsev, Yu., Rapp, Ö., and Andersson, M. (1998) Empirical scaling of the vortex glass line above 1 T for high-T_c superconductors of varying anisotropy, *Phys. Rev. B*, in press

Lundqvist, B., Larsson, J., Herting, A., Rapp, Ö., Andersson, M., Ivanov, Z. G., and Johansson, L.-G., submitted to *Phys. Rev. B*

Safar, H., Rodriguez, E., and de la Cruz, F. (1992) Observation of two-dimensional vortices in $Bi_2Sr_2CaCu_2O_x$, *Phys. Rev. B.* **46**, pp. 14238–14241

Safar, H., Gammel, P. L., Huse, D. A., Majumdar, S. N., Schneemeyer, L. F., Bishop, D. J., Lopez, D., Nieva, G., and de la Cruz, F. (1994) Observation of a nonlocal conductivity in the mixed state of $YBa_2Cu_3O_{7-\delta}$: Experimental evidence for a vortex line liquid, *Phys. Rev. Lett.* **72**, pp. 1272–1275

VORTEX LATTICE MELTING AND VISCOSITY IN $Y_{0.6}Dy_{0.4}Ba_2Cu_3O_{7-x}$ SUPERCONDUCTOR STUDIED BY ELECTRICAL RESISTIVITY

M. PEKALA AND W. GADOMSKI
Department of Chemistry, University of Warsaw,
Al. Zwirki i Wigury 101,
PL-02-089 Warsaw, Poland

I. NEDKOV
Institute of Electronics, Bulgarian Academy of Sciences,
Tzarigrazdsko Chaussee Boulevard 72
1784 Sofia, Bulgaria

AND

H. BOUGRINE AND M. AUSLOOS
SUPRAS, Institute of Physics,
Sart Tilman, B5
University of Liège , B-4000 Liège, Belgium

1. INTRODUCTION

Among the scientific problems raised by High Critical Temperature (T_c) Superconductors (HTcS) extensive experimental research has clearly shown the complexity of these perovskite cuprate compounds both because of their crystallographic structure (layered structures with various phases and defects), as well as the basic microscopic electronic interactions. Experiments have now produced a well established empirical knowledge of many properties of these materials. However, a more fundamental and unified understanding of the nature and mechanisms of the phenomenon of HTcS is still lacking in particular with respect to the symmetry of the order parameter and the system effective dimensionality.

One of the most popular themes is that pertaining to the phenomenology of the mixed state. The investigation of the mixed state in HTcS changes the

301

M. Ausloos and S. Kruchinin (eds.), Symmetry and Pairing in Superconductors, 301–316.

simple notions about some superconductivity aspects as derived from low T_c (LTcS) superconductors. It appears that superconductor physical parameters present a quite complicated behavior especially in high magnetic fields, but also at low fields. The thermal fluctuations, vortex dissipation, quasiparticle scattering "interferences" are in fact the main obstacles for well controlling HTcS properties and subsequent applications. Thus some fine investigations are still needed.

The mixed state of HTcS has been investigated in many papers during the last decade, but the quality of the HTcS samples was pretty bad [1]. Only in the last few years has the quality of HTcS sample begun to reach the quality of the low T_c (LTcS) samples. Comparison of experimental results on LTcS and HTcS has then helped in the understanding of phenomena and peculiarities of HTcS [2]. Such an understanding on good materials allow us to go back to "not-so-good-ones" and propose a better way of understanding their properties. The main feature of HTcS is their highly anisotropic (layered) structure. As a result, the dissipation contributions in the mixed state of HTcS with respect to those in LTcS are to be well defined operationally speaking. Due to quasi two dimensional in-plane electronic motion, strong fluctuation renormalization of the one-electron density of states manifests itself in a very wide range of temperatures above and below T_c. This peculiarity evokes such "hot" problems as the effect of the role of density of states fluctuations, vortex-antivortex interactions in layered materials, features of in-plane and out-of-plane transport in model S-I-S and S-N-S superconducting superlattices, etc. Crossover between 2D and 3D regimes is also important for the understanding of the phenomena in layered superconductors [3]

One reason for the failure of recent mixed state investigations in HTcS is not in the understanding of the nature of the Abrikosov state. A large majority of scientists consider the Abrikosov state to be the vortex lattice itself. Therefore the main results of the thermal fluctuation effects should be exemplified in the vortex lattice melting. This point of view is expressed in almost all papers (see for example the review by Blatter et al. [4]). But this point of view is not entirely correct. The Abrikosov state ought be considered as the state with a long-range phase coherence. A majority of theoretical works seems not to have considered that the long-range phase coherence does not depend on the symmetry or order of the vortex distribution. Therefore attempts to describe the transition widely known as the "vortex lattice melting" seem to consider only some order symmetry disappearance in the vortex distribution. A more correct interpretation of this transition was given in [5, 6], where the transition is interpreted as a transition from the Abrikosov state into the normal state with superconducting fluctuations [7]. The phase coherence determines the interesting superconductor property which is the most important one for applications, i.e. a zero electric resistance. Fluctuation effect investigations have shown that the existence of the superconducting electrons does not exclude the absence of the phase coherence and, as consequence, nonzero electric resistance [8]. This state is perfectly realized only below the so-called percolation temperature T_p. Thus the interest to consider data between T_p and T_c in presence of a (low or not) magnetic field.

The absence of phase coherence is the main obstacle for HTcS applications in a high magnetic field by the way. In fact, the state with superconducting electrons without phase coherence exists in a wide region for HTcS in an applied magnetic field. This state is widely known now as the "vortex liquid". As shown by Nikulov et al. [6] this state can be observed over a wider region in LTcS thin films than in HTcS ones in fact. Investigations have shown that the width of this region depends on different causes. One of them will be taken into account below, i.e. a distribution of vortex lattice temperatures.

The stability of vortex configurations in superconducting bulk, thin or thick films and in layered Josephson - coupled superconductors in presence of an applied current has been investigated by various authors, e.g. [9] - both analytically and numerically. In fact, as the current increases, transitions occur between a pinned state, a plastic-flow state, and next an elastic-flow state. A much more complex behavior has been predicted for layered superconductors due to the anisotropic self-field. The vortex state phase diagram can contain a topologically ordered Bragg-glass phase with quasi-long-range translational order at low magnetic fields, and another glassy phase with dislocations or a pinned liquid at high magnetic fields. The symmetry of the vortex lattices in presence of a set of random pinning potential has been also intensively discussed elsewhere [10, 11, 12]. However it is not easy to define a vortex lattice melting transition from variations of physical features, even though the magnetization jump is considered to be the signature of such a three dimensional first order transition [9].

The present paper is aimed at an advanced analysis of this expected vortex lattice melting transition. The mixed state of a polycrystalline High-T_c Superconductor like $Y_{0.6} Dy_{0.4} Ba_2 Cu_3 O_{7-x}$ was precisely investigated through the electrical resistivity. No jump is found, but it will be argued that the jump is masked due to a distribution of melting temperatures, say one for each grain. A Gaussian distribution for the vortex lattice melting temperatures T_M is assumed. The variations of the mean melting temperature T_{MA}, the melting temperature distribution width and the dB_{c2}/dT parameter in a magnetic field are calculated from electrical resistivity data fits. The viscosity of the vortex liquid is deduced at various magnetic field strengths from electrical resistivity data taken between the vortex lattice melting temperature T_M and the critical temperature T_c. The behavior of the viscosity is shown to reproduce the (B, T) map. Differences between a meaningful vortex lattice melting, irreversibility and percolation transition temperatures T_p are discussed. In conclusion, this analysis of the electrical resistivity variation in a broad temperature and magnetic field ranges does supply some information complementary to the studies of the vortex lattice melting by magnetostatic and magnetodynamical investigations.

2. SAMPLES

A Dy-substituted 123-YBCO based ceramics with nominal composition $Y_{0.6}Dy_{0.4}Ba_2Cu_3O_{7-x}$ was synthesized, starting from commercial Y_2O_3, Dy_2O_3, CuO, $BaCO_3$. Stoichiometric amounts of initial materials were homogenized by wet milling. The wet milling was carried in a ball-mill type "Fritsch" for four hours in ethanol medium. Thus prepared powders were dried and then milled in a vibration mill - KM for 30 min. The powders were pressed into 15 mm diameter pellets under a 5 MPa pressure. The pellets were sintered at 950 C with isothermal delay of 24 hours. A vibration mill was used for a second milling and the grain size was controlled. For the last pressing we used a powder with grain size of about 1 micron. The powder was pressed into 10 mm diameter pellets under a 3.5 MPa pressure. The pellets were sintered at 950 C during 50 hours. They were cooled to 650 C with isothermal delay for 24 hours. Then the pellets were cooled down to room temperature.

Using an optical microscope, after metallographic treatment of the sample surfaces, the grain distribution was seen to roughly range between 10 and 50 μm. The particle size in the polycrystalline structure for four different isothermal treatments at 950 C were investigated. An X-ray diffraction analysis proved that the samples were almost single phase with only traces of Dy_2BaCuO_5, CuO and $BaCuO$ phases. The presence of these minority phases was also confirmed by scanning electron microscopy. Raman spectra showed that the oxygen content was very close to seven.

3. ELECTRICAL RESISTIVITY MEASUREMENTS

The four-probe method was used for precise measurements of the electrical resistivity of the samples. A dc current up to 30mA was injected as one second pulses with successively reversed direction in order to avoid Joule and Peltier effects [13, 14]. The electrical resistivity was calculated taking into account the voltage drop and the sample dimensions. The electrical resistivity measurements were performed at several magnetic field strengths between 0 and 0.6 T and the temperature was varied between 20 and 300

K in a closed cycle refrigerator with stabilization by a DRC 91 controller with accuracy of 0.02 K. Further details are described elsewhere [15].

The electrical resistivity versus temperature variation measured in a broad temperature range is shown in Fig. 1.

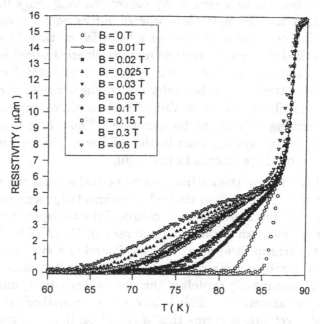

Figure 1. Electrical resistivity vs temperature of a polycrystalline $Y_{0.6}Dy_{0.4}Ba_2Cu_3O_{7-x}$ superconductor measured at various magnetic field strengths between 0 and 0.6 T.

The absolute value of the resistivity is 18 $\mu\Omega$m at 100 K. In the normal state the resistivity increases approximately proportionally to the temperature with an average rate of 0.04 $\mu\Omega$m /K. The transition width defined as the difference between the normal state and the resistivity percolation temperature T_p [16, 17] is narrower than 3 K in absence of external magnetic field. The transition width becomes gradually broader up to 37 K when the magnetic field of 0.6 T. Even in a non-zero magnetic field the electrical resistivity drops down to about 45 % of the normal state value in a remarkably narrow interval between 85 and 88 K which is typical for intragrain superconductivity. The intergrain processes are seen to be spread out over a much broader temperature range rising gradually up to 25 K with the magnetic field strength (Fig. 1).

4. VORTEX LATTICE MELTING PROCESS

The vortex lattice melting transition should concern both static and moving vortex structures. In an idealized picture the unperturbed solid vortex lattice should obey both short and long range ordering conditions, e.g. be a triangular or square lattice. The vortex system may respond in quite different manners when subjected to external forces, depending on the solid or liquid vortex state. In the first case the vortices take part in a so called "elastic" motion, i.e. as a rigid body conserving long range intervortex ordering. In contrast, the much less ordered vortex liquid, revealing no long range ordering of vortices, undergoes a "plastic" motion, in which various groups (so called clusters) of vortices move at different velocities and change their mutual coordination [18]. Vortices in the "liquid state" may be entangled. This is related to a vortex deformation phenomenon parametrised by changes in the tilt and shear moduli of vortices which become meaningful when approaching a "melting temperature". Thus there is more freedom in the liquid vortex system than in the solid one. This in turn causes the dissipation to be more intense in the liquid.

Beyond the purely theoretical and numerical models the vortex lattice melting process is most often studied experimentally using magnetometric and calorimetric methods. An application of the transport methods to this question is not so much popular though see [10, 11, 12]. The distinguishing feature of the transport approach is that it probes a system of the moving vortices in the mixed state of HTcS. The transport studies of the vortex melting processes mainly exploited the electrical resistivity and critical current density measurements. This is in contrast to methods sensitive to the static vortex systems. It seems that a report on the thermoelectric power requires a more cautious interpretation [12, 19].

In an ideally homogeneous superconductor the vortex lattice melting transition should occur abruptly in a very narrow temperature interval. However, various forms of inhomogeneities, like at least grain boundaries, chemical and structural defects, spread out the temperature interval where a vortex lattice melting transistion should occur, just like it is expected at a liquid-solid transition. In order to consider realistic conditions of the vortex lattice melting in HTcS an unsophisticated model was successfully applied to Y123 thin films by Casaca et al. [20] and to Bi2212 superconducting tapes by Pekala et al. [21]. Assumptions of this model seem not to be oversimplified however since it *a posteriori* supplies reasonable values of physical quantities. The main model assumption is that the all "small enough" local regions, which may be identified with grains or so, are homogeneous. This in turn allows one to approximate an abrupt electrical resistivity drop from the vortex flow value $\rho_{VF}(T,B)$, occuring in each region of a sample, as a step like function at the local melting temperature T_M characteristic *for each region*. The inhomogeneities present in the superconductor sample composed of a large number of small and locally homogeneous regions lead to a broadened distribution of the melting temperatures. Thus in order to avoid further too much sophisticated asumptions a simple Gaussian distribution for the local melting temperatures T_M is introduced

$$G(T_M, B) = \frac{1}{(2\pi)^{1/2}\Delta T_M} exp[\frac{-(T_M - T_{MA}(B))^2}{2(\Delta T_M)^2}] \tag{1}$$

with the average melting temperature being T_{MA} and the width being ΔT_M.

The link between the electrical resistivity data and the model is based on two approximations. The vortex flow resistivity $\rho_{VF}(T, B)$ in the mixed state of HTcS is taken as

$$\rho_{VF}(T, B) = \rho_N(T)(B/B_{c2}(T)) \tag{2}$$

where $\rho_N(T)$ is the normal state resistivity, and B the applied magnetic field. In the vicinity of the critical temperature T_c, the second critical field $B_{c2}(T)$ may be written as:

$$B_{c2}(T) = (T - T_c)(dB_{c2}/dT) \tag{3}$$

The next dilemma pertains to the *magnitude* of the electrical resistivity drop at T_M. One may imagine that that the electrical resistivity should not be necessarilly reduced down to zero in the moving vortex solid system below T_M. This condition distinguishes the melting transition from the irrevesibility line separating regions of the (B,T) phase diagram, where the vortex solid or liquid is free to move on one side and is frozen or blocked on the opposite one. Thus a drop to a zero resistivity should be observed at the irreversibility temperature. Unfortunately no theoretical estimation of the finite resistivity value below T_M is really available [8]. For this reason the calculations consider the electrical resistivity to be given by Eq.(2) and by zero above and below T_M, respectively. This question is discussed in details elsewhere [21]. The model described above results in the expression

$$\rho(T, B) = \frac{\rho_N(T)B}{(T - T_c)(dB_{c2}/dT)(2\pi)^{1/2}\Delta T_M} * \tag{4}$$

$$\int_0^T exp[-\frac{(T_M - T_{MA}(B))^2}{2(\Delta T_M)^2}]dT_M$$

describing the mixed state electrical resistivity around the vortex lattice melting temperature.

5. DISTRIBUTION OF THE VORTEX LATTICE MELTING TRANSITION IN A PHASE DIAGRAM

The model described in the previous section has been applied to the electrical resistivity data taken on the above $Y_{0.6}Dy_{0.4}Ba_2Cu_3O_{7-x}$ sample in the low temperature range of the mixed state, where the vortex lattice melting process strongly affects the electrical resistivity. The temperature variation of electrical resistivity is correctly reproduced numerically for values roughly below 5 $\mu\Omega m$ (Fig. 2).

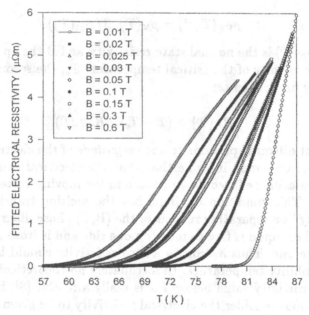

Figure 2. Electrical resistivity data of Fig.1 vs temperature at various magnetic fields fitted to the model described in the text.

The additional test for the high quality of the applied fitting procedure is supplied by the precise coincidence of the position and shape of the $d\rho(T)/dT$ maxima calculated from the modelled $\rho(T)$ curves from the experimental $d\rho(T)/dT$ data. It was observed that deviations between the fitted curves and the experimental data remarkably increased above 5 $\mu\Omega m$. Thus the fitting was performed up to a temperature 3 to 8 K higher than T_{MA}, depending on the magnetic field. This condition reveals the highest temperature at which the vortex lattice multi melting temperature model is applicable. Values of T_{MA}, ΔT_M and dB_{c2}/dT used as free fitting parameters were calculated with high accuracy for various magnetic fields.

Values of the mean melting temperature T_{MA} are plotted in a phase diagram of Fig. 3. They shift monotonically from 84 down to 62 K when the magnetic field rises from 0.01 to 0.6 T.

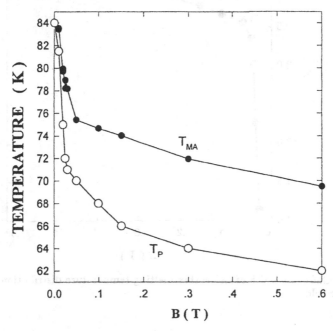

Figure 3. Phase diagram of the discussed sample showing the average vortex lattice melting temperature T_{MA} derived from model calculation and the percolation temperature T_p of the electrical resistivity as a function of external magnetic field.

The phase diagram in Fig. 3 shows also the field dependence of the electrical resistivity percolation temperature T_p, at which the resistivity vanishes. It is worth noticing that both the $T_{MA}(B)$ and $T_p(B)$ exhibit qualitatively very similar behaviors as a function of the magnetic field. Two intervals may be distinguished: the abrupt drop in a first interval below 0.05 T is followed by a much slower variation at high fields.

The shape of the melting curve was numerically analysed in both intervals. For magnetic fields below 0.05 T this curve is well approximated by $B_{MA}(T) = (1 - T/T_c)^n$ with n = 1.98 (0.2). Such a behavior corresponds perfectly with the dependence $B_{MA}(T) = (1 - T/T_c)^2$ predicted in [4, 22, 23, 24, 25]. It is observed that above B = 0.05 T the exponent n equals 5.2 (0.3). A similar procedure applied to the percolation line gives a value of n equal to 1.6 (0.3) and 7.9 (0.6) below and above B = 0.05 T respectively. The high field percolation curve exponent found here is not very different from the values 4.48 and 6.25 derived for the Bi2212 tapes [21] and the Hg1223 HTcS close to T_c [26], respectively.

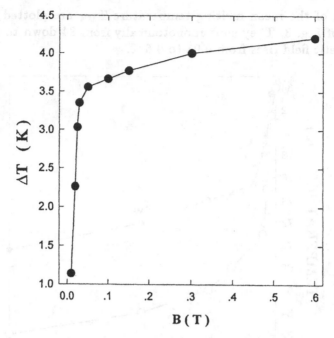

Figure 4. Gaussian width of the vortex melting temperature distribution as a function of magnetic field.

Values of the magnetic field dependence of the melting temperature distribution width ΔT_M increase smoothly from 1.1 K at 0.01 T up to 4.2 K at 0.6 T and seem to approach some saturation at high fields (Fig.4). In analogy to $T_{MA}(B)$ and $T_p(B)$, the $\Delta T_M(B)$ plot reveals around B = 0.05 T a crossover between some abrupt and moderate variations. A comparison of Fig. 4 and 5 shows that the prevailing vortex lattice melting temperatures occur in a ribbon shape phase diagram area located around the $T_{MA}(B)$ curve with a width increasing to more than 4 K for a 0.6 T magnetic field. This region lies remarkably above the percolation temperature T_p pointing out that the vortex lattice melting transition should not be confused with the vortex blocking/pinning processes revealed by the percolation line in electrical resistivity or by the irreversibility line determined from magnetic studies. Such a behavior seems to be characteristic for HTcS since a qualitatively very similar phase diagram in the vicinity of $T_{MA}(B)$ was also reported for Bi2212 polycrystalline tapes[21].

6. SECOND CRITICAL FIELD AND VORTEX VISCOSITY

The calculated parameter dB_{c2}/dT is an increasing function of the magnetic field (Fig. 5). The dB_{c2}/dT values extrapolate to zero at B = 0. Moreover, a change in the slope of the dB_{c2}/dT vs B plot is seen at B = 0.05 T but the high magnetic field variation is very close to the linear one. Often B_{c2} is an average of $B_{c2,c}$ and $B_{c2,ab}$ and is a function of the angle between the field and the c-axis crystallographic direction. Therefore the magnetic phase diagram should precisely pin point the experimental conditions. The vortex viscosity parameter η is calculated as:

$$\eta = \Phi_0 B_{c2}(T)/\rho(T) \tag{5}$$

where Φ_0 denotes the flux quantum.

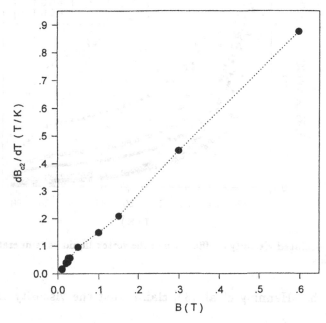

Figure 5. Temperature derivative of the second critical magnetic field dB_{c2}/dT vs magnetic field as derived from the model.

Fig. 6 shows the temperature variation of $\eta(T)$ for fixed magnetic field strengths corresponding to various vortex densities. Values of η are somewhat smaller than the magnitude $10^{-9} Js/m^3$ reported in [21] for Bi2212 tapes and about two orders of magnitude smaller than those reviewed by Golosovsky et al. [29] for Y123 HTcS. Generally $\eta(T)$ is found to be a monotonically decreasing function of temperature. Such a behavior is similar to that observed for various classical and polymeric liquids. All these curves exhibit the so called "knee" located at T_{kn} a few K below T_c. One may see that T_{kn} shifts downwards when the magnetic field rises up to 0.6 T. The T_{kn} temperature seems to reveal a remarkable change in the strength of intervortex interactions and/or in the structure of vortices. It is hard to compare this "knee" effect to that in other HTcS systems, since the analysis of data in terms of vortex viscosity is surprisingly rather rare in the "classical" HTcS literature known to the authors. Usually the viscosity is listed as a simple number characterizing a system at chosen temperatures [29].

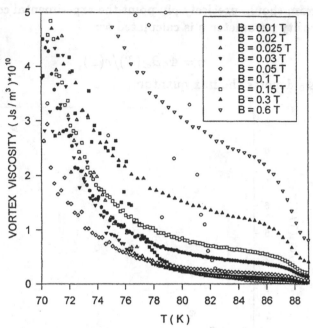

Figure 6. Calculated viscosity coefficient η of the vortex liquid vs temperature at various magnetic fields.

Notice that Henning et al. [30] claims that the viscosity should follow the formula

$$\eta(T) = \eta(0)(1 - (T/T_c)^2) \qquad (6)$$

where $\eta(0)$ denotes the amplitude of the vortex viscosity. The vortex viscosity amplitude extrapolates to zero at B = 0 as should be expected for the decaying mixed state of HTcS. However an Arrhenius plot of $ln(\eta(T))$ vs. $(1/T)$ as done in order to determine an activation energy rather points to such an exponential behavior. It is known from the theory of liquids that it is rather hard to distinguish the true temperature dependence of the viscosity indeed, and further arguments should be in order.

In fact the viscosity is an indirect measurement of the different pinning regimes, "shape" of the vortices and their relative motion (or better relative momentum) influence. Moreover the critical current density is usually related to the viscosity by

$$j_c = \eta c v_c / B \qquad (7)$$

313

where v_c is usually taken as a decaying exponential as due to thermal (de)activation over the pinning barrier. Phenomenologically the critical current density has different behaviors for the various pinning regimes which can also be distinguished by different characterisitc lengths which can in turn be expressed in terms of the characterisitc field strengths [4]. The behavior of the viscosity displayed here as a function of field and temperature is in fact reminiscent of the more classical (B, T) plane or phase diagram itself crossed by transition lines, as illustrated in Fig. 18 and 39 of ref. [4]. This is *in fine* rather obvious from the definition of the viscosity of Bardeen and Stephen [27], and also from Eq.(3.87) in [4].

Any anomaly in the field dependence of the viscosity should be the signature of some anomalous field dependence or by extension that of the different regimes. For example, a smooth temperature behavior expected at small field is followed by the small bundle pinning regime and next at high field and higher temperature by the large bundle pinning regime. The viscosity is seen to have a different temperature and field dependences in these three regimes. Of course, one might ask whether the three regimes are indeed bounded above by some glass regime transition line, and still further by the melting field line. According to Blatter et al. [4] (see their Fig. 42) the glass and melting lines are identical at small field and large temperature. It is most likely the region of investigation here, but the anomalies are likely to be too small to be observed.

A formula with physical parameters for the melting field as a function of temperature based on the Lindemann criterion may be found in the review by Blatter et al. [4] (table XIV; Eq. (4.107)). It can be simplified near T_c (see Eq. (5.5) in [4]). By comparison the B_{c2} line is quite steep $\sim exp(T^2)$ (see Eq. (8.125) in [4]) for $B < B_{2D}$. This should remind anyone that the effective dimensionality plays a role in discussing the phase diagram and that the origins of the various lines at zero field are controlled by the Ginzburg Levanyuk number, which depends on the anisotropy parameter in general. It depends also on both the zero temperature thermodynamic critical field and the coherence length (and is scaled by T_c, of course). By extension one could consider that different regimes are measured by this number as well, to be field and temperature dependent. Klemm [28] has found a dimensional cross-over for the magnetic field equal to

$$B_0 = \zeta^2 \Phi_0 / (4\pi s^2) \qquad (8)$$

where $\zeta = 0.34589/(0.09133\gamma)$, γ is the anisotropy parameter, s is the distance between copper oxyde planes.

In order to inspect differences below and above the T_{kn} temperature, a thermally activated viscosity mechanism was assumed and the activation energies E_L and E_H were calculated for both ranges respectively.The activation energy E_H for the temperature interval above T_{kn} varies between 450 and 350 meV (Fig. 7). This shows that the viscosity mechanism in this range depends weakly on the magnetic field. The activation energy E_L goes down from 310 meV for B = 0.01 T to 35 meV for B = 0.6 T with a sharp turn near B = 0.05 T, a value also seen as "special" for other parameters in this system. One may get the impression from our data that the parabolic-like depinning line $B_{dp}(T)$ of Blatter et al. Fig. 18 in [4] should be flattened out allowing for T_{kn} to lie between $B_m(T)$ and $B_{c2}(T)$ lines. For the weak field regimes, e.g 0.01 T, T_{kn} may thus coincide with the single/small bundle pinning temperature T'_{dp}.

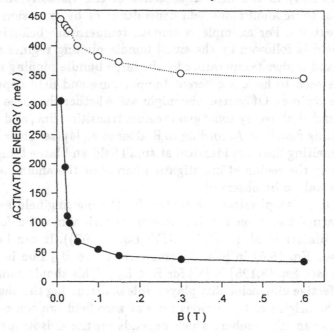

Figure 7. Activation energy E_L and E_H vs magnetic field limited to 0.6 T as calculated from the vortex viscosity (black dots) below (L) and (open circles) above (H) the "knee" temperature T_{kn} respectively.

7. CONCLUSIONS

Scientific progress is expected to depend on the combination of experimental and theoretical investigations that provide deep understanding of physical phenomena from high reliability of results. The investigation reported here above has proposed a simple model to discuss (i) the pecularities of the electrical resitivity behavior in the mixed state in a highly-anisotropic superconductor, (ii) the conditions of the phase coherence transition in the mixed state of a type-II superconductors and (iii) information about microscopical parameters and mechanisms of HTcS. These results are useful for applications, because they rely on a simple assumption, i.e. a distributon of vortex melting temperatures, but they are also of fundamental interest. From the mixed state electrical resistivity value of a polycrystalline $Y_{0.6}Dy_{0.4}Ba_2Cu_3O_{7-x}$ HTcS and assuming a Gaussian distribution for the vortex lattice melting temperatures, the variations of the melting temperature distribution width and the dB_{c2}/dT parameter in a magnetic field were obtained. The viscosity of the vortex liquid was determined between the vortex lattice melting temperature T_M and T_c at various magnetic field strengths.

Acknowledgements

Work supported in part by Polish - Belgian Scientific Exchange Program and Actions de recherche concertées (ARC) program (ARC 94-99/174) of the Communauté Française de Belgique, DG Enseignement non obligatoire et recherche scientifique, Direction de la recherche scientifique through the University of Liège Research Council.

References

1. C. Hannay, R. Cloots, H.W. Vanderschueren, P.A. Godelaine, G.J. Tatlock, D.G. McCartney, and M. Ausloos, Supercond. Sci. Technol. **5**, S296 (1992)
2. M. Pękala, R. Cloots, H. Bougrine, E. Maka, and M. Ausloos, Z. Phys. B **97**, 67 (1995)
3. S. Dorbolo, M. Ausloos, and M. Houssa, Phys.Rev.B **57**, 5401 (1998)
4. G. Blatter, M.V. Feigelman, V.B. Geshkenbein, A.I. Larkin, and V.M. Vinokur, Rev. Mod. Phys. **46**, 1125 (1994)
5. V.A.Marchenko and A.V.Nikulov, JETP Lett. **34**, 17 (1981)
6. A.V.Nikulov, D.Yu.Remisov, and V.A.Oboznov, Phys.Rev.Lett. **78**, 2586 (1995)
7. A. A. Varlamov and M. Ausloos, in *Fluctuation Phenomena in High Temperature Superconductors*, M.Ausloos and A. A. Varlamov, Eds., vol. **32** in NATO ASI Partnership Sub-Series 3: High Technology (Kluwer, Dordrecht, 1997) p. 3
8. J. Mannhart, in *Ultimate Limits of Fabrication and Measurement*, M.E. Welland and J. K. Gimzewski, Eds. (Kluwer, Dordrecht, 1995) p.241
9. E. Zeldov, D. Majer, M. Konczykowski, V. B. Geshkenbein, V. M. Vinokur, and H. Shrikman, Nature **375**, 373 (1995)
10. W.K. Kwok, J. Fendrich, S. Fleshker, U. Welp, J. Downey, and G.W. Crabtree, Phys. Rev. Lett. **72**, 1092 (1994)
11. U. Welp, J.A. Fendrich, W. K. Kwok, G.W. Crabtree, and B.W. Veal, Phys. Rev. Lett. **76**, 4809 (1996)
12. H. Ghamlouch, M. Aubin, R. Gagnon, and L. Taillefer, Phys. Rev. B **54**, 9070 (1996)
13. Ch. Laurent, S.K. Patapis, M. Laguesse, H.W. Vanderschueren, A. Rulmont, P. Tarte, and M. Ausloos, Solid State Commun. **66**, 445 (1989)
14. H. Bougrine and M. Ausloos, Rev. Sci. Instrum. **66**, 199 (1995)
15. M. Pękala, A. Tampieri, G. Celotti, M. Houssa, and M. Ausloos, Supercond. Sci. Technol. **9**, 644 (1996)
16. E.A. Jagla and C.A. Balseiro, Phys. Rev. B **53**, R538 (1996)
17. M. Pękala, E. Maka, D. Hu, V. Brabers, and M. Ausloos, Phys. Rev. B **52**, 7647 (1995)

18. D.R. Nelson, Phys. Rev. Lett. **60**, 1973 (1988)
19. M. Ausloos, H. Bougrine, M. Houssa, and M. Pękala, Phys. Rev. B (1998)
20. A. Casaca, G. Bonfait, M. Lenkens, G. Mller, K. Lander, and J.A. Edwards, Supercond. Sci. Technol. **10**, 75 (1997)
21. M. Pękala, H. Bougrine, W. Gadomski, C.G. Morgan, C.R.M. Grovenor, R. Cloots, and M. Ausloos, Physica C **303**, xxx (1998)
22. G. Blatter and B.I. Ivlev, Phys. Rev. B **50**, 10272 (1994)
23. E.H. Brandt, Rep. Progr. Phys. **58**, 1465 (1995)
24. D.J.C. Jackson and M.P. Das, Supercond. Sci. Technol. **9**, 713 (1996)
25. P.H. Kes, H. Pastoriza, T.W. Li, R. Cubitt, E.M. Forgan, S.L. Lee, M. Konczykowski, B. Khaykovich, D. Majer, D.T. Fuchs, and E. Zeldov, J. Physique (France) I **6**, 2327 (1996)
26. K.C. Hung, Supercond. Sci. Technol. **10**, 837 (1997)
27. J. Bardeen and M.J. Stephen, Phys. Rev. **140**, A1197 (1965)
28. R. A. Klemm, in *Fluctuation Phenomena in High Temperature Superconductors*, M.Ausloos and A. A. Varlamov, Eds., vol. **32** in NATO ASI Partnership Sub-Series 3: High Technology (Kluwer, Dordrecht, 1997) p.377
29. M. Golosowsky, M. Tsindlekht, and D. Davidov, Supercond. Sci. Technol. **9**, 1 (1996)
30. T. Henning, H. Kliem, A. Weyers, and W. Bauhofer, Supercond. Sci. Technol. **10**, 721 (1997).

IS FLUCTUATION CONDUCTIVITY A PROBE TO DETERMINE THE PRESENCE OF S OR D -WAVE PAIRING IN HIGH-T_C SUPERCONDUCTORS?

S. K. PATAPIS

Solid State Section, Physics Department, University of Athens
Panepistimiopolis, Zografos, Athens 157 84 Greece

Abstract. Can the fluctuation conductivity(or paraconductivity) be used as an experimental probe for the detection of s or non-s wave (d wave?) pairing in high temperature superconductors? As it can be shown Maki - Thompson effect turns out to be absent for pure d-wave superconductors. In other words, the scattering of the electrons to impurities has the effect of destroying the pairing, if this is not of the s-wave kind, since the out-coming electrons do not satisfy any more the pairing relation in the momentum space. Thus, the presence or absence of Maki - Thompson terms in the fluctuation conductivity of a high T_c superconducting sample can be used to determine the presence or absence of s-wave pairing. Experimental results concerning fluctuation measurements on different superconductors are presented.

1. Introduction

Since the discovery of the new high temperature superconductors an enormous amount of work has been done and still remains to be done to reach a better theoretical understanding of the complex properties of this cuprate material. However, the problems are still unsolved.

Among the above problems is that of pairing. The determination of the pairing state has been a fundamental issue for both experiment and theory. This is a basic problem for clarifying the microscopic mechanism of superconductivity.

The pairing state influences basic properties such as penetration depth,

M. Ausloos and S. Kruchinin (eds.), Symmetry and Pairing in Superconductors, 317–326.
© 1999 *Kluwer Academic Publishers. Printed in the Netherlands.*

optical response, sensitivity to impurities and the macroscopic Ginzburg-Landau behavior. The matter may even influence commercial application (e.g. a d-wave superconductor will have a finite resistivity at microwave frequencies even without weak links present[1]).

Concerning the YBaCuO there are many experiments indicating a highly anisotropic energy gap and furthermore there is strong evidence that it changes sign as expected in a predominantly $d_{x^2-y^2}$ pairing state and unlike some highly anisotropic s-wave states[2, 3, 4]. Although there are experiments indicating s or s+d pairing[1, 5, 6, 7]. Generally the problem of pairing is not explicitly solved for all the high temperature superconductors[8] and specially for YBaCuO material. Recently, we have discussed about s,d-wave pairing[6, 9]. Note that the orthorombic structure of YBaCuO predicts a mixed pairing state proceeding from group theoretical consideration[1, 10].

As it will be shown below among different ways of experimental indicative probes for the kind of pairing one may consider the Maki - Thompson effect in the fluctuation conductivity.

2. Theoretical Background

Among the main features of the high temperature superconductors are the large fluctuation effects observable well above the transition temperature T_c. The higher transition temperature T_c, the shorter coherence length ξ and the higher anisotropy ($\xi_c \ll \xi_{a,b}$) favor these fluctuations for temperatures further enough from the transition temperature[11]. The coherence length of the order of unit cell has as result a small coherence volume containing few Cooper pairs. This implies that fluctuations have a higher influence here and can be easier detected. Fluctuations manifest themselves in many properties such as electrical conductivity thermolectric power specific heat and magnetic susceptibility as well (for a review see Refs[12, 13]). Specially for fluctuation conductivity effect, the fluctuations have as result higher conductivity near the transition temperature as we approach it from above. Experimental data obtained from this excess conductivity or paraconductivity measurements are of particular interest for studying fluctuation behavior.

Although electron pairing above T_c is not favored because of the higher energy there are always fluctuations of pairing. These fluctuation pairings at temperatures above T_c increase the conductivity to higher values as T_c is approached from above. This excess conductivity or paraconductivity can be expressed by the Aslamazov - Larkin[14] relation (as it will be shown below) and related theoretical approaches such as those by Maki and Thompson[15, 16, 17] (Maki-Thomson effect).

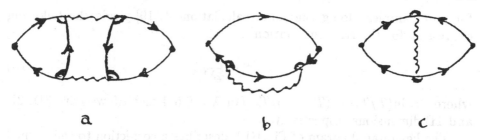

Figure 1. Different Feynman digrams contributing to the fluctuation resistivity.

The Aslamasov - Larkin conribution is not the only one to fluctuation conductivity, the Maki - Thompson effect and the influence of density of states play also a decisive -depending on temperature- role. So the expression for the full fluctuation conductivity consists of the three main contributions as follows

$$\delta\sigma = \delta\sigma_{AL} + \delta\sigma_{DOS} + \delta\sigma_{MT} \tag{1}$$

where $\delta\sigma_{AL}$ is the Aslamazov - Larkin contribution, $\delta\sigma_{DOS}$ is the contribution from the density of states and $\delta\sigma_{MT}$ is the Maki - Thompson contibution. These above main contributions to the fluctuation conductivity come out of three distinct classes of Feynman diagrams shown in Fig.1 (a: is the Aslamazov - Larkin diagram, b: is a diagram giving corrections to the normal density of states and c: is a Maki - Thompson diagram)

The Aslamazov - Larkin (AL) term[14], which corresponds to the Feynman diagram of Fig.1(a), describes the extra contribution to the conductivity that exists above T_c due to the virtual production of Cooper pairs. In this process a Cooper pair is spontaneously formed and after some time τ_{GL} is destroyed. However, during the time of its existence it carries a supercurrent that changes the conductivity of the whole medium. Assumming[13] that Cooper pairs can be treated as carrriers of charge 2e and life-time $\tau_{GL} = \pi/8(T - T_c)$ in the usual Drude formula, we obtain that the AL contribution to the fluctuation conductivity should be

$$\delta\sigma^{AL} \propto \frac{N_{Cp}(2e)^2\tau_{GL}}{2m} = \frac{\pi e^2}{4m}\frac{1}{(T - T_c)}N_{Cp}, \tag{2}$$

where m is the effective mass and N_{Cp} is the number of fluctuating Cooper pairs. Very close to the transition temperature[18], the AL term is the most singular term and dominates the fluctuation conductivity. At this region, for pure samples, doing the exact calculation[14, 19] yields the following expression for the AL contribution

$$\delta\sigma^{AL} \propto \frac{1}{\epsilon^\lambda}, \tag{3}$$

where $\epsilon = \ln(T/T_c) \approx (T - T_c)/T_c$. For $\lambda = 0.5, 1$ and 1.5 we have 3D, 2D and 1D dimensions respectively.

The Feynman diagram of Fig.1(b), describes a correction to the normal phase conductivity above the transition temperature due to the fact that some of the electrons of the system contribute to the fluctuating Cooper pairs. Thus, the effective number of carriers taking part in the one-electron charge tranfer diminishes, and a renormalization in the density of states (DOS) is needed. This leads to a decrease of conductivity[13]

$$\delta\sigma^{DOS} \propto -\frac{2N_{Cp}e^2\tau}{m}, \tag{4}$$

where τ is the normal state quasi-particle life-time. According to the expressions (2) and (4), the singularity in $\delta\sigma^{AL}$ is ϵ times larger than the respective singularity in $\delta\sigma^{DOS}$,

$$\delta\sigma^{AL} \propto \frac{\delta\sigma^{DOS}}{\epsilon}, \tag{5}$$

since generally $\epsilon \ll 1$.

In order to justify the high value of the excess conductivity $\delta\sigma$ observed mainly in films Maki and Thompson interpreted this effect as due to a decay of the fluctuation pairs into pairs of quasi-particles which continue to be much accelerated. This leads to an increase of the electrical conductivity which is otherwise limited by strong inelastic scatterers and other pair-breaking interactions. For the two dimensions gives an additional term as follows

$$\delta\sigma_{MT} = \frac{e^2}{8k\hbar d} \frac{1}{(\epsilon - \delta_T)} \ln \frac{\epsilon}{\delta_T} \tag{6}$$

with d the film thickness and $\delta_T = 2\xi(0)/d^2\delta$ where $\xi(0)$ is the coherence length at $T = 0$, and δ is a pair-breaking parameter varying with field and temperature. A similar more elaborated but logarithmic relation has been recently derived by Maki and Thompson[20]. From the above relation it is clear that a Maki - Thompson contribution increases logarithmically for $\epsilon \gg \delta_T$. Indeed the influence of a Maki - Thompson term is greater further from the transition temperautre while closer to T_c an Aslamsov - Larkin term dominates as has been shown by Hikita and Suzuki[21].

Finally, the Feynman diagram of Fig.1(c) corresponds to a Maki - Thompson (MT) contribution[15, 16, 17]. Such a contribution to the fluctuation conductivity is related to the coherent electron pair scattering[13]. It is crucially dependent on the one-electron state phase-breaking time τ_ϕ and on the presence of diffusion motion of the electrons[19]. It also plays a dominant role in the fluctuation conductivity far from the transition temperature[21]. In this region the MT contribution is proportional to $\ln \epsilon$, while the AL contibution is proportional to ϵ [18]. In this region, using the expression of Eq.(6) we can see that the MT correction is less singular than the AL correction a fact that is depicted in the graphs of the fluctuation resistivity as a smaller inclination of $d\Delta R/dT$ far from the transition temperature. Thus, the presence or absence of the MT correction can easily be found if the experimental curves of $d\Delta R/dT$ versus ϵ are given.

Returning now, to our original problem, on the application of the fluctuation conductivity as a probe for detecting pure non-s wave pairing in HTS, we note that the MT term turns out[22] to be generally absent in for pure d-wave superconductors in three-dimensions. For this pairing (d) in two dimensions, it also results[22] that the MT term is negligible compared to the AL term. This conclusion can also be generalized[22] for every pairing interaction that averages to zero over the Fermi surface. The reason for this result is that the scattering of electrons to impurities has the effect of destroying the pairing, if this is not of the s-wave kind. The out-coming electrons don't satisfy the pairing relation in the momentum space after the scattering. Thus, the presence or absence of MT terms in the fluctuation conductivity of a HTS sample can be used to determine the presence or absence of s-wave pairing.

3. Experimental Results and Discussion

Fluctuation conductivity behavior has been studied in different samples, bulk and thin films, of YBaCuO composition where the Maki - Thompson effect has emerged for temperatures further of the critical temperature indepedently of the behavior close to it. There will be also mentioned fluctuation measurement in BSSCO compounds where the Maki - Thompson effect is not observed.

Firstly let's start with the bulk material of YBaCuO. For a sample of this material a logarithmic behavior (corresponding to critical exponent $\lambda = 0$ and consequently to a two dimensional Maki-Thompson effect) has been observed[23, 24, 25]. Further from T_c, for $\ln \epsilon$ between -1.5 and -4 or for 2K to 30K above the transition temperature, a critical exponent equal to 0 is extracted [23, 24] (Fig.2). In an other sample the region is extended down to $\ln \epsilon = 0$ (it means 90K above the T_c) as it is depicted in Fig.3,[25]. A similar logarithmic behavior corresponding to a two-dimensional (2D) Maki - Thompson relation and for the same about region ($\ln \epsilon$ between -2

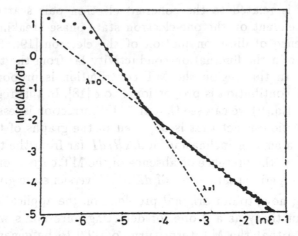

Figure 2. Log-log plot for a bulk YBaCuO material from which the critical exponents are determined. The value $\lambda = 0$ leads to a two dimensional Maki-Thomson effect, $\epsilon = (T - T_c)/T_c$.(The curves come from Refs.[23] and [24].

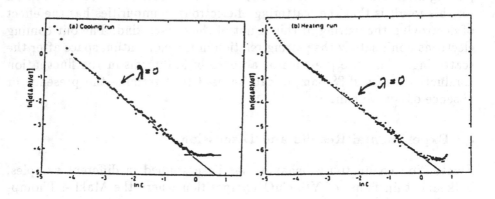

Figure 3. Same as Fig.2 for a different bulk material of YBaCuO (after Ref.[25]).

and -4.2) is also observed[26] in another bulk sample, Fig.4.

Fluctuation conductivity, in zero magnetic field, is also analysed in two high quality c-oriented epitaxial thin films grown on different substrates[27, 28] (e.g. LaAlO$_3$, KTaO$_3$). The critical exponents extracted from the temperature dependence of fluctuation lead to the same value $\lambda \approx 0$ which corresponds to a logarithmic behavior and consequently to a 2D Maki - Thompson effect. This is observed for a similar temperature region (between -2.3 and -4.5) before a cross-over closer to T_c as is depicted in Fig.5.

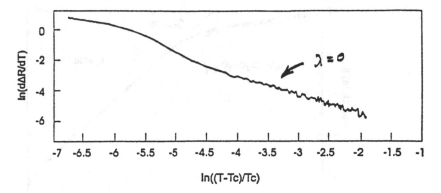

Figure 4. Same as Fig.2 for a different bulk material of YBaCuO (after Ref.[26]).

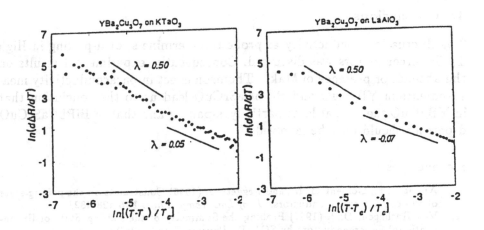

Figure 5. Same as Fig.2 for epitaxial films of YBaCuO grown on two different substrates(Refs.[27,28]). The value $\lambda \approx 0$ emerges for higher temperatures.

On the contrary fluctuation measurements in BSSCO compound indicate no Maki - Thompson effect. These indications come from measurements in three BiPbCaSrCuO bulk samples[29, 30] where no critical exponent with the value zero is observed up to $\ln \epsilon = 0$(Fig.6). This absence of Maki - Thompson effect is also observed in thin films of the same composition.

324

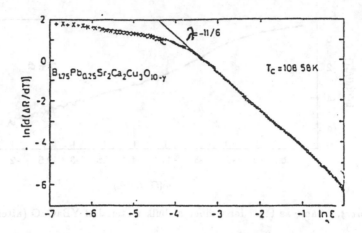

Figure 6. Same as Fig.2 for a bulk material of BSCCO type. Note that no $\lambda = 0$ is observed.

4. Conclusions

The fluctuation conductivity as probe to determine s- or d-pairing in High T_c Superconductors was discussed. Consequently, experimental results on the absence or presence of Maki - Thomson effect in paraconductivity measurements in YBaCuO and BiPbCaSrCuO lead us to the conclusion that in YBaCuO there is, at least partially, s-pairing and that in BiPbCaSrCuO d-pairing should only be involved.

References

1. Annet, J. F., Goldenfeld, N., and Leggett, J. (1996) Constraints on the pairing state of the cuprate superconductors, *J. of Low Temp. Phys.* **105**, 428-482.
2. Van Harlingen, D. J. (1997) Probing the Symmetry of the Pairing State of Unconventioanl Superconductors by SQUID, *Physica C* **282**, 128-131.
3. Junod, A., Roulin, M., Revar, B., Mirmelstein, A., Genard, J.Y., Walker, E., and Erb, A. (1997) Specific Heat of High Temperature Superconductors in High Magnetic Fields, *Physica C* **282**, 1399-1400.
4. Aubin, H., Behnia, K., Ribourt, M., Gagnon, R., and Taillefer, L. (1997) In-plane symmetry of the superconducting gap in YBCO probed by the angular dependence of thermal magnetoresistance, *Physica C* **282**, 1505-1506.
5. Miller Jr, J.H., Zou, Z.G., Lin, J., Zheng, J.S., and Chu, N.K. (1997) Possible evidence for d+s pairing symmetry in YBCO using tricrystal microbridges, *Physica C* **282**, 1515-1516.
6. Patapis, S.K. and Kruchin, S.K. (1997) Fluctuation conductivity behavior in thin films of YBaCuO indicates s-wave pairing, *Physica C* **282**, 1509-1510.
7. Béal-Monod, M.T. and Maki, K. (1997) Ginzburg-Landau equation and vortex state in d+s superconductivity, *Physica C*, **282**, 1841-1842.

325

8. Klemm. R.A. (1998) In Bose, S.M. and Gang, K.B. editors, *International workshop of High Temperature superconductivity: Ten years after the discovery*, Jaipur, India, 1997 (to be published).
9. Kruchinin, S.P. and Patapis, S.K. (1996) Thermodynamics of d-wave pairing in cuprate superconductors, *J. Low Temp. Phys.* **105**, 717-721.
10. Scalapino, D.J. (1969) *Phys. Rev.*, **105**, 473.
11. Fischer, D.S., Fischer, M.P.A., and Huse, D.A. (1991) Thermal fluctuations, quenched disorder, phase transitions, and transport in type-II superconductors, *Phys. Rev. B* **43**, 130-159.
12. Ausloos, M., Patapis, S.K., and Clippe, P. (1993) Superconductivity fluctuation effects on electrical and thermal transport phenomena. H=0 $T > T_c$, In Kossowsky, R., Raveau, B., Wohlleben, W., and Patapis, S.K. editors, *Physics and Materials Science on High Temperature Superconductors II*, Kluwer Academic Publishers, Dodrecht, 755-785.
13. Varlamov, A.A., Balestrino, G., Milani, E., and Livanov, D.V. (1996) The role of density of states fluctuations in the normal state properties of higt T_c superconductors. *Electronic preprint cond-mat Los Alamos National Laboratory*, No.9710175.
14. Aslamazov, L.G. and Larkin, A.I. (1968) Influence of fluctuations on the properties of superconductor above critical temperature, *Sov. Phys. - Solid State* **10**, 875-880.
15. Maki, K. (1968) The critical fluctuation of the order parameter in type II superconductors, *Progr. Theoret. Phys.* **39**, 897-906.
16. Maki, K. and Thompson, R.S. (1989) Fluctuation-induced pair breaking in High T_c superconductors, *Physica C* **162**, 1441-1442.
17. Thompson, R.S. (1970) Microwave Flux Flow, and Fluctuation Resistence of Dirty Type-II Superconductors, *Phys. Rev. B* **1**, 327-333.
18. Riezer, M.Yu. (1992) Fluctuation conductivity above the superconducting transition: Regularization of the Maki-Thompson term, *Phys. Rev. B* **45**, 12949-12958.
19. Dorin, V.V., Klemm, R.A., Varlamov, A.A., Buzdin, A.I., and Livanov, D.V. (1993) Fluctuation conductivity of layered superconductors in a perpendicular magnetic field, *Phys. Rev. B* **48**, 12951-12965.
20. Maki, K. and Thompson, R.S. (1989) Fluctuation conductivity of high-T_c superconductors *Phys. Rev. B* **39**, 2767-2770.
21. Hikita, M. and Suzuki, M. (1990) Fluctuation conductivity and normal resistivity in $YBa_2Cu_3O_y$, *Phys. Rev. B* **41**, 834-837.
22. Yip, S.K. (1990) Fluctuations in an impure uncoventional superconductor, *Phys. Rev. B* **41**, 2612-2615.
23. Ausloos, M., Clippe, P., and Laurent, Ch. (1990) Homogeneous and fractal behavior of superconducting fluctuations in the electrical resistivity of granular ceramic superconductors, *Phys. Rev. B* **41**, 9506-9509.
24. Clippe, P., Laurent, Ch., Patapis, S.K., and Ausloos, M. (1990) Superconductivity fluctuations in electrical and thermoelectrical preperties of granular ceramic superconductors: Homogeneous versus fractal behavior, *Phys. Rev. B* **42**, 8611-8614.
25. Ausloos, M., Gillet, F., Laurent, Ch., and Clippe, P. (1991) High temperature crossover in paraconductivity measurements of granular YBaCuO, *Z. Phys. B* **84**, 13-16.
26. Patapis, S.K., Moraitakis, E., Vekinis, G., Niarchos, D., and Clippe, P. (1997) Percolation conductivity and fractal behavior in an YBaCuO sample, *Mod. Phys. Lett. B* **11**, 517-519.
27. Patapis, S.K., Jones, E.C., Phillips, J.M., Norton, D.P., and Lowndes, D.H. (1995) Fluctuation conductivity of YBaCuO epitaxial thin films grown on various substrates, *Physica C* **244**, 198-206.
28. Patapis, S.K., Jones, E.C., Phillips, J.M., Norton, D.P., Lowndes, D.H., and Somekh, R.E. (1994) Fluctuations and the Maki-Thompson model in YBaCuO thin films, *Physica C* **235**, 1965-1966.
29. Laurent, Ch., Patapis, S.K., Green, S.M., Luo, H.L., Politis, C., Durczewski, W.,

and Ausloos, M. (1989) Fluctuation conductivity effects on the thermoelectric power of granular Bi(Pb)CaSrCuO superconductor, *Mod. Phys. Lett. B* **3**, 241-248.

30. Ausloos, M., Laurent, Ch., Patapis, S.K., Politis, C., Luo, H.L., Godelaine, P.A., Gillet, F., Dang, A., and Cloots, R. (1991) Superconductivity fluctuations in Bi(Pb) based granular ceramic superconductors: evidence for fractal behavior *Z. Phys. B* **83**, 355-359.

PSEUDO-GAP AND CROSSOVER FROM THE 2D HEISENBERG TO THE EVEN-LEG SPIN-LADDER REGIME IN UNDERDOPED CUPRATES

V. V. MOSHCHALKOV, L. TRAPPENIERS AND J. VANACKEN
Laboratorium voor Vaste-Stoffysica en Magnetisme,
Katholieke Universiteit Leuven, Celestijnenlaan 220D,
B-3001 Heverlee, Belgium

1. INTRODUCTION

The temperature dependent resistivity $\rho(T)$ of high-T_C cuprates at temperatures above the pseudogap regime has recently been interpreted in the framework of the two-dimensional (2D) Heisenberg model [1]. The approach proposed in Ref. 1 is based on three basic assumptions: (i) the dominant scattering mechanism in high-T_C's in the whole temperature range is of *magnetic origin*; (ii) the specific temperature dependence of the resistivity $\rho(T)$ can be described by the *inverse quantum conductivity* σ^{-1} with the inelastic length L_ϕ being fully controlled, via a strong interaction of holes with Cu^{2+} spins, by the *magnetic correlation length* ξ_m, and, finally, (iii) the proper *2D expressions* should be used for calculating the quantum conductivity with $L_\phi \sim \xi_m$.

In the present work we have analysed the temperature dependence of the resistivity $\rho(T)$ of the novel $Sr_{2.5}Ca_{11.5}Cu_{24}O_{41}$ spin-ladder compound under hydrostatic pressure of up to 8 Gpa. We have explained the latter by assuming that the relevant length scale for electrical transport is given by the magnetic correlation length related to the opening of a spin-gap in a one-dimensional (1D) even-chain spin-ladder (1D-SL). Comparing the $\rho(T)$ curves of SL compounds and underdoped cuprates we have found that the $\rho(T)$ dependencies of $YBa_2Cu_4O_8$ and underdoped $YBa_2Cu_3O_x$ demonstrate a remarkable scaling with the $\rho(T)$ of the 1D-SL compound

M. Ausloos and S. Kruchinin (eds.), Symmetry and Pairing in Superconductors, 327–336.

$Sr_{2.5}Ca_{11.5}Cu_{24}O_{41}$. This scaling implies that the pseudo-gap below T^* in underdoped $YBa_2Cu_4O_8$ and $YBa_2Cu_3O_x$ is the spin-gap in the even-chain 1D-SL formed at $T < T^*$ in these materials.

2.   DEVELOPMENT OF THE MODEL

With a few straightforward modifications, these ideas can also be applied to the 1D quantum spin-ladder (SL) systems [2]. In this case, instead of 2D expression [1], we rely upon the 1D quantum resistance. The latter is a linear function of the inelastic length L_ϕ [3] and $\rho(T)$ can be represented as

$$\rho_{1D}^{-1}(T) = \sigma_{1D}(T)\big|_{L_\phi = \xi_m} = \frac{1}{b^2}\frac{e^2}{\hbar}\xi_{m1D}(T) \tag{1}$$

Here b is the diameter of the 1D wire and $L_\phi \sim \xi_{m1D}$ is given by the spin-correlation length ξ_{m1D} in the 1D regime. For 1D even-chain Heisenberg spin-ladder compounds this length has been determined in recent Monte Carlo simulations [4]:

$$(\Delta\xi_{m1D})^{-1} = \frac{2}{\pi} + A\left(\frac{T}{\Delta}\right)\exp\left(\frac{-\Delta}{T}\right) \tag{2}$$

where $A \sim 1.7$ and Δ is the spin-gap. The combination of Eq. 1 and Eq. 2 leads to

$$\rho(T) = \rho_0 + CT\exp\left(-\frac{\Delta}{T}\right) = \frac{\hbar b^2}{e^2 a}\left\{\frac{2\Delta}{\pi J_{//}} + A\frac{T}{J_{//}}\exp\left(-\frac{\Delta}{T}\right)\right\} \tag{3}$$

where $J_{//}$ is the intra-chain coupling and a is the spacing between the 1D wires. Eq. 3 can also be written as

$$\ln\left\{\frac{\rho(T) - \rho_0}{T}\right\} = const - \frac{\Delta}{T} \tag{4}$$

3. EXPERIMENTAL VERIFICATION

To verify the validity of the proposed 1D SL model $\sigma_{1D}\sim\xi_{m1D}$, we will use it in the present work first for the description of the resistivity data obtained on the novel even-chain spin ladder compound $Sr_{2.5}Ca_{11.5}Cu_{24}O_{41}$ [5]. This compound, due to its specific crystalline structure, definitely contains a two-leg ($n_c = 2$) Cu_2O_3 ladder and therefore its resistivity along the ladder direction should indeed obey Eq. 1 with ξ_{m1D} given by the recent Monte Carlo calculations [4], through the admixture of the linear temperature dependence of ξ_{m1D}^{-1} with the exponential term containing the spin gap D and the constant A (see Eq. 2). The results of the $\rho(T)$ fit with Eq. 3 are shown in Fig. 1. This fit demonstrates a remarkable quality over the whole temperature range $T \sim 25$-300 K, except for the lowest temperatures where the onset of the localisation effects, not considered here, is clearly visible in the experiment [5]. Moreover, the used fitting parameters ρ_0, C and Δ in Eq. 3 all show very reasonable values.

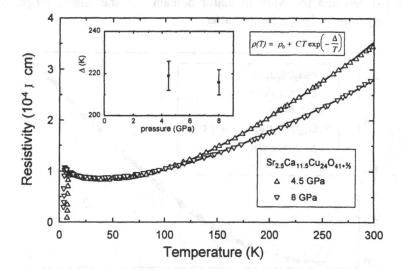

Figure 1. Temperature dependence of the resistivity for a $Sr_{2.5}Ca_{11.5}Cu_{24}O_{41}$ even-chain spin-ladder single crystal at 4.5 GPa and 8 GPa (experimental data points after Ref. 5). The solid line represents a fit using Eq. 3 describing transport in 1D SL's.

The expected residual resistance $\rho_0 = \hbar b^2\, 2\Delta/e^2 a\,\pi\,J_{//}$ for $b \sim 2a \sim 7.6$ Å, $\Delta \sim 200$ K and $J_{//} \sim 1400$ K (the normal value for the CuO_2 planes) is $\rho_0 \sim 0.5\cdot10^{-4}$ Ωcm which is in good agreement with $\rho_0 \sim 0.83\cdot10^{-4}$ Ωcm found from the fit. The fitted gap $\Delta \sim 216$ K (at 8 GPa) (Fig. 1) is close to $\Delta \sim 320$ K determined for the undoped SL $SrCu_2O_3$ from inelastic neutron scattering experiments [6]. In doped systems it is natural to expect a reduction of the spin gap. Therefore the difference between the fitted value (216 K) and the one measured in an undoped system (320 K) seems to be quite fair. Finally the calculated fitting parameter $C = \rho_0 A\pi/2\Delta = 0.0103$ (in units of 10^{-4} Ωcm/K) is to be compared with $C = 0.013$ (from the 8 GPa fit on Fig. 1). Using the fitting procedure for the two pressures 4.5 GPa ($\Delta \sim 219$ K) and 8 GPa ($\Delta \sim 216$ K), we have obtained a weak suppression of the spin-gap under pressure $d\Delta/dp \sim$ -1 K/GPa.

A rapidly growing experimental evidence [7,8,9,10,11] indicates that the 1D scenario might be also relevant for the description of the underdoped high-T_C cuprates where 1D stripes can be eventually formed. Since mobile carriers in this case are expelled from the surrounding Mott insulator phase into the stripes, they provide then *the lowest resistance paths. This makes the transport properties very sensitive to the formation of the stripes (both static and dynamic).* Moreover, due to the proximity effects [8] between the Mott-Insulator domain and the charge stripe, the former may impose $L_\phi \sim \xi_m$ for the latter.

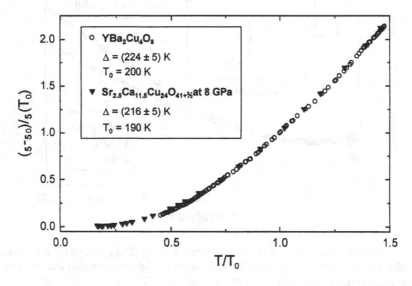

Figure 2. Scaled temperature dependence of the resistivity for a $YBa_2Cu_4O_8$ single crystal (open circles) and a $Sr_{2.5}Ca_{11.5}Cu_{24}O_{41}$ even-chain SL single crystal at 8 GPa (filled down-triangles).

To investigate the possibility of using the 1D scenario for describing transport properties of the 2D Cu-O planes, we directly compare the temperature dependencies of the resistivity of a typical underdoped high-T_C material $YBa_2Cu_4O_8$ with that of the even-chain SL compound

$Sr_{2.5}Ca_{11.5}Cu_{24}O_{41}$. The crystal structure of the $YBa_2Cu_4O_8$ compound
('124') differs substantially from that of the more common $YBa_2Cu_3O_7$
('123'), since 124 contains double CuO chains stacked along the c-axis and
shifted by *b/2* along the b axis [12]. These chains are believed to act as
charge reservoirs, therefore they may have a strong influence on the
transport in the CuO_2 planes themselves. In the 124 case, the 1D features of
this double CuO chain can be expected to induce an intrinsic doping
inhomogeneity in the neighbouring CuO_2 planes thus enhancing in a natural
way the formation of the 1D stripes. But even in pure 2D planes, without
their coupling to the 1D structural elements, the formation of the 1D stripes
is possible. Using a simple scaling parameter T_0, we have found a perfect
overlapping of the two sets of data: *(ρ-ρ₀)/ρ(T₀)* versus *T/T₀* (with ρ_0 being
the residual resistance) for $YBa_2Cu_4O_8$ and $Sr_{2.5}Ca_{11.5}Cu_{24}O_{41}$. Note that
ρ_0 should be subtracted from *ρ(T)* since ρ_0 may contain contributions from
several scattering mechanisms depending on the sample quality.

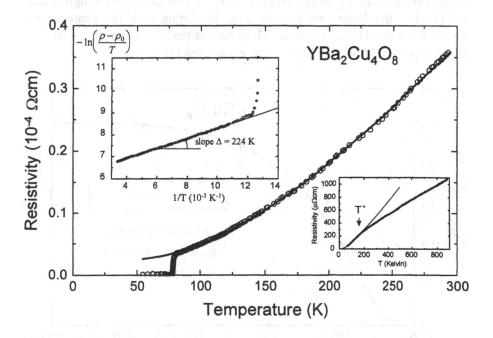

Figure 3. Temperature dependence of the resistivity for a $YBa_2Cu_4O_8$ single crystal (open
circles); the solid line represents the fit using Eq. 3. The following fit parameters have been
obtained: $\rho_0 = 0.024\ 10^{-4}\ \Omega cm$, $C = 0.00242\ 10^{-4}\ \Omega cm/K$ and $\Delta = 224\ K$. The high-
temperature data taken on another crystal (after Ref. 13), shown in the inset, illustrate the
crossover at T^* to 2D (linear behaviour). Insert (upper left): fit of experimental data using
Eq. 4.

The remarkable scaling behaviour (Fig. 2) convincingly demonstrates that resistivity vs. temperature dependencies of underdoped cuprates in the pseudo-gap regime at $T < T^*$ and even-chain SL with the spin-gap D are governed by *the same underlying mechanism*. Since the 1D quantum conductivity model (Eq. 3) works very well for even-chain SL's (see Fig. 1), then it should be also applicable for the description of the $\rho(T)$ behaviour in the underdoped cuprates. The fitting of the $\rho(T)$ curve for $YBa_2Cu_4O_8$ with Eq. 3 resulted in a spin-gap $\Delta = (224 \pm 5)\,K$ (Fig. 3) and the slope of $ln[(\rho-\rho_0)/T]$ versus $1/T$ (see insert fig. 3) defined the spin gap value. Therefore, the resistivity of underdoped cuprates below T^* simply reflects the temperature dependence of the magnetic correlation length $\xi_{m1D} = 1/\rho_{1D}$ in the even-chain SL's (see Fig. 5 in Ref. 4) and the pseudo-gap is the spin-gap formed in the 1D stripes. It seems to be quite important to support these observations and conclusions by using similar ideas in the analysis of other physical properties. Since in underdoped cuprates the scaling temperature T_0 works equally well for resistivity and Knight-shift data [13,14], the latter can also be used for fitting with the expressions derived from the 1D SL models. Then the temperature dependence of the Knight shift should obey the following expression [15]:

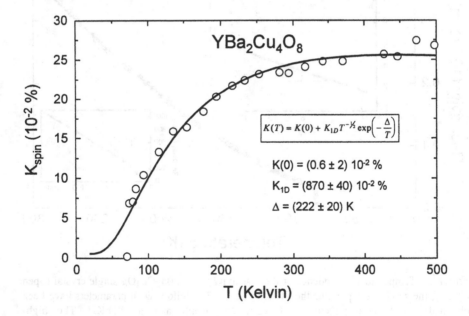

Figure 4. Knight shift data (open circles) (after Ref. 13) fitted with Eq. 5 (solid line). A gap of $\Delta = (222 \pm 20)\,K$ was obtained.

$$K(T) \sim T^{-\frac{1}{2}} \exp\left(-\frac{\Delta}{T}\right) \tag{5}$$

Fitting the $K(T)$ data [13] for YBa$_2$Cu$_4$O$_8$ with Eq. 5 gives an excellent result (Fig. 4) with the spin-gap $\Delta = (222 \pm 20) K$ which is very close to the value $\Delta = (224 \pm 5) K$ derived from the resistivity data (Fig. 3).

More experimental evidence for this scaling behavior can be found in the study of YBa$_2$Cu$_3$O$_x$ in which the doping is continuously changed by tuning the oxygen content x. We systematically varied the oxygen content in several c-axis-oriented YBa$_2$Cu$_3$O$_x$ films and measured the temperature dependence of the resistivity. It becomes clear that although the reduction of the oxygen content in YBa$_2$Cu$_3$O$_x$ gives rise to different temperature dependencies of the transport properties, the changes take place in a very systematic way, where the T-linear part systematically shifts towards higher temperatures when x is lowered and simultaneously a pronounced S shape develops at lower temperatures. In order to quantify this evolution we can define a characteristic temperature $T^* \sim T_0$ as the temperature above which $\rho(T)$ is linear in T. Using the characteristic temperature T_0, we may now scale the temperature axis and simultaneously scale the resistivity with the value of ρ at T_0.

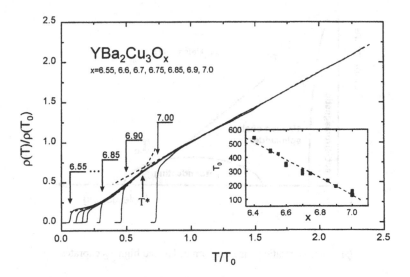

Figure 5. Scaled in-plane resistivity $\rho/\rho(T_0)$ versus scaled temperature T/T_0 for 7 sets of YBCO thin film data. Dashed lines are theoretical curves for 1D ($T<T^*$) and 2D ($T>T^*$) regimes.

The result of this operation for all the available thin film resistivity data (7 data sets in total) is shown in Fig. 5. Very interestingly, we see in this figure that to a good approximation all the $\rho(T)$ curves collapse onto one universal curve. Given the considerable overlap in the reduced-

334

temperature scale between the 7 resistivity curves, it is very unlikely that this scaling result is a fortuitous coincidence or an artificial effect of splicing separate curves together. The universal resistivity curve consists of an super-linear part for $T/T_o<1$, followed by a linear part for $T/T_o>1$, which approximately extrapolates to zero for $T=0$. Hence this scaling curve clearly demonstrates the limited validity of the often-cited T linearity of the resistivity.

The proposed approach to the description of the scaling behaviour of the resistive properties of cuprate high-T_C's and the existence of the pseudo-gap is based on the assumption that DC transport in high-T_C's is caused by the quantum conductivity with the inelastic length L_ϕ fully governed by the magnetic correlation length ξ_m. Depending upon the effective dimensionality - 2D $(T > T^*)$ or 1D (stripes at $T < T^*$) (Fig. 5) - conductivity is found to change from $\sigma_{2D} \sim ln(\xi_{m2D})$ to $\sigma_{1D} \sim \xi_{m1D}$ (Eq. 1), respectively. The validity of the $\sigma_{1D} \sim \xi_{m1D}$ approach has been checked directly using resistivity data for real SL systems.

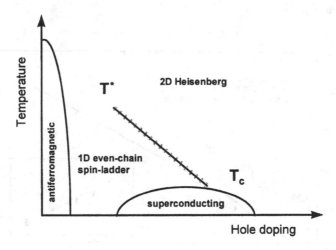

Figure 6. Schematic phase diagram of layered high-T_C cuprates

The temperature dependences $\rho(T)$ of single crystals of the novel $Sr_{2.5}Ca_{11.5}Cu_{24}O_{41}$ ladder compound under hydrostatic pressure of up to 8 GPa [5] and the underdoped cuprate $YBa_2Cu_4O_8$ have been explained in the framework of the model assuming the opening of a spin-gap in the 1D even-chain spin-ladder (1D-SL) and a pseudo-gap at $T < T^*$ in $YBa_2Cu_4O_8$. The $\rho(T)$ dependence of the underdoped cuprate $YBa_2Cu_4O_8$ also demonstrates a remarkable scaling with the $\rho(T)$ of the 1D-SL compound. This scaling implies that the pseudo-gap below $T^*(p)$ (p being the hole concentration) in underdoped $YBa_2Cu_4O_8$ is the spin-gap in the even-chain SL's formed at $T < T^*(p)$ in these materials (Fig. 6) and strongly reflects the 1D nature of their microscopic electronic structure in this regime. The proposed approach considers the pseudo-gap temperature $T^*(p)$ as the crossover between the high temperature 2D Heisenberg and the 1D quantum even-chain SL regime. The latter is established as temperature decreases and charge expulsion from the Mott insulator phases [16] gets stronger and finally leads to the formation of well defined stripes which can be modelled by the SL with an even number of chains. The effective 1D case at $T < T^*(p)$ makes all considerations of the non-Fermi liquid behaviour [8,17,18] and possible charge/spin separation [19,20] important for the physics of underdoped high-T_c's in general. The weakly doped even-chain SL's have the spin-gap, show hole-hole pairing (though rather short-range along the chain) with an "approximate" $d_{x^2-y^2}$ symmetry [11]. These materials, being lightly doped insulators, show also features of a large metal-like Fermi surface. Since the 1D SL phase precedes the superconducting transition, then it is quite reasonable to consider similar mechanisms being responsible for superconductivity in both underdoped cuprates and in SL with an even number of chains [11]. The short-range correlation for the pairs along the chains in even-leg SL's [21] contradicts ARPES and corner-junction SQUID measurements carried out at $T < T_c(p)$. This can be explained by a substantial low-temperature modification of these correlations due to the onset at $T \sim T_c(p)$ of the Josephson-like coupling between the stripes [8, 22], resulting eventually in an effective recovery of the 2D-character of the CuO system at $T < T_c(p)$. The Josephson-like coupling between the 1D SL causes the onset of superconductivity at $T_c(p)$ with the $T_c(p)$ value increasing with the hole doping. Isolated 1D even-chain SL's, most probably, have no chance to develop along the chain a real macroscopic superconducting coherence of bosons, formed along rungs in the SL. From this point of view, recent pulsed field data [23] can be interpreted as an experimental evidence of the insulating ground state of field-decoupled SL's in underdoped cuprates.

The 1D SL phase of the CuO planes becomes unstable at high temperatures, when entropy effects lead to stripe meandering and eventually destroy the 1D stripes [9], and at high doping levels, when the distance between the stripes becomes so small that the Mott insulator phase between stripes collapses, thus recovering the 2D regime. As a result, in optimally doped and overdoped regime the superconducting transition takes place when the CuO_2 layer is in the regime of the 2D doped Heisenberg system. Therefore, the rather "fragile" 1D SL phase seems to exist only in underdoped cuprates in the temperature window between T^* and T_c.

4. CONCLUSIONS

We have described the transport properties of he even-chain SL compound $Sr_{2.5}Ca_{11.5}Cu_{24}O_{41}$ under pressure in terms of the 1D quantum transport model. Taking into account a remarkable scaling behaviour of the resistivity of $YBa_2Cu_4O_8$ and $YBa_2Cu_3O_x$ at $T < T^*$ and $Sr_{2.5}Ca_{11.5}Cu_{24}O_{41}$ we have assumed that the former is also in the 1D regime at temperatures $T_C < T < T^*$ and therefore *the pseudo-gap of underdoped high-T_C cuprates is simply the spin-gap in the even-chain SL's associated with the stripe formation in the CuO planes at $T < T^*$*.

Acknowledgements. This work was supported by the GOA and FWO-Vlaanderen programs. L.T. is a research fellow supported by the Flemish institute for the stimulation of the scientific-technological research in industry (IWT); J.V. is a research fellow from the Flemish FWO. The authors are thankful to Y. Bruynseraede for useful discussions.

5. REFERENCES

[1] Moshchalkov V.V., *Sol. St. Comm.* **86** (1993) 715.

[2] MoshchalkovV.V., *cond-mat / 9802281*

[3] Abrikosov A.A., *Fundamentals of the Theory of Metals*, (North-Holland, Amsterdam) 1988.

[4] Greven M., Birgeneau R.J. and Wiese U.-J., *Phys. Rev. Lett.* **77** (1996) 1865.

[5] Nagata T . et al., *Physica C* **282-287** (1995) 153.

[6] Takano M. et al, *Physica C* **282-287** (1995) 149.

[7] Batlogg B., *Physica C* **282-287** (1997) XXIV.

[8] Emery V.J., Kivelson S.A., and Zachar O., *Physica C* **282-287** (1997) 174.

[9] Zaanen J. and van Saarloos W., *Physica C* **282-287** (1997) 178.

[10] Tranquada J.M., *Physica C* **282-287** (1997) 166; Tranquada J.M. et al., *Nature* **375** (1995) 561.; Tranquada J.M. et al., *Phys. Rev. Lett.* **73** (1997) 338.

[11] Dagotto E. and Rice T.M., *Science* **271** (1996) 618.

[12] Karpinski J., Kaldis E., Jilek E., Rusiecki s., Bucher B., *Nature* **336** (1988) 660.

[13] Bucher B., Steiner P., Wachter P., *Physica B* **199-200** (1994) 268.

[14] Wuyts B., Moshchalkov V.V. and Bruynseraede Y., *Phys. Rev. B* **53** (1996) 9418.

[15] Troyer M., Tsunetsugu H. and Würtz D., *Phys. Rev. B* **50** (1994) 13515.

[16] Schrieffer J.R., Zhang S.C. and Wen X.G., *Phys. Rev. Lett.* **60** (1988) 944.

[17] Maekawa S., Tohyama T., *Physica C* **282-287** (1997) 286.

[18] Ding M. et al., *Phys. Rev Lett.* **76** (1996) 1533.

[19] Anderson P.W., *Science* **235** (1987) 1196.

[20] Fukuyama H. and Kohno H., *Physica C* **282-287** (1997) 124.

[21] Chakravarty S., *Science* **278** (1997) 1412.

[22] Kashine J. and Yonemitsu K., Journ. Phys. Soc. Jap. **66** (1997) 3725.

[23] Ando Y. et al., *Phys. Rev. Lett.* **77** (1996 2065.

THE INFLUENCE OF GRAIN BOUNDARY ROUGHNESS ON TRICRYSTAL SYMMETRY TESTS

J.R. KIRTLEY AND C.C. TSUEI
IBM T.J. Watson Research Center
P.O. Box 218, Yorktown Heights, NY 10598, USA

K.A. MOLER
Dept. of Physics
Princeton University, Princeton, NJ 08544, USA

AND

J. MANNHART AND H. HILGENKAMP
Exp. Phys. VI, Center for Electronic Correlations and Magnetism,
Institute of Physics, Augsburg University, D-86135 Augsburg,
Germany

1. Introduction

Phase sensitive experiments in tricrystal grain boundary samples[1, 2, 3, 4, 5, 6] have demonstrated that a number of the high-T_c cuprate superconductors have predominantly d-wave symmetry. The reproducibility of these experiments is remarkable, when one considers the roughness of the grain boundary interfaces. Here we briefly describe how this roughness affects the magnetic properties of grain boundaries in high-T_c superconductors. We then describe experiments designed and performed to make conclusive symmetry tests despite this roughness.

2. Grain boundary roughness

In tricrystal phase sensitive pairing symmetry experiments, a single crystal of $SrTiO_3$ is cut, polished, reoriented, and fused together to make three joined [001] oriented crystals with controlled misorientation angles between them, as diagrammed in Figure 1. Thin film high-T_c cuprate superconductors are grown epitaxially on these tricrystal substrates. The substrate

337

M. Ausloos and S. Kruchinin (eds.), Symmetry and Pairing in Superconductors, 337–346.

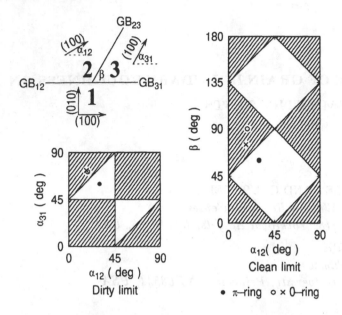

Figure 1. General tricrystal geometry, and diagrams showing values of grain boundary angle β and crystalline axis misorientation angles α in which the tricrystal point is frustrated (π-ring, open) or non-frustrated (0-ring, shaded) for a $d_{x^2-y^2}$ superconductor, in both the clean and dirty limits.

geometry is chosen to either have, or not have, frustration at the central, tricrystal point, for a $d_{x^2-y^2}$ superconductor. This frustration, involving an odd number of sign changes in the components of the superconducting order parameter normal to the grain boundary interfaces, results in spontaneous magnetization at the tricrystal point with a total flux of half the superconducting flux quantum $\Phi_0 = h/2e$. To use the presence or absence of the half-flux quantum as a test of the symmetry of the pairing wave function, it is necessary to understand the influence of grain boundary interface roughness.

It is well known from transmission electron microscopy (TEM) imaging that high-T_c thin film grain boundaries are generally composed of facets having typical dimensions of the order of 10-100 nm[7, 8, 9]. The combination of these facets with submicron-sized structural defects at the fusion boundary of the SrTiO$_3$ epitaxial growth substrates can cause the grain boundary to wander over lengths up to a micron[10].

An example of the effects of faceting on the grain boundary is shown in Figure 2, which is an atomic force microscope (AFM) image of an asymmetric 45 degree grain [001]-tilt boundary in a ~150 nm thick YBa$_2$Cu$_3$O$_{7-\delta}$ (YBCO) film. The grain boundary orientation can be locally significantly different from the average value.

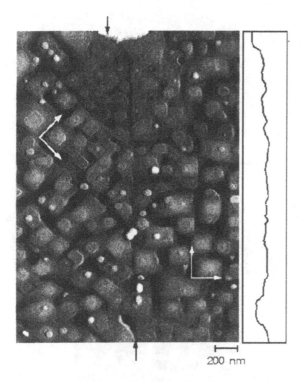

Figure 2. AFM image of the surface of a $\sim$ 150 nm thick $YBa_2Cu_3O_{7-\delta}$ film with an asymmetric 45 deg [001]-tilt boundary. The orientation of the grains is clearly visible by the orientation of the facets of the growth islands. The location of the grain boundary is indicated by the arrows. At the boundary, the film thickness is reduced by 1-2 nm compared to the areas between the growth islands. The meandering path of the boundary has been replotted to the right of the figure. (Reprinted from Ref. 14)

3. Spontaneous flux generation in grain boundaries

The combination of grain boundary roughness with $d_{x^2-y^2}$ symmetry of the high-T_c order parameter has profound effects on the transport properties of high-T_c cuprate grain boundaries through the spontaneous generation of magnetic flux[11]. This can be understood as follows: The Josephson critical current across an interface can be written as $I_c \propto \Delta_1(\hat{n})\Delta_2(\hat{n})$, where $\Delta_1(\hat{n})$ and $\Delta_2(\hat{n})$ are the normal projections of the superconducting gaps onto the two sides of the interface[12]. A d-wave superconductor is characterized by a momentum dependent order-parameter that can change sign. Therefore, it is possible for a rough interface in a d-wave superconductor to have sections with π relative phase shifts: $I = I_c sin\phi = -\mid I_c \mid sin\phi = \mid I_c \mid sin(\phi + \pi)$. This is called a π-junction; one without such a sign change is a 0-junction. A grain boundary with alternating 0- and π-regions can

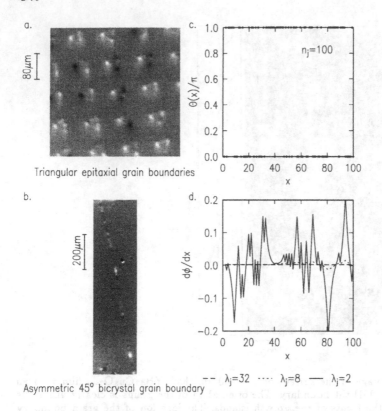

a. Triangular epitaxial grain boundaries

b. Asymmetric 45° bicrystal grain boundary

c. $n_j = 100$

$-- \lambda_j = 32 \quad \cdots \lambda_j = 8 \quad - \lambda_j = 2$

Figure 3. a. SSM image of triangular shaped biepitaxial grain boundaries in thin film $YBa_2Cu_3O_{7-\delta}$. The triangles are $30\mu m$ on a side. b. SSM image of an asymmetric 45 degree [001] bicrystal grain boundary. c. Randomly assigned 0 or π phase shifts of a linear array of Josephson junctions. d. Solution of the linear array of Josephson junctions model described in the text, for different values of the Josephson penetration depth λ_j. As the array is "cooled" by decreasing the value of the Josephson penetration depth, magnetic flux is spontaneously generated in the array.

spontaneously generate supercurrents along the grain boundary, with a corresponding magnetic flux through the grain boundary.

While such spontaneous magnetization can be inferred from current-voltage characteristics of narrow bridges fabricated across the grain boundaries[11], a more direct way is to magnetically image the sample[13]. The magnetic images in this paper were made with a high spatial resolution scanning SQUID microscope(SSM)[14]. In our microscope, a small diameter (between 4 and 10 microns) thin film pickup loop, which is an integral part of a low-T_c Superconducting Quantum Interference Device (SQUID), is scanned relative to a sample with the plane of the pickup loop nearly parallel to, and less than a pickup loop diameter above, the surface of the sample. The grey-scale intensity in the images in this paper is proportional to the total magnetic flux threading the pickup loop.

An example of spontaneous flux generation in grain boundaries is shown in Figure 3a, an SSM image of an array of triangular biepitaxial grain boundaries[15]. These grain boundaries were produced by photolithographically patterning a MgO seed layer on a single crystal substrate of $LaAlO_3$, followed by an epitaxial layer of CeO_2 and a 250nm thick epitaxial film of YBCO[16]. In this type of sample, the crystalline orientation inside the triangles is rotated by 45 degrees relative to that outside the triangles. The sample was cooled in nominally zero field, and imaged at 4.2K. Such images show that the grain boundaries have localized regions of magnetic flux, with the flux lines both entering and exiting the films. These magnetic images can be modelled as arising from isolated monopole sources of magnetic flux, each with much less than a superconducting quantum of flux in them, but summing to integer multiples of a flux quantum over each enclosed grain boundary path[15].

A second example is shown in Figure 3b. This is an SSM image of a 1mm long section of an asymmetric 45 degree [001] tilt angle grain boundary in YBCO[13]. The sample was cooled and imaged in nominally zero external field, and imaged at 4.2K. The only flux visible in this image is that spontaneously generated in the grain boundary. As in the triangular grain boundary, the flux lines both enter and exit the film, and are consistent with sources of flux less than a superconducting flux quantum. It has been proposed that such "fractional" vortices are evidence for time reversal symmetry breaking in the bulk[17] or at the grain boundary interfaces[18]. However, it appears that a mechanism of spontaneous flux generation arising from interface roughness and d-wave symmetry can adequately explain our measurements, without invoking time-reversal symmetry breaking.

Spontaneous flux generation in a rough grain boundary between d-wave superconductors can be simply modelled as a linear array of inductively coupled Josephson junctions, with randomly assigned 0 and π relative phases[19]. For a set of n_j junctions uniformly spaced by Δx, the magnetic behavior is described by the difference equation

$$\phi_{n+1} - 2\phi_n + \phi_{n-1} = sin(\phi_n + \theta_n)/\alpha, \tag{1}$$

where ϕ_n is the superconducting phase of the nth Josephson junction, θ_n is randomly assigned the values 0 or π, $\alpha = \lambda_j^2/(\Delta x)^2$, and λ_j is the Josephson penetration depth. In the absence of an applied magnetic field or current, suitable boundary conditions are $\phi_{n_j} = \phi_{n_j-1}$ and $\phi_2 = \phi_1$. The magnetic field threading between a particular pair of junctions is given by $\Phi_n = (h/2e)(\phi_n - \phi_{n-1})/2\pi$.

An example of such a calculation in shown in Figure 3. Figure 3c shows a set of 100 randomly assigned values for $\theta(n)$. We simulate cooling by making an initial guess for $\phi(n)$ and solving Eq. 1 for a value of $\lambda_j \gg \Delta x$. This solution is used as the initial condition for the next iteration with a reduced value of λ_j, continuing until the desired value of λ_j is reached. Figure 2d plots $d\phi(x)/dx$ (proportional to the magnetic flux), for various values of λ_j during the simulated cooling process.

This simulation confirms our intuition: When the penetration depth is much longer than the spacing between switches in θ, the sign changes in I_c are averaged out, and little spontaneous flux generation occurs. On the other hand, when λ_j is comparable to the spacing between sign changes, there is appreciable flux, with both signs, trapped in the array. However, even for this small value of λ_j, the maximum values of the trapped flux are smaller than a half-flux quantum. The generation of flux as large as a half-flux quantum only occurs when the spatial extent between sign changes is appreciably longer than λ_j[3, 19].

Both examples of spontaneous magnetization in Figure 3 are for grain boundaries with misorientation angles of 45 degrees. Experiments with a number of other misorientation angles in YBCO failed to observe any spontaneous magnetization[13]. The tricrystal experiments on pairing symmetry were done on samples with crystalline misorientations of 30 degrees or less: we have never seen such spontaneous magnetization in YBCO films in the tricrystal samples. Spontaneous magnetization is sometimes seen in tricrystal samples with other cuprates. We associate this effect with poor sample morphology. When spontaneous magnetization in the grain boundaries occurs, the samples are discarded as tests of the pairing symmetry. As the samples improve, this spontaneous magnetization disappears, and symmetry tests become possible.

4. Design of tricrystal symmetry tests

In the design of tricrystal symmetry tests, it is crucial to take into account the effects of interface roughness. Assuming that the critical current across a grain boundary in a $d_{x^2-y^2}$ superconductor, oriented in the x direction, locally follows the Sigrist-Rice formula[12]:

$$I_c(x) = I_0 cos(2\theta_1(x))cos(2\theta_2(x))sin(\phi(x)), \qquad (2)$$

where $\theta_1(x)$ and $\theta_2(x)$ are the angles between the local surface normal and the crystalline principle axis on the two sides of the grain boundary, and we have explicitly included the spatial dependence of the phase difference ϕ across the grain boundary. In general, $\theta_1(x)$ and $\theta_2(x)$ will be characterized by some distribution function $f(\delta\theta)$ of deviations of the grain boundary orientation from its average value The clean limit of a perfectly smooth interface is given by $f(\delta\theta) = \delta(\delta\theta)$. Since a $d_{x^2-y^2}$ superconductor has 4-fold symmetry, the dirty limit is given by $f(\delta\theta) = 2/\pi, 0 < \delta\theta < \pi/2$. In

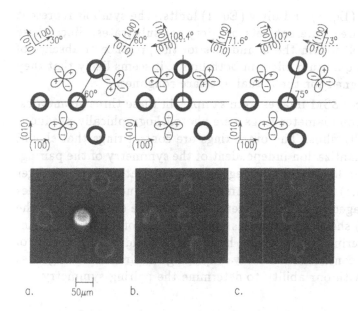

a.   ⊢50μm⊣   b.   c.

Figure 4. Scanning SQUID microscope images of YBCO ring samples grown on three different SrTiO₃ tricrystal substrates. The samples were all cooled and imaged in nominally zero field, and imaged at 4.2K. The ring circling the tricrystal point in geometry (a) has a half-flux quantum of spontaneously generated magnetic field; all the other rings have no magnetic flux included in them. These results are consistent with YBCO having $d_{x^2-y^2}$ pairing symmetry.

the limit where λ_j is much longer than a characteristic length for $I_c(x)$ to switch sign, $\phi(x)$ is independent of x, and in the clean limit the Sigrist-Rice formula is recovered:

$$I_c = I_0 cos(2\theta_1)cos(2\theta_2) \qquad \text{Clean limit} \qquad (3)$$

In the dirty limit

$$I_c = I_0 cos(2(\theta_1 + \theta_2)) \qquad \text{Dirty limit} \qquad (4)$$

In contrast, if the Josephson penetration depth is shorter than the lengths over which I_c switches sign, magnetic flux will be spontaneously generated at the location of the sign switch[19]. Our scanning SQUID microscope has a pickup loop flux sensitivity of about $10^{-3}\Phi_0$: spontaneous flux is detected if present. Therefore such sign switching does not occur for the samples we use for tests of pairing symmetry, and we need only consider the clean and dirty limit cases described above.

Figure 1 shows parameter space diagrams for the presence (open) or absence (hatched) of the half-integer flux quantum effect for the tricrystal geometry as a function of the misorientation (α) and grain boundary (β)

angles for the clean (Eq. 3) and dirty (Eq. 4) limits. The symbols represent design points for three different sets of tricrystal substrates. Since these design points are well within the boundaries for the presence or absence of the half-integer flux quantum effect in both limits, it seems likely that they will perform as designed for the actual interface roughness.

Figure 4 compares SSM images from samples in these three geometries. In each case 50 micron diameter rings were photolithographically produced from epitaxial YBCO films. The outer rings are control rings that should show integer flux quantization independent of the symmetry of the pairing. The central ring should show half-integer flux quantization (a) or integer flux quantization (b,c) for a $d_{x^2-y^2}$ symmetry superconductor. The samples were cooled and imaged in nominally zero field: only the central ring in the π-ring geometry (a) shows spontaneous magnetization of a half-flux quantum. All of our experiments, involving hundreds of cooldowns of dozens of samples, have been consistent with d-wave pairing: the interface roughness has not interfered with our ability to determine the pairing symmetry.

5. A model independent test

The tricrystal experiments described above depend for their interpretation on the assumption that the pair transport across the grain boundaries follows, at least qualitatively, the Sigrist-Rice angular dependence. The results of such experiments are self-consistent: the half-integer flux quantum effect can be turned on and off in agreement with the predictions from this formula.

However, Walker and Luettmer-Strathmann[20] have designed a geometry that relies only on symmetry considerations for its interpretation. In their original proposal (Figure 5a) a high-T_c ring crosses the grain boundaries of two c-axis oriented tetragonal cuprate grains that are rotated about the c-axis by an angle of $\pi/4$ with respect to each other. The symmetrical placement of the two grain boundary junctions assure that for a pair state with $d_{x^2-y^2}$ or d_{xy} symmetry the currents across the two junctions are equal in magnitude but opposite in sign. As a result of this π phase shift, the ring will generate a half-flux quantum. However, for s symmetry, there is no sign reversal, and the ring should show the standard integer flux quantization. It is extremely difficult to make a bicrystal $SrTiO_3$ substrate with the required $\pi/4$-rotated wedge as well as two identical grain boundaries that are smooth and free of microscopic voids. To overcome these technical difficulties, we achieved the bicrystal wedge configuration by effectively fusing two bicrystals along the dividing line MM' in Fig. 5a. We have demonstrated that in the tricrystal symmetry experiments it is not necessary to pattern a ring: a half-flux quantum is still produced upon cooling if required by the

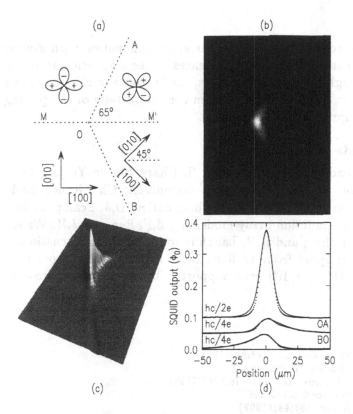

Figure 5. (a) Schematic diagram of the tetracrystal geometry. (b) Top view of an SSM image of a SrTiO₃ tetracrystal with tetragonal Tl-2201 epitaxially grown on it, cooled in nominally zero field and imaged at 4.2K. (c) 3-dimensional rendering of the same data. (d) Detailed modelling of a bulk Abrikosov vortex (top) imaged in the same cooldown, and of two cross-sections through the vortex in the directions indicated, show that it has a half-flux quantum of total flux.

sample symmetry, but in this case it appears as a Josephson vortex at the tricrystal point[21].

Figure 5b shows an SSM image of the central point of such a tetracrystal with an epitaxial c-axis oriented 200nm thick film of $Tl_2Ba_2CuO_{6+\delta}$ (Tl-2201) [5]. The sample was cooled in nominal zero field, and imaged at 4.2K. While Tl-2201 can take on orthorhombic symmetry, extensive experiments were performed to determine that the films on which our experiments were done had tetragonal symmetry[22, 23]. The presence of the half-integer Josephson vortex at the central point in the crystal, demonstrates that Tl-2201 has pure d-wave symmetry. Since this experiment depends only on symmetry considerations for its interpretation, it should not be influenced by the details of the interface roughness.

346

6. Conclusions

Although grain boundary interface roughness, in combination with d-wave pairing symmetry, can have profound influences on the magnetic and transport properties of high-T_c grain boundaries, the influence of this roughness can be accounted for to design and perform conclusive tests of the pairing symmetry of the cuprate superconductors.

Acknowledgements

We thank P. Chaudhari, M.B. Ketchen, N. Khare, Shawn-Yu Lin, and T. Shaw for work on the biepitaxial grain boundaries; Ch. Gerber, and M. Sigrist for work on the bicrystal grain boundaries; D.J. Scalapino for work on the Josephson junction array modelling; Z.F. Ren and J.H. Wang for growth of Tl-2201 films; and R.B. Laughlin for stimulating discussions. KAM acknowledges support from an R.H. Dicke Fellowship and from the NSF. The work of JM and HH was supported by the BMBF (project 13N6918/1).

References

1. C.C. Tsuei et al., Phys. Rev. Lett. 73:593(1994).
2. J.R. Kirtley et al., Nature 373:225(1995).
3. J.H. Miller, Jr., et al. Phys. Rev. Lett. 74:2347(1995).
4. C.C. Tsuei et al., Science 272:329(1996).
5. C.C. Tsuei et al., Nature 387:481(1997).
6. J.R. Kirtley et al., Czech. J. Physics 46:3169(1996).
7. J.A. Alarco et al. Ultramicroscopy 51:239(1993).
8. C.L. Jia et al., Physica C 196:211(1992).
9. S.J. Rosner, K. Char, and G. Zaharchuk, Appl. Phys. Lett. 60:1010(1992).
10. E.B. McDaniel et al., Appl. Phys. Lett. 70:1882(1997).
11. H. Hilgenkamp, J. Mannhart, and B. Mayer, Phys. Rev. B 53:14586(1996).
12. Manfred Sigrist and T.M. Rice, J. Phys. Soc. Japan 62:4283(1992).
13. J. Mannhart et al. Phys. Rev. Lett. 77:2782(1996).
14. J.R. Kirtley et al. Appl. Phys. Lett. 66:1138(1995).
15. J.R. Kirtley et al. Phys. Rev. B 51:12057(1995).
16. K.Char et al. Appl. Phys. Lett. 72:1084(1994).
17. M. Sigrist, D.B. Bailey, and R.B. Laughlin, Phys. Rev. Lett. 74:3249(1995).
18. M. Fogelström and S.-K. Yip, preprint.
19. J.R. Kirtley, K.A. Moler, and D.J. Scalapino, Phys. Rev. B 56:886(1997).
20. M.B. Walker & J. Luettmer-Strathmann, Phys. Rev. B 54:598(1996).
21. J.R. Kirtley, et al. Phys. Rev. Lett. 76:1336(1996).
22. C.A. Wang et al. Physica C 262:98(1996).
23. Z.F. Ren, J.H. Wang, & D.J. Miller, Appl. Phys. Lett. 69:1798(1996).

LOCALIZED CLUSTER STATES STUDIED BY EPR IN R123 COMPOUNDS

N. GUSKOS[1,2], V. LIKODIMOS[1], M. WABIA[2], J. TYPEK[2]
[1]Solid State Physics, Department of Physics, University of Athens,
Panepistimiopolis, 15 784 Zografos, Athens, Greece
[2]Institute of Physics, Technical University of Szczecin, Al.Piastow 17,
70-310 Szczecin, Poland

1. Introduction

Since the discovery of high-T_c cuprates, there has been a large number of experiments attesting that doping with holes of the antiferromagnetic (AFM) CuO_2 layers in the corresponding parent compounds leads to phase separation into hole-enriched and hole-depleted AFM domains [1, 2]. Magnetic susceptibility and resistivity experiments in $La_2CuO_{4+\delta}$ and $La_{2-x}Sr_xCuO_{4+\delta}$ compounds[3-5] have provided strong evidence for the presence of electronic phase separation of percolative type as previously predicted on a theoretical basis [6]. According to the latter theoretical model the introduction of holes in the AFM CuO_2 planes induces the formation of spin-polarized ferromagnetic clusters, also denoted as magnetic polarons, which have only low mobility, whereas the hole inside the cluster is highly mobile [6]. It was shown that such a process is favored by the lowering of the kinetic energy compared to the magnetic energy required for turning the copper spins from the AFM configuration [6, 7]. At higher hole concentration and due to cluster diffusion a percolation network is built, leading to the transition from the AFM insulating to the conducting and below T_c superconducting, state. Theoretical calculation of the structure of such spin-polarized localized states arising in the charge transfer gap of the undoped antiferromagnet, indicated that for low doping the holes preferably create separate clusters [7, 8]. Recently, the formation of spin-polarized one-dimensional stripes within the context of the microscopic percolative phase model [9] and other approaches [10-12] has been suggested in order to explain the charge and spin ordering in the CuO_2 planes and many experimental observations in the high-T_c superconductors [13-15].

According to the experimental results on $La_2CuO_{4+\delta}$ [3, 4] phase separation proceeds via diffusion of the spin-polarized clusters in the temperature range of 150-200 K leading to the formation of the metallic network. On the other hand, fast quenching from the latter temperatures effectively suppresses cluster diffusion and eventually superconductivity, enhancing the paramagnetic contribution of the dissociated clusters. The latter effect was exploited in EPR experiments on quenched $La_2CuO_{4+\delta}$ samples revealing the correlation between the suppression of superconductivity with the observation of intense

347

M. Ausloos and S. Kruchinin (eds.), Symmetry and Pairing in Superconductors, 347-357.
© 1999 Kluwer Academic Publishers. Printed in the Netherlands.

EPR signals at low-temperatures which were attributed to isolated spin-polarized Cu^{2+} clusters containing one delocalized oxygen hole [16].

High-T_c cuprates and their insulating AFM parent compounds have been the subject of numerous EPR investigations as recently reviewed [17], in view of the possibility of direct access to the static and dynamic properties of intrinsic Cu^{2+} ions. However, soon it was inferred that many of the observed EPR signals were due to impurity phases and that no EPR signal corresponding to the bulk divalent copper ions could be detected even at temperatures well above the Neel temperature of the AFM materials [18]. The Cu^{2+} EPR silence has been interpreted in terms of the very fast relaxation rates invoked either in the metallic or the antiferromagnetic phase [17, 18]. However, EPR studies of oxygen-deficient $YBa_2Cu_3O_{6+x}$ (Y123) have shown the presence of weak intrinsic Cu^{2+} EPR spectra [19-23] which for the heavily doped Y123 compounds were attributed to the paramagnetic chain fragments in the Cu(1) plane and further were analyzed in terms of bottleneck of Cu(1) moments through the CuO_2 planes, thus reflecting the spin-gap in the dynamic susceptibility of the CuO_2 planes [22]. On the other hand, EPR studies of oxygen-deficient Y123 and Gd123 materials prepared by quenching of the corresponding high-T_c materials from high-temperature (500-800 °C) revealed the presence of intense EPR signals at low-temperature (T<40 K) with temperature and frequency dependent g-factors, which have been attributed to the ferromagnetically ordered clusters predicted within the model of percolative phase separation [23-25]. Similar intense EPR signals at low-temperatures have been also independently detected by other groups in oxygen-deficient R123 (R=rare earth) compounds [26-29], while earlier reports in high-T_c and non-high T_c materials [30-33] reported a strong correlation of the intensity of low-temperature EPR signal with the $BaCuO_2$ impurity phase content and thus considered an extrinsic origin of the EPR signal, at least in the high-T_c phase [32, 33]. However, intensity considerations of the low-temperature EPR signals in oxygen-deficient R123 argue against any correlation with impurity inclusions [23-25, 29, 34], while they are consistent with the concentration dependence of paramagnetic centers with a maximum near the metal-insulator transition ($x\sim0.4$), as deduced from NMR studies of the spin-lattice relaxation of ^{169}Tm nuclei in Tm123 compounds [10, 15]. Recently, the temperature dependence of the low-temperature EPR signal intensity in oxygen-deficient Y123 compounds has been accessed within the framework of spin-polarized clusters with ground state of spin S=2 [34], consistently with theoretical calculations of the low energy spectrum of pentanuclear-based spin-polarized clusters in the CuO_2 planes [35].

In this context, we present the results of EPR investigations revealing the presence of ferromagnetic localized clusters which are associated with spin-polarized copper clusters induced by oxygen holes in the low-doped R123 compounds.

2. Low-temperature EPR signal in oxygen-deficient R123

All samples used in the present study were prepared from the corresponding high quality high-T_c polycrystalline samples of $RBa_2Cu_3O_{6+x}$ and mixed $R_{0.5}R'_{0.5}Ba_2Cu_3O_{6+x}$

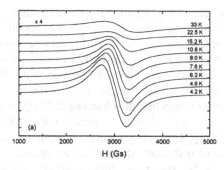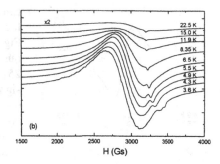

Figure 1. Temperature dependence of the EPR spectra of (a) Nd123, (b) (La,Eu)123 after annealing at $T \sim 1100$ K in He followed by fast cooling at room temperature

($x \sim 0.95$, $T_c \sim 90\text{-}95$ K) where R and R' are rare earth ions, which have been prepared by the solid state reaction method taking special care in order to eliminate the formation of any impurity phases. In particular, part of the latter samples has been subjected to thermal annealing in He at temperatures of 650-850 °C for time intervals of 4 up to 24 h, followed by fast cooling at room temperature in the reducing atmosphere. The samples were characterized with powder X-ray diffraction using a D5000 Siemens diffractometer employing CuKa radiation. XRD characterization confirmed the formation of the R123 crystal structure and further showed only traces of barium cuprate, $BaCuO_{2+x}$ for R=La and the green phase R_2BaCuO_5 for R=Yb materials. The latter procedure is similar to that used in earlier magnetic resonance studies of R123 where the samples were quenched from high temperatures by dropping into water or liquid nitrogen [23-25, 33]. EPR measurements were carried out using standard X-band spectrometers with 100 KHz modulation using fine powdered samples, sealed in quartz tubes.

Figs. 1a, b show the temperature dependence of the X-band EPR spectra for oxygen-deficient $NdBa_2Cu_3O_{6+x}$ and mixed $La_{0.5}Eu_{0.5}Ba_2Cu_3O_{6+x}$ samples prepared from the corresponding high-T_c materials after annealing at $T \sim 1100$ K in He followed by fast cooling at room temperature. Below $T \sim 50$ K in both cases an intense, almost symmetric EPR line is observed, which narrows rapidly with decreasing temperature and below $T \sim 10$ K shifts substantially towards lower fields. For the (La,Eu)123 sample a much weaker monomer Cu^{2+} EPR signal can be also traced from the peak at $H \sim 3250$ Gs, while some additional satellite structure appears at $H \sim 3500$ Gs below 4 K. The intensity of the EPR signal is comparable or even larger to that recorded for bulk $BaCuO_{2+x}$ samples while the linewidth is substantially smaller than in $BaCuO_{2+x}$ [36], thus excluding its relation to small inclusions of the impurity barium cuprate phase. Analysis of the EPR lineshape shows that it can be accurately fitted by a single Lorentzian line, indicative of strong "motional" averaging, at temperatures above 8 K and 12 K for Nd123 and (La,Eu)123 samples, respectively. However, below the latter temperatures a deviation from the single Lorentz lineshape occurs which can be taken into account by using a two-Lorentz lineshape. Representative examples of such fits

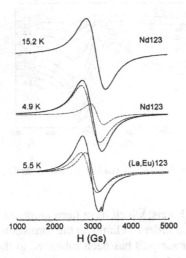

Figure 2. Fitting of the EPR spectra for Nd123 and (La,Eu)123 samples, with single or two Lorentz lineshape (dotted-dashed lines).

can be seen in Fig. 2, while the corresponding g-factors as a function of temperature are plotted in Fig. 3a. Comparison with the EPR spectra previously reported for quenched R123 ceramics [24-25] reveals close similarity with the present EPR signals. In that case, single-crystal EPR measurements showed that the observed EPR signal at X-band is slightly anisotropic with $g_\parallel(\mathbf{H}\|c) < g_\perp(\mathbf{H}\perp c)$ and angular dependence inconsistent with that expected for an anisotropic $S=1/2$ EPR spectrum, thus making powder simulation analysis of the present EPR spectra with $S=1/2$ inappropriate. Therefore, the observed deviation of the EPR signal from the single Lorentz lineshape can be associated with the small g-anisotropy which becomes more prominent at $T<10$ K due to the divergent behavior of the g-factors. The linewidth (ΔH=half-width at half height) of the EPR signal decreases rapidly with decreasing temperature reaching a minimum value at $T\sim7$ K and then increases down to the lowest investigated temperature (Fig.3b). Furthermore, analysis of the integral intensity $I(T)$ which is proportional to the static susceptibility shows that $I(T)$ increases with decreasing temperature faster than $1/T$ which is expected for a simple $S=1/2$ paramagnetic centers as Cu^{2+} ions, indicative of ferromagnetic interactions (Fig. 4).

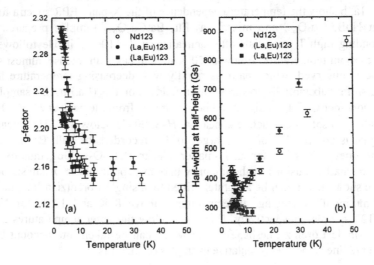

Figure 3. Temperature dependence of the EPR parameters (a) g-factor, (b) ΔH half-width at half-height for the oxygen-deficient Nd123 and (La,Eu)123 samples.

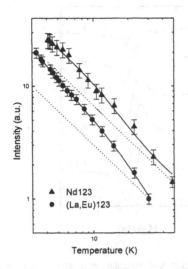

Figure 4. Temperature dependence of the integral intensity I(T) for the oxygen-deficient Nd123 and (La,Eu)123 samples. Solid lines represent the fitting curves according to Eq.1, while dotted lines show the corresponding $1/T$ curves.

Re-annealing of the oxygen-deficient Nd123 and Sm123 samples at 850 °C for 24 h in He results in the suppression of the intense EPR signal which is replaced by an anisotropic Cu^{2+} EPR spectrum with substantially reduced intensity, at least an order of magnitude smaller (Fig.5). The latter powder EPR spectrum can be simulated with a rhombic g-tensor only slightly temperature dependent, anisotropic linewidths slightly increasing below 10 K by approximately 25 MHz~10 G and integral intensity $I(T)$ varying as $T^{-0.82}$ for most of the investigated temperature range [29]. On the other hand, the EPR spectra for Yb123 samples annealed in He at 850 °C for 4 h (sample B) and 900 °C for 24 h (sample C), revealed a different behavior than in R123 (R=Nd, Sm). In particular, for Yb123 B the low-temperature EPR signal coexists with the non-resonant microwave absorption (MWA) signal, which, though weak, persists up to high temperatures indicating the presence of high-T_c superconducting phase (Fig. 6). However, the intensities of the EPR signal and MWA anticorrelated at low-temperatures implying a competing effect. On the other hand, for Yb123 sample C the MWA signal is drastically reduced indicating the suppression of the superconducting properties, while the intensity of the EPR signal is greatly enhanced (Fig. 6).

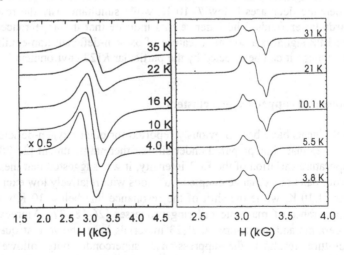

Figure 5. Temperature dependence of the EPR spectra for Sm123 after (a) annealing in He at 1123 K for 4 h, (b) annealing in He at 1123 K for 24 h. Dotted lines show the simulated EPR spectra.

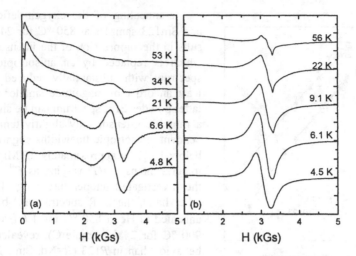

Figure 6. Temperature dependence of the EPR spectra for Yb123 (a) after annealing in He at 1123 K for 4 h, (b) annealing in He at 1173 K for 24 h.

At low temperatures ($T<10$ K) the EPR spectrum is dominated by an intense, almost isotropic EPR line which, as the temperature increases above $T≈10$ K, gradually evolves to an anisotropic powder Cu^{2+} EPR spectrum which becomes clearly resolved at the highest investigated temperatures. The isotropic EPR signal is similar to the corresponding low-temperature EPR signal in Yb123 B, though its intensity is at least an order of magnitude larger, implying an analogous increase of the concentration for the underlying paramagnetic centers. The linewidth of the EPR line increases considerably as the temperature decreases below $T≈10$ K, while simultaneously the resonance field shifts towards lower fields. The latter results indicate that the appearance of the low-temperature EPR signal is a common feature of oxygen-deficient non-equilibrium $R123$ compounds, though it can be affected by the particular $R123$ environment.

3. Spin-polarized copper-oxygen clusters

Similar EPR signals have been previously reported for other oxygen-deficient $R123$ and mixed $(R,R')123$ materials prepared under similar thermal treatment [26-28, 37]. Based on the temperature variation of the EPR intensity, it was suggested that these EPR spectra result from pairs of exchange coupled Cu^{2+} ions with relatively low excitation energy of the order of 10 K, while the shift of the resonance line below 10 K was associated with the appearance of magnetic ordering processes [26-28, 37]. Moreover, previous studies of ceramic and single-crystal $R123$ materials have shown that quenching from high-temperatures results in the suppression of superconductivity followed by the appearance of intense magnetic resonance signals which have been attributed to magnetic clusters formed in hole localization regions [24, 25]. Based on the divergent behavior of the g-factors which becomes more pronounced at Q-band, it has been suggested that the clusters have a quasi-one-dimensional structure with magnetic ordering in the CuO_2 planes [25]. Recently, it has been suggested that the intense low-temperature R123 EPR signals arise from spin-polarized clusters through oxygen-holes in the CuO_2 planes possessing a ground state with spin $S=2$ [10, 34].

Theoretical calculations have shown that the ground state of a CuO_2 layer with a single hole is described by a spin-polarized cluster with 5-8 parallel copper spins, due to the gain of kinetic energy for the delocalized hole in the ferromagnetic environment [7]. In the case where the motion of a hole is confined around a central copper site in a pentanuclear copper cluster, the low energy spectrum is predicted to be described by an effective hamiltonian of the form $H_{eff} = -F_1S(S+1)+F_2(S_z)^2$ with S being the total spin [35]. The interaction parameter F_1 results from the spin-spin coupling of the oxygen hole with the copper spins and it is of the order of 4 meV, while F_2 (~0.05-0.2 meV) results from the anisotropic exchange involving the contribution of excited states to the $d(x^2-y^2)$ copper state through spin-orbit coupling [35]. The ground state thus determined has total spin $S=2$, three excited states with $S=1$ and two with $S=0$ are predicted at energies F_1 and $4F_1$ respectively, while the degeneracy of the $S=2$ and $S=1$ cluster multiplets is further raised by F_2. The $S=2$ ground state is predicted to be split in one singlet $M_S=0$ as the ground state and two *non-Kramers* doublets with quantum numbers M_S equal to ±1 and ±2 at energies 0.6-2.3 K and 2.3-9.3 K, respectively. It has been also shown that the contribution of the copper spins in the second and higher coordination sphere of the central copper site on the basis of the AFM exchange coupling does not alter the low-energy spectrum, while higher excited states occur at $J/2$~750 K [34]. In this respect, allowing for the low-energy spectrum of the cluster, the temperature intensity of the EPR intensity for $g\beta H<<kT$ can be described by

$$I = \frac{A}{T(\sum_i \exp(-\varepsilon_i/kT)+9\exp(-4F_1/kT)+2\exp(-6F_1/kT))} \tag{1}$$

where ε_i (i=1-5) are the Zeeman levels of the $S=2$ ground state with zero-field splitting given by F_2, while for powder EPR spectra the angular average of Eq. (1) should be used [34]. Fitting of the EPR intensity with Eq.(1) (solid lines in Fig.4) results in the values $F_1=0.95(8)$ meV and $F_2=0.061(4)$ meV for (La,Eu)123, while for Nd123 only the value of $F_1=0.7(2)$ meV can be fairly accurately derived. These values are consistent with the theoretical predictions [35], while they can be compared with the values of $F_1=0.8$ meV and $F_2=0.08$ meV reported for an oxygen deficient Y123 sample with oxygen content 6.25 [34]. For the latter value of F_2 denoted as D, the usual axial zero-field splitting constant, it has been shown that the corresponding X-band EPR powder pattern consists of a single resonance line at g~2.1, while another resonance is predicted to occur at H~9000 Gs, which however was not experimentally accessible [34]. Furthermore, using the frequency independent diagrams of the EPR spectra for $S=2$ [38] it can be shown that for zero-field splitting D in the range of 0.06-0.12 meV the corresponding

EPR powder pattern in X-band comprise a dominant spectral line at $g \sim 2$ with a weaker shoulder at the low-field side, while much weaker lines are expected to occur at higher magnetic fields $H > 8000$ Gs. However, the corresponding Q-band EPR spectra should exhibit a rich fine structure which has not been observed in the corresponding Q-band experiments reported so far [24, 25]. In this case, collapse of the fine structure lines should occur through motional narrowing in a similar way as in the exchange narrowing of the Gd^{3+} ($S = 7/2$) crystal-field split EPR transitions in metallic hosts through the Barnes-Plefka mechanism [39]. It should be also noted that consideration of the tetranuclear copper cluster with one hole confined in the b_{2g} symmetry, which has a ground state with total spin $S = 3/2$ further split by anisotropic exchange in two Kramers doublets with an energy gap of 5-20 meV [35], can not account for the temperature dependence of the EPR signal intensity which requires a rich energy spectrum with lower excitation energies (Fig. 4).

The other well established feature of the EPR signal's temperature variation is the divergent behavior of both the g-factor and the EPR linewidth below 10 K (Figs. 4 and 5). In particular, the g-shift which has been found to increase remarkably at Q-band in opposite sense for the two orientations of magnetic field, has been interpreted as due to the presence of quasi-one dimensional ordering of the magnetic clusters [24, 25]. Phenomenologically, the observed behavior may be explained assuming the presence of fluctuating internal fields which add to the applied magnetic field altering the resonance condition, while at the same time slowing down of the fluctuation rate would produce a progressive broadening of the resonance line as the temperature decreases. The occurrence of such fluctuating fields might be traced to short range order among the spin clusters. The temperature dependence of the EPR frequency has been treated theoretically for one-dimensional Heisenberg magnets where it was shown that the g-shift originates from the anisotropic magnetic susceptibility induced by the dipole-dipole interaction or single-ion anisotropy [40]. If the presence of dissociated spin-polarized clusters in the CuO_2 planes is responsible for the intense EPR signal, then the frequency dependent g-shift of the resonance signal induced by the short-range correlations should be considerably enhanced by the zero-field splitting of the ground state. However, the presence of short range correlations at low-temperature does not comply with the isolated copper-oxygen cluster model inferred from the intensity considerations (Eq. 1), while theoretical calculations show that two clusters slightly attract each other but only up to a certain distance (5-9 lattice constants) [7]. An alternative explanation emerges if we allow for the contribution in the EPR signal of the $S = 1$ excited states which occur at approximately the same energy range ($T \sim 10$ K) where the g-factor and linewidth start to deviate.

On the other hand, for the EPR signals observed in quenched $La_2CuO_{4+\delta}$ neither shift of the g-factor nor broadening of the linewidth were reported [16]. The latter system can be discriminated from the $R123$ one by the presence of the $Cu(1)$ chains which for low oxygen doping are expected to maintain oxygen holes in short copper-oxygen chain fragments, while further increase of the oxygen content results in doping of the AFM CuO_2 planes[41]. When no charge transfer to the CuO_2 planes occurs, chain fragments with n oxygens atoms are predicted to comprise $2n$ holes, namely $n+1$ d holes due to Cu^{2+} ions and $n-1$ p holes due to O^-, each having spin $S = 1/2$ which through AFM coupling maintain a singlet ground state [42]. Paramagnetically active chain fragments with odd number of holes leading to $S=1/2$ ground state, occur when charge transfer of holes to the CuO_2 planes takes place. The minimal length for a fragment to be paramagnetic is $n=3$ with one O^- hole transferred, while the concentration dependence of such clusters is calculated to have a maximum at oxygen content $x\approx0.8$, which however shifts and broadens towards lower x values as the quenching temperature increases [42]. The latter paramagnetic contribution has been directly observed in polarized neutron diffraction studies of oxygen-deficient Y123 single-crystals which also indicated a anti-correlation of the Cu(1) magnetization with the chain ordering and non-zero magnetization in the apical oxygen sites connecting the Cu(1) chains with CuO_2 planes indicative of magnetic coupling between chains and planes, as well [43, 44]. Based on the close correlation of hole doping in the CuO_2 planes with the oxygen-ordering of the adjacent Cu(1) planes, it may be argued that quenching $R123$ from high-temperature resulting in oxygen disorder in the Cu(1) planes would lead to a weak but disordered hole doping of the CuO_2 planes, as well. In this case, the resulting paramagnetic chain fragments, most likely of minimal length, coupled with localized spin-polarized clusters in the CuO_2 planes may serve as a possible candidate for the low-temperature EPR signal.

The disappearance of the intense EPR signal after increasing the high-temperature annealing time, are indicative of a reduction of the cluster concentration which is normally expected to occur for a high degree of oxygen deficiency. The latter inhibits the formation of Cu(1) and mostly Cu(2) clusters by decreasing sufficiently the number of available oxygen holes, and thus only the EPR spectra of more "isolated" Cu^{2+} magnetic defects may be observed. The difference between the EPR spectra of the various $R123$ (R=Nd, Sm, Yb) compounds might be rationalized in terms of the effect that the R ions have on the 123 structure. In particular the shift of the metal-insulator transition towards lower oxygen content values for the smaller R ions [44], indicate that the conditions under which localized clusters producing the intense EPR signals can be different along the $R123$ series.

4. Conclusions

The intense EPR signals observed in oxygen-deficient $R123$ fast cooled from high-temperature can be associated with the formation of localized ferromagnetic copper clusters. Their origin can be related to the spin-polarization of divalent copper ions by delocalized holes in the CuO_2 planes, though the contribution of paramagnetic chain fragments in the oxygen-deficient Cu(1) planes can not be excluded.

References

1. Muller, K.A. and Benedek, G., Eds. (1993) *Phase Separation in Cuprate Superconductors*, World Scientific, Singapore.
2. Sigmund, E. and Muller, K.A., Eds. (1994) *Phase Separation in Cuprate Superconductors*, Springer-Verlag, Berlin, Heidelberg.

356

3. Kremer, R.K., Sigmund, E., Hizhnyakov, V., Hentsch, F., Simon A., Muller, K.A., and Mehring, M. (1992) Percolative phase separation in $La_2CuO_{4+\delta}$ and $La_{2-x}Sr_xCuO_4$, Z. Phys. B 86, 319-324.
4. Kremer, R.K., Hizhnyakov, V., Sigmund, E., Simon, A., and Muller, K.A. (1993) Electronic phase separation in La-cuprates, Z. Phys. B 91, 169-174.
5. Sigmund, E., Hizhnyakov, V., Kremer, R.K., and Simon, A. (1994) On the existence of phase separation in high-T_c cuprates, Z. Phys. B 94, 17-20.
6. Hizhnyakov, V. and Sigmund, E. (1988) High-T_c superconductivity induced by ferromagnetic clustering, Physica C 156, 655-666.
7. Klemm, D., Letz, M., Sigmund, E., and Zavt, G.S. (1994) Electronic phase separation: Extended mean-field calculations for CuO_2 layers in high-T_c superconductors, Phys. Rev. B 50, 7046-7055.
8. Seibold, G., Sigmund, E., and Hizhnyakov, V. (1993) Spin cluster states in CuO_2 planes, Phys. Rev B 48, 7537-7544.
9. Seibold, G., Seidel, J., and Sigmund, E. (1996) Inhomogeneous spin structures in high-T_c cuprates, Phys. Rev B 53, 5166-5169.
10. Bakharev, O.N., Eremin, M.V., and Teplov, M.A. (1995) Ordering of magnetic clusters in CuO_2 planes of 1-2-3 superconductors, JETP Lett. 61, 515-519.
11. Emery, V.J. and Kivelson, S.A (1996) Charge ordering in high-temperature superconductors, Physica C 263, 44-48.
12. Salkola, M.I., Emery, V.J., and Kivelson, S.A (1996) Implications of charge ordering for single-particle properties of high-T_c superconductors, Phys. Rev. Lett. 77, 155-158.
13. Cho, J.H., Chou, F.C., and Johnston, D.C. (1993) Phase separation and finite size scaling in $La_{2-x}Sr_xCuO_{4+\delta}$, Phys. Rev. Lett. 70, 222-225.
14. Tranquada, J.M., Sternlieb, B.J., Axe, J.D., Nakamura, Y., and Uchida, S. (1995) Evidence for stripe correlations of spins and holes in copper oxide superconductors, Nature 375, 561.
15. Teplov, M.A., Bakharev, O.N., Dooglav, A.V., Egorov, A.V., Krjukov, E.V., Mukhamedshin, I.R., Sakharatov, Y.A., Brom, H.B., and Witteveen, J. (1997) Tm NMR and Cu NQR studies of phase separation in TmBaCuO compounds, in Kaldis E. et al. (eds.), High-T_c Superconductivity 1996:Ten Years after the Discovery, Kluwer Academic Publishers, Dordrecht, The Netherlands, pp. 531-562.
16. Wubbeler, G. and Schirmer, O.F. (1992) Phase separation in $La_2CuO_{4+\delta}$ superconductors as studied by paramagnetic resonance, Phys. Stat. Sol. (b) 174, K21-K25.
17. Punnoose, A. and Singh, R.J. (1994) EPR studies of high-T_c superconductors and related systems, Int. J. Mod. Phys. B 35, 1123-1156.
18. Simon, P., Bassat, J.M., Oseroff, S.B., Fisk, Z., Cheong, S.W., Wattiaux, A., and Schultz, S. (1993) Absence of Cu^{2+} electron-spin resonance in high-temperature superconductors and related insulators up to 1150 K, Phys. Rev. B 48, 4216-4218.
19. Shaltiel, D., Bill, H., Fischer, P., Francois, M., Hagemann, H., Peter, M., Ravisekhar, Y., Sadowski, W., Scheel, H.J, Triscone, G., Walker, E., and Yvon, K. (1989) Single crystal ESR studies on tetragonal $YBa_2Cu_3O_{6+x}$, Physica C 158, 424-432.
20. Alekseevskii, N.E., Garifullin, I.A., Garif;yanov, N.N, Kochelaev, B.I., Mitin, A.V., Nizhanovski, V.I., Tagirov, L.R., Khaliullin, G.G., and Khlybov, E.P. (1989) EPR study of polycrystalline superconductors with $YBa_2Cu_3O_6$ structure, J. Low Temp. Phys. 77, 87-118.
21. Garifullin, I.A., Garif;yanov, N.N., Alekseevskii, N.E., and Kim, S.F. (1991) EPR data on the evolution of the oxygen distribution in single crystals of $YBa_2Cu_3O_{7-\delta}$, Physica C 179, 9-14.
22. Sichelschmidt, J, Elschner, B., Loidl, A., and Kochelaev, B.I. (1995) EPR study of the dynamic spin susceptibility in heavily doped $YBa_2Cu_3O_{6+\delta}$, Phys. Rev. B 51, 9199-9207.
23. Badalyan, A.G., Baranov, P.G., Aleksandrov, V.I., Borik, M.A., and Osiko, V.V (1989) Magnetic resonance and relaxation in $GdBa_2Cu_3O_x$ single crystals below T_c, JETP Lett. 49, 697-701.
24. Baranov, P.G. and Badalyan, A.G. (1993) Magnetic resonance and magnetic ordering in the oxygen-deficient $RBa_2Cu_3O_x$ superconductors, Solid State Commun. 85, 987-990.
25. Baranov, P.G., Badalyan, A.G., and Il'in, I.V. (1995) Phase separation in cuprate superconductors: magnetic resonance studies, Phys. Solid State 37, 1811-1816.
26. Guskos, N., Triberis, G.P., Calamiotou, M., Trikalinos, C., Koufoudakis, A., Mitros, C., Gamari-Seale, H., and Niarchos, D. (1990) Temperature dependence of the EPR spectra of $GdBa_2Cu_3O_{7-\delta}$ Compounds in orthorhombic and tetragonal phases, Phys. Stat. Sol (b) 162, 243-249.

27. Guskos, N., Triberis, G.P., Likodimos, V., Kondos, A., Koufoudakis, A., Mitros, C., Gamari-Seale, H., and Niarchos, D. (1991) Electron spin resonance of Cu^{2+} ions in $Pr_{0.5}Dy_{0.5}Ba_2Cu_3O_{7-\delta}$, $Sm_{0.5}Gd_{0.5}Ba_2Cu_3O_{7-\delta}$ compounds in the tetragonal phase, *Phys. Stat. Sol (b)* **164**, K105-K108.

28. Gamari-Seale, H., Guskos, N., Koufoudakis, A., Kruk, I., Mitros, C., Likodimos, V., Niarchos, D. and Psyharis, V. (1992) A possible new magnetic phase transition in tetragonal $RBa_2Cu_3O_{6+x}$ (R=Nd, Sm, Tm), *Phil. Mag. B* **65**, 1381-1387.

29. Likodimos, V., Guskos, N., Gamari-Seale, H., Koufoudakis, A., Wabia, M., Typek, J., and Fuks, H. (1996) Copper magnetic centers in oxygen deficient $RBa_2Cu_3O_{6+x}$ (R=Nd, Sm): An EPR and magnetic study, *Phys.Rev.B* **54**, 12342-12352.

30. Shaltiel, D., Genossar, J., Grayvesky, A., Kalman, Z.H., Fisher, B., and Kaplan, N. (1987) ESR in new high temperature superconductors, *Solid State Commun.* **63**, 987-990.

31. Rettori, C., Davidov, D., Belaish, I., and Felner, I. (1987) Magnetism and critical fields in the high-T_c superconductors $YBa_2Cu_3O_{7-x}S_x$ ($x=0$, 1): An ESR study, *Phys. Rev. B* **36**, 4028-4031.

32. Vier, D.C., Oseroff, S.B., Salling, C.T., Smyth, J.F., Schultz, S., Dalichaouch, Y., Lee, B.W., Maple, M.B., Fisk, Z., and Thompson, J.D. (1987) Precautions when interpreting EPR and dc magnetization measurements of high-T_c $RBa_2Cu_3O_{9-x}$-phase superconducting materials, *Phys. Rev. B* **36**, 8888-8892.

33. Genossar, J., Shaltiel, D., Zevin, V., Grayvesky, A., and Fisher, B. (1989) Comparison of the ESR spectra in ceramic $YBa_2Cu_3O_{7-y}$ ($1>y>0$) and related phases, *J. Phys.: Condens. Matter* **1**, 9471-9482.

34. Eremina, R.M., Gafurov, M.R., and Kurkin, I.N. (1997) Intensity of the EPR spectrum in quenched samples of $YBa_2Cu_3O_x$ compounds, *Phys. Solid State* **39**, 374-377.

35. Eremin, M.V. and Sigmund, E. (1993) On the importance of anisotropic exchange coupling on the structure of spin-polarized clusters in layered cuprates, *Solid State Commun.* **90**, 795-797.

36. Guskos, N., Likodimos, V., Psycharis, V., Mitros, C., Koufoudakis, A., Gamari-Seale, H., Windsch, W., and Metz, H. (1995) Structural, magnetic and EPR studies of $BaCuO_{2+x}$, *J. Solid State Chem.* **119**, 50-61.

37. Guskos, N., Likodimos, V., Londos, C.A., Windsch, W., Metz, H., Koufoudakis, A., Mitros, C., Gamari-Seale, H., and Niarchos, D. (1992) Low-temperature dependence of the EPR spectra of $Gd_{0.5}R_{0.5}Ba_2Cu_3O_{7-\delta}$ compounds in the tetragonal phase, *Phys. Stat. Sol (b)* **170**, 597-607.

38. Mabbs, F.E. and Collison, D. (1992) *Electron Paramagnetic Resonance of d Transition Metal Compounds*, Elsevier, Amsterdam.

39. Barnes, S.E. (1981) Theory of electron spin resonance of magnetic ions in metals, *Adv. Phys.* **30**, 801-938.

40. Karasudani, T. and Okamoto, H. (1977) Temperature dependence of EPR frequencies in pure- and pseudo-one dimensional Heisenberg magnets, *J. Phys. Soc. Japan* **43**, 1131-1136.

41. Uimin, G. and Rossat-Mignod, J. (1992) Role of Cu-O chains in the charge transfer mechanism in $YBa_2Cu_3O_{6+x}$, *Physica C* **199**, 251-261.

42. Uimin, G. (1994) Order and disorder in the ensemble of Cu-O chain fragments in oxygen-deficient planes of $YBa_2Cu_3O_{6+x}$, *Phys. Rev. B* **50**, 9531-9547.

43. Boucherle, J.X., Henry, J.Y., Papoular, R.J., Rossat-Mignod, J., Schweizer, J., Tasset, F., and Uimin, G. (1993) Polarised neutron study of high-T_c superconductors, *Physica B* **192**, 25-38.

44. Henry, J.Y, Papoular, R.J., Schweizer, J., Tasset, F., Uimin, G., and Zobkalo, I. (1994) Polarised neutron study of the high-T_c superconductor $YBa_2Cu_3O_{6+x}$ *Physica C* **235-244**, 1659-1660.

FLUCTUATION CONTRIBUTIONS
TO THE TRANSPORT PROPERTIES
OF $Hg_{1.4}Ba_2Ca_2Cu_3O_{8+\delta}$ AND $Bi_2Sr_{1.8}Ca_{1.2}Cu_2O_8$
POLYCRYSTALLINE SUPERCONDUCTORS

H. BOUGRINE
S.U.P.R.A.S., Montefiore Institute of Electricity B28
University of Liège, B-4000 Liège, Belgium

S. DORBOLO
S.U.P.R.A.S., Institute of Physics B5
University of Liège, B-4000 Liège, Belgium

R. CLOOTS
S.U.P.R.A.S., Chemistry Institute B6
University of Liège, B-4000 Liège, Belgium

AND

M. PEKALA
Departement of Chemistry, University of Warsaw,
Al. Zwirki i Wigury. 101 Pl-02089 Warsaw, Poland

Abstract. After some discussion on the precision and originality of the measurement methods used in order to measure the electrical resistivity and simultaneously the thermopower and the thermal conductivity as a function of temperature (20-300K), we report on a study of the superconductivity fluctuation contributions of these transport properties for Hg-BaCaCuO (1223) and BiSrCaCuO (2212) polycrystalline superconductors. The normal-state background of the thermal conductivity and the electrical resistivity is found to obey a linear function of temperature, while the thermopower is found to obey a cubic function of temperature above the critical temperature. After substracting the background, the fluctuation regimes are extracted. The results are discussed and compared to the literature with respect to the system effective dimensionality.

M. Ausloos and S. Kruchinin (eds.), Symmetry and Pairing in Superconductors, 359-370.
© *1999 Kluwer Academic Publishers. Printed in the Netherlands.*

1. Introduction

It is best to recall first that Varlamov and Livanov have recently calculated the fluctuation contribution $\kappa_{e,fl}$ to the thermal conductivity and the thermoelectric power in the clean limit-model for layered superconductors [1]. This contribution is given by

$$\frac{\kappa_{e,fl}}{\kappa_{e,n}(T_c)} = A \left(\epsilon(\epsilon + \delta^2)\right)^{-1/2} \qquad (1)$$

with

$$\epsilon = \frac{T - T_c}{T_c} \qquad (2)$$

$$A = \frac{9\pi^5 \hbar}{128(7\zeta(3))^2 \epsilon_F \tau_{tr}(T_c)} \qquad (3)$$

and

$$\delta = \left(\frac{7\zeta(3)J_c^2}{8\pi^2(k_B T_c)^2}\right)^{1/2} \qquad (4)$$

where T_c is the critical temperature, A the amplitude of the fluctuation contribution, $\zeta(3) = 1.022$ where $\zeta(x)$ is the Riemann Zeta function, ϵ_F the Fermi energy, τ_{tr} the transport relaxation time, δ_{VL} the Varlamov-Livanov parameter, and J_c is the interlayer coupling energy between the superconducting layers. The Varlamov-Livanov expression predicts a crossover from 2D behavior to 3D behavior, i.e for

$$\delta^2 >> \epsilon \rightarrow (3D) \rightarrow \frac{\kappa_{e,fl}}{\kappa_{e,n}(T_c)} = A(1/\delta)\epsilon^{-1/2} \qquad (5)$$

$$\delta^2 >> \epsilon \rightarrow (2D) \rightarrow \frac{\kappa_{e,fl}}{\kappa_{e,n}(T_c)} = A\epsilon^{-1} \qquad (6)$$

This effective dimensionnality crossover occurs at the T_{VL} temperature given by

$$T_{VL} = T_c + T_c \delta_{VL}^2 \qquad (7)$$

We can notice that all those expressions can also be valid for the electrical conductivity. It is of interest to examine whether they formally hold for all transport coefficients, bearing in mind that much discussion has already taken place on the so-called background in earlier theoretical and experimental works [2,3].

2. Experiment

The samples have a nominal composition $Hg_{1.4}Ba_2Ca_2Cu_3O_{8+\delta}$ and Bi_2-$Sr_{1.8}Ca_{1.2}Cu_2O_8$ respectively. The $Bi_2Sr_{1.8}Ca_{1.2}Cu_2O_8$ polycrystal was prepared using the method described in Ref. 4. While the $Hg_{1.4}Ba_2Ca_2Cu_3O_{8+\delta}$ was made by a solid state reaction from stoichiometric mixtures of HgO, BaO, CaO, and CuO powders. The starting mixtures were ground together under dry nitrogen and pressed into pellets of 10mm diameter. The superconducting phase was synthesized in sealed quartz ampoules at 860C for 5 hours [5]. A small (length $10mm$, cross section $S_e = 1mm^2$) bar was cut from each pellet, glued on the sample holder and introduced into a cryogenic system made of a compressor and a double stage cold head CTI-20 model. Electrical and thermal transport properties have been measured from room temperature to 20 K. During the experimental run, the cooling power is obtained through a Gifford-Mac Mahon cycle. The electrical resistivity ρ was measured using the four-point method (Fig.1(a)), and following a slow temperature T heating run, considering the following equation :

$$\rho = (V/I)(S_e/L) \tag{8}$$

where I is the dc current, V the voltage drop across the sample, S_e the cross section, and $L(= 1mm)$ the distance between the voltage leads on the sample. The voltage drop V across the sample is measured carefully since the dc current I is small (1mA), lasts one second and is injected successively on both sides of the sample in order to avoid Joule and Peltier effects. The details and the accuracy of the technique are described in ref. [6]. The thermal conductivity and the thermopower were measured simultaneously using the steady state longitudinal heat flux method and the differential method respectively (Fig. 1(b)). In order to generate a heat flow, a small heater ($R = 150\Omega$) is attached to one end of the sample. Two calibrated chromel-constantan thermocouples are used to measure the temperature difference between two points on the sample. We take into account the residual temperature, losses by radiation, conduction and convection. A typical temperature difference across the sample is $0.5 < \Delta T < 1.5K$, but it midly depends on the temperature of the sample. The thermopower S is deduced from the following equation :

$$S(T) = S_{Au}(T) - (V_s(T)/\Delta T(T)) \tag{9}$$

where $S_Au(T)$ is the thermopower of the gold wires used to measure the voltage drop V_s at the hot junctions of both thermocouples.

The differential Fourier equation for the thermal conductivity $\kappa(T)$ leads to the steady state equation:

$$[P(T)/S_e] = \kappa(T)[\Delta(T)/L] \tag{10}$$

where $P(T)$ is the power dissipated by Joule effect in the sample heater for a one-dimensional heat flux through a sample with cross section area S_e and length L measured between the thermocouples for a temperature difference $\Delta T(T)$. The details and the accuracy of the technique taking into account the thermal losses, are fully described in refs. [7,8]. The relative error on the thermal conductivity κ is estimated to be ca. 8% . It is due mainly to the applied power, the thermal losses, the temperature gradient, the distance between thermocouples, and the cross section area of the sample. The estimated error for measuring the thermopower is estimated to be ca. 2%, while the relative error $\Delta\rho/\rho$ is estimated to be 3% for the electrical resistivity .

3. Results and Discussion

3.1. ELECTRICAL CONDUCTIVITY

In Figs. 1a and 1b the measured electrical conductivity is shown together with the extrapolated fit of the normal state background σ_n obtained with the expression $\sigma_n = 1/(AT + B)$. The parameters of the normal state background for the $Hg_{1.4}Ba_2Ca_2Cu_3O_{8+\delta}$ and $Bi_2Sr_{1.8}Ca_{1.2}Cu_2O_8$ are $A = 9.6310^{-2}(\mu\Omega.m/K)$, $B = 1.6210^{-1}(\mu\Omega.m)$ and $A = 4.810^{-2}(\mu\Omega.m/K)$, $B = 2.584(\mu\Omega.m)$ respectively.

For clarity we compare in Figs. 2(a) and 2(b) the $log[\sigma_{fl}/\sigma(T_c)]$ as a function of $log[(T - T_c)/T_c]$ for the corresponding normal state backgrounds illustrated above. Here $\sigma(T_c) = 0.12(\mu\Omega.m)^{-1}$ and $0.44(\mu\Omega.m)^{-1}$ for $Hg_{1.4}Ba_2Ca_2Cu_3O_{8+\delta}$ and $Bi_2Sr_{1.8}Ca_{1.2}Cu_2O_8$ respectively as determined from the experimental data, where the critical temperature T_c is determined from the maximum of the electrical resistivity derivative. We notice that both curves manifest a crossover from the 3D Gaussian $\sigma_{fl} \approx \epsilon^\alpha$ with $\alpha = -1/2$, to the critical fluctuations, $\sigma_{fl} \approx \epsilon^\alpha$ with $\alpha = -1/3$ at the Ginzburg Landau temperature $T_{GL} = 136.9K$ and $80.62K$ respectively corresponding to $Hg_{1.4}Ba_2Ca_2Cu_3O_{8+\delta}$ and $Bi_2Sr_{1.8}Ca_{1.2}Cu_2O_8$, and a crossover from the 3D gaussian fluctuations to 2D behavior, $\sigma_{fl} \approx \epsilon^\alpha$ with $\alpha = -1$, at the Varlamov-Livanov temperature $T_{VL} = 141.7K$ and $82.5K$ for $Hg_{1.4}Ba_2Ca_2Cu_3O_{8+\delta}$ and $Bi_2Sr_{1.8}Ca_{1.2}Cu_2O_8$ respectively.

We present in Table 1 the corresponding slopes and the amplitudes for the normal state background σ_n, the interlayer coupling energy J_c calculated from equations (4) and (7), the Ginzburg Landau constant K_{GL} estimated from $(T_{GL} - T_c) = 1.0710^{-9}[(K_{GL}^4 T_c^3)/B_{c2}(0)]$, where we take $B_{c2}(0) = 100T$ [17], the relaxation time τ_{tr} and the mean free path l which can be estimated from the Eqs.(3) and (6) and $(l = v_F \tau_{tr})$ respectively, and where $v_F = (2\epsilon_F/m^*)^{1/2}$ is the Fermi velocity. We have taken $\epsilon_F = 0.1eV$ and $m^* = 10m_0$.

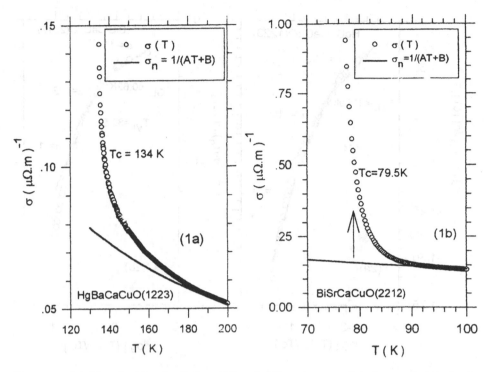

Figure 1. (a) Electrical conductivity $\sigma(T)$ and different extrapolated normal state backgrounds versus temperature T for a polycrystalline HgBaCaCuO (1223). (b) Electrical conductivity $\sigma(T)$ and different extrapolated normal state backgrounds versus temperature T for a polycrystalline BiSrCaCuO(2212).

$\sigma_n(T)$	$T_{GL}(K)$	$T_{VL}(K)$	$J_c(meV)$	K_{GL}	$\tau_{tr}(s)$	$l(nm)$
Hg(1223)	136.9	141.7	8.5	183.2	$2.4 \ 10^{-13}$	14.2
Bi(2212)	80.62	82.5	3.89	217.3	$2.49 \ 10^{-13}$	13.2

TABLE 1. Summary of electrical conductivity results

3.2. THERMAL CONDUCTIVITY

For the thermal conductivity, it is difficult to know the normal state background κ_n behavior at this time from theoretical investigation [9-11]. Several fits can be tested. We find that a linear fit is suitable for our data, i.e $\kappa_n = CT + D$. We report in Figs 3(a) and 3(b) the best extrapolated background curves with respect to the total thermal conductivity as a function of

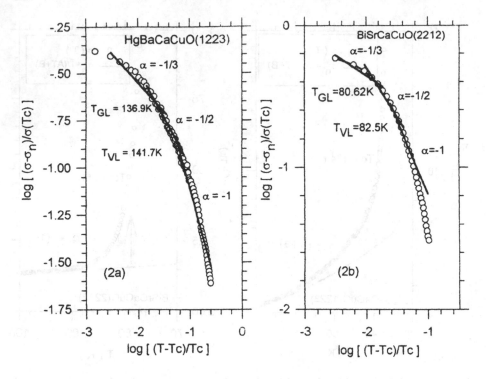

Figure 2. (a) Comparison between two normalized fluctuation contributions $log[\sigma\Pi_{fl}/\sigma(T_c)]$ for the excess electrical conductivity near T_c as a function of $log[(T - T_c)/T_c]$ for a polycrystalline HgBaCaCuO (1223) (b) Comparison between two normalized fluctuation contributions $log[\sigma_{fl}/\sigma(T_c)]$ for the excess electrical conductivity near T_c as a function of $log[(T - T_c)/T_c]$ for a polycrystalline BiSrCaCuO(2212).

temperature. The slopes and the amplitudes are $C = -2.8310^{-4}(W/m.K^2)$, $D = 8.08(W/m.K)$, and $C = 1.1710^{-2}(W/m.K^2)$, $D = 3.04(W/m.K)$ for $Hg_{1.4}Ba_2Ca_2Cu_3O_{8+\delta}$ and $Bi_2Sr_{1.8}Ca_{1.2}Cu_2O_8$ respectively.

In Figs.4(a) and 4(b) we present a comparison of $log[\kappa_{fl}/\kappa_{e,n}(T_c)]$ as a function of $log[(T - T_c)/T_c]$ for the normal state backgrounds illustrated in Figs. 3(a) and 3(b), when $\kappa_{e,n}(T_c)$ is fixed to $0.448W/m.K$ and to $1.1W/m.K$ for $Hg_{1.4}Ba_2Ca_2Cu_3O_{8+\delta}$ and $Bi_2Sr_{1.8}Ca_{1.2}Cu_2O_8$ respectively as estimated with the variational method [9].

Even with a normal state background fit as used in this report, the curves plotted in Figs. 4a and 4b are quite similar to those reported by Houssa et al. [9-11] who have calculated the normal state background behavior from $\kappa_n = \kappa_{e,n}(T) + \kappa_{ph}(T)$, using the explicit Tewordt-Wolkhausen theory [12] and the variational method, in terms of the the phonon thermal conductivity $\kappa_{ph}(T)$. Similar parameters are obtained when we use a linear

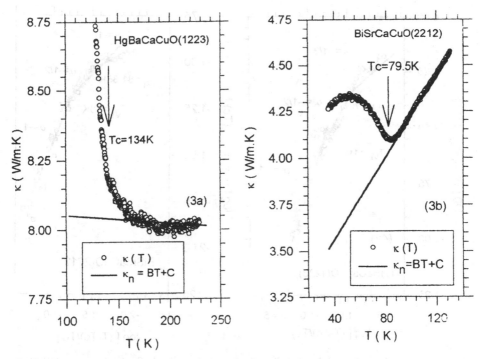

Figure 3. (a) Thermal conductivity $\kappa(T)$ and different normal state extrapolated backgrounds versus temperature T for a polycrystalline HgBaCaCuO (1223). (b) Thermal conductivity $\kappa(T)$ and different normal state extrapolated backgrounds versus temperature T for a polycrystalline BiSrCaCuO(2212).

$\kappa_n(T)$	$T_{GL}(K)$	$T_{VL}(K)$	$J_c(meV)$	K_{GL}	$\tau_{tr}(s)$	$l(nm)$
Hg(1223)	137.7	143.1	9.2	194.7	$1.6\ 10^{-13}$	9.55
Bi(2212)	81.35	83	4.4	246.3	$3.9\ 10^{-13}$	20.6

TABLE 2. Summary of thermal conductivity results

fit for the normal state background.

Obviously similar expressions, as discussed here above, can be used to determine, J_c, K_{GL}, τ_{tr}, and l as function of the normal state background slope. A summary of the collected values for T_{GL}, T_{VL}, J_c, K_{GL}, τ_{tr} and l is presented in Table 2.

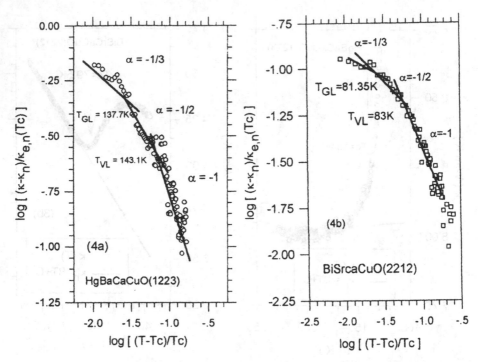

Figure 4. (a) Comparison between two normalized fluctuation contributions $log[\kappa_{fl}/\kappa_{e,n}(T_c)]$ for the excess thermal conductivity near T_c as a function of $log[(T - T_c)/T_c]$ for a polycrystalline HgBaCaCuO (1223). (b) Comparison between two normalized fluctuation contributions $log[\kappa_{fl}/\kappa_{e,n}(T_c)]$ for the excess thermal conductivity near T_c as a function of $log[(T - T_c)/T_c]$ for a polycrystalline BiSrCaCuO(2212).

3.3. THERMOPOWER

The above mentioned Varlamov-Livanov theory predicts the fluctuation contribution to the thermopower as well. The background term relevance is however here much more crucial, since for this coefficient one cannot linearly superpose various contributions [13]. By analogy with the above, however the Varlamov-Livanov expression is used here to analyse the fluctuation contribution to the thermopower (S) data expecting similar singular behavior irrespective of the background term. Figs. 5(a) and 5(b) show the extarapolated fit of $ET^3 + FT^2 + GT + H$ and $LT^4 + MT^3 + NT^2 + OT + P$ behavior of the normal state background S_n [13].

In Figs. 6(a) and 6(b) we report the $log[(S_{fl})/S(T_c)]$ dependence on $log[(T - T_c)/T_c]$. The curve behavior is quite similar to that observed on the electrical conductivity and thermal conductivity. A summary of the corresponding values for J_c, K_{GL} and l is given in Table 3.

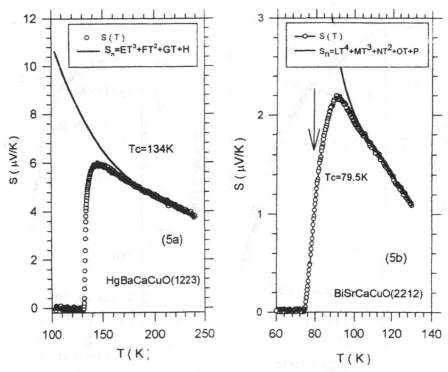

Figure 5. (a) Thermopower $S(T)$ and the best extrapolated normal state background versus temperature T for a polycrystalline HgBaCaCuO (1223). (b) Thermopower $S(T)$ and the best extrapolated normal state background versus temperature T for a polycrystalline BiSrCaCuO(2212).

$S_n(T)$	$T_{GL}(K)$	$T_{VL}(K)$	$J_c(meV)$	K_{GL}	$\tau_{tr}(s)$	$l(nm)$
Hg(1223)	140.2	146.7	10.9	220	$2.0 \ 10^{-13}$	11.87
Bi(2212)	81.65	84.5	5.25	255.7	$1.0 \ 10^{-13}$	6

TABLE 3. Summary of thermopower results

4. Conclusion

One can see that the $Hg_{1.4}Ba_2Ca_2Cu_3O_{8+\delta}$ and $Bi_2Sr_{1.8}Ca_{1.2}Cu_2O_8$ samples presented here have similar parameters and dimensionality near the critical temperature. For $Hg_{1.4}Ba_2Ca_2Cu_3O_{8+\delta}$, let us recall that Kim et al. [14] estimated the Ginzburg Landau parameter K_{GL} to be 100, by analyzing the experimental results of the magnetization versus T near T_c,

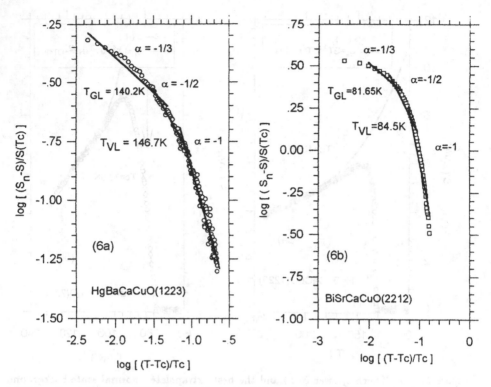

Figure 6. (a) Normalized fluctuation contribution $log[S_{fl}/S(T_c)]$ to the thermopower near T_c as a function of $log[(T - T_c)/T_c]$ for a polycrystalline HgBaCaCuO (1223). (b) Normalized fluctuation contribution $log[S_{fl}/S(T_c)]$ to the thermopower near T_c as a function of $log[(T - T_c)/T_c]$ for a polycrystalline BiSrCaCuO(2212).

in various applied magnetic fields. The interlayer coupling energy J_c can also be estimated from the theoretical expression $J_c = (2\hbar/\gamma d)(\epsilon_F/m^*)^{1/2}$, where γ is the anisotropy parameter, and d the distance between the superconducting layers. For $\gamma = 16$ [15], $d = 0.5nm$, $\epsilon F = 0.1eV$ and $m^* = 10m_0$, this yields the reasonable result $J_c = 7meV$. For the $Bi_2Sr_{1.8}Ca_{1.2}Cu_2O_8$ sample we estimate $K_{GL} = 200 - 250$ which is in good agreement with values found in the literature i,e., K_{GL} is between $100 - 300$ [16,17]. The interlayer coupling energy J_c in the sample is estimated to be $4.5meV$, in quite good agreement with values found by other studies [18,19]. Finally the mean free path is found to be $l_e = 9.5 - 14nm$ and $l_e = 6 - 20nm$ respectively for $Hg_{1.4}Ba_2Ca_2Cu_3O_{8+\delta}$ and $Bi_2Sr_{1.8}Ca_{1.2}Cu_2O_8$, in reasonable agreement with the values $l_e = 10nm$ recently estimated from the analysis of the thermal Hall conductivity measurement by Krishana et al. [20]. Therefore it is essential that normal state properties be studied further in order to obtain fine background values. This is certainly the case for the thermoelectric power and thermal conductivity even though much work on that, as recalled here above, has already been presented.

Acknowledgements

Part of this work has been financially supported by Actions de Recherche Concertées (ARC) 94/99-174, Communauté Française de Belgique Direction de la Recherche Scientifique. S. Dorbolo acknowledges financial support from FRIA. We thank Prof. H.W. Vanderschueren for allowing as to use the M.I.E.L equipement and for his constant interest. M.Pekala acknowledges financial support from KBN through 3T09A 00811 grant. This work has also been supported by Grant Agency of Slovak Academy of Sciences Project No. 2/1323/94 (I.S.). Thanks to NATO for financially supporting this workshop and the organizers for providing such a fine setting for a meeting.

References

[1] Varlamov A A and Livanov D V 1990 Sov. Phys. JETP **71** 325

[2] Ausloos M, Patapis S K and Clippe P 1991 in *Physics and Materials Science of High Temperature Superconductors II Proc. NATO-ASI on High Temperature Superconductors*, Porto Carras 1991 Ed. by Kossowsky R, Raveau B, Wohlleben D and Patapis S K (Dordrecht : Kluwer) p 755

[3] Houssa M, Bougrine H, Stassen S, Cloots R and Ausloos M 1996 Phys. Rev. B 54 R6885

[4] Maeda A, Hase M, Tsukada I, Noda K, Takebayashi S and Uchinokura K 1990 Phys. Rev. B **41** 6418

[5] Sargankova I, Diko P, Kavecansky V, Kovac J, Konig W, Mair M, Gritzner G and Longauer S 1997 Superlattices and Microstructures **21** 3

[6] Laurent Ch, Patapis S K, Laguesse M, Vanderschueren H W, Rulmont A, Tarte P and Ausloos M 1989 Solid State Commun. **66** 445

[7] Bougrine H 1994 Ph.D.Thesis University of Liege, Belgium

[8] Bougrine H and Pekala M 1997 Supercond. Sci. Technol. **10** 621

[9] Houssa M and Ausloos M 1996 Physica C. **257** 321

[10] Houssa M, Ausloos M, Cloots R and Bougrine H 1997 Phys.Rev B**56** 802

[11] Houssa M Ph.D. Thesis 1996 University of Liege, Belgium

[12] Tewordt L and Wolkhausen Th 1989 Solid State Commun. **70** 83

[13] Durczewski K and Ausloos M 1996 Phys. Rev. B **53** 1762

[14] Kim Y C, Thompson J R, Ossandon J G, Christen D K, Pranthaman M 1995 Phys.Rev. B **51** 11767

[15] Couach M, Khodar A F, Calemczuk R, Marcenat Ch, Tholence J L, Capponi J J and Gorius M F 1994 Phys.Lett.A **188** 85

[16] Kleiner R and Müller P 1993 Phys. Rev. B **49** 1327

[17] Dhard C P, Bhatia S N, Sastry P V, Yakhari J V and Nigam A K 1994 J. Appl. Phys. **76** 6944

[18] Houssa M, Ausloos M and Durczewski K 1996 Phys. Rev. B **54** 6126

[19] Reggiani L, Vaglio R and Varlamov A A 1991 Phys. Rev. B **44** 9541

[20] Krishana K, Harris J M and Ong N P 1995 Phys. Rev. Lett. **75** 3529

PART III.

.... with respect to other reports.

TRANSFORMATION OF THERMAL ENERGY INTO ELECTRIC ENERGY IN AN INHOMOGENEOUS SUPERCONDUCTING RING

A.V.NIKULOV

Institute of Microelectronics Technology and
High Purity Materials,
Russian Academy of Sciences,
142432 Chernogolovka, Moscow District, Russia.

Abstract. Let an inhomogeneous superconducting ring (hollow cylinder) be placed in a magnetic field. It is shown that a direct voltage appears across sections with the lowest critical temperature when the ring is switched periodically from the normal state to the superconducting state and back, if the magnetic flux contained within the ring is not divisible by the flux quantum. The superconducting transition can be first order in this case. In the vicinity of this transition, thermal fluctuations can induce the voltage if the ring has a rather small size.

1. Introduction

Superconductivity is a macroscopic quantum phenomenon. One of the consequences of this is the periodical dependence of energy of a superconducting ring on a magnetic flux within this ring. This dependence is caused by a quantization of the velocity of the superconducting electrons v_s, according to ref.[1].

$$v_s = \frac{1}{m}(\hbar\frac{d\phi}{dr} - \frac{2e}{c}A) = \frac{2e}{mc}(\frac{\Phi_0}{2\pi}\frac{d\phi}{dr} - A) \qquad (1)$$

the velocity along the ring (or tube) circumference must have fixed values dependent on the magnetic flux because

M. Ausloos and S. Kruchinin (eds.), Symmetry and Pairing in Superconductors, 373–382.
© *1999 Kluwer Academic Publishers. Printed in the Netherlands.*

$$\int_l dl v_s = \frac{2e}{mc}(\Phi_0 n - \Phi) \tag{2}$$

and $n = \int_l dl(1/2\pi)d\phi/dr$ must be an integer number since the wave function $\Psi = |\Psi|\exp(i\phi)$ of the superconducting electrons must be a simple function. Where ϕ is the phase of the wave function; $\Phi_0 = \pi\hbar c/e = 2.07 \ 10^{-7} G \ cm^2$ is the flux quantum; A is the vector potential; m is the electron mass and e is the electron charge; $l = 2\pi R$ is the ring circumference; R is the ring radius; $\Phi = \int_l dl A$ is the magnetic flux contained within the ring.

The energy of the superconductor increases with the superconducting electron velocity increasing. Therefore the $|v_s|$ tends towards a minimum possible value. If Φ/Φ_0 is an integer number, the velocity is equal to zero. But v_s cannot be equal to zero if Φ/Φ_0 is not an integer number. Consequently, the energy of the superconducting state of the ring depends in a periodic manner on the magnetic field value. The energy of the superconducting ring changes for two reasons : because the kinetic energy of the superconducting electrons $n_s s l m v_s^2/2$ changes and because the energy of a magnetic field $LI_s^2/2$ induced by superconducting current $I_s = sj_s = s2en_s$ changes. Here L is the inductance of the ring; n_s is the superconducting pair density; s is the area of cross-section of the ring wall.

The former is the cause of the Little-Parks effect [2]. Little and Parks discovered that the critical temperature, T_c, of a superconducting tube with narrow wall depends in a periodic way on the magnetic flux value within the tube. This effect has been explained by M.Tinkham [3]. According to ref.[3], the critical temperature of the homogeneous ring (which we consider now) is shifted periodically in the magnetic field:

$$T_c(\Phi) = T_c[1 - (\xi(0)/R)^2(n - \Phi/\Phi_0)^2] \tag{3}$$

because the kinetic energy changes periodically with the magnetic field. Here $\xi(0)$ is the coherence length at T = 0. The value of $(n - \Phi/\Phi_0)$ changes from -0.5 to 0.5. The T_c shift is visible if the tube radius is small enough (if $R \simeq \xi(0)$).

The magnetic field energy $F_L = LI_s^2/2 = Ls^2 2e^2 v_s^2 n_s^2$ does not influence the critical temperature value because it is proportional to n_s^2. A value of this energy depends on the temperature because the n_s value depends on the temperature. The n_s value change causes the superconducting current change and a voltage appaears as a consequence of the electromagnetic induction law. Consequently the superconducting ring can be used for the transformation of thermal energy into electric energy if the velocity $v_s \neq 0$. A temperature change will induce a voltage in the ring if $v_s \neq 0$. In the present paper a most interesting case - an induction of direct voltage in an inhomogeneous ring - is considered.

A ring (tube) with a narrow wall (the wall thickness $w \ll R, \lambda$) is considered. In this case $\Phi = BS \simeq HS$, because the magnetic field induced by the superconducting current in the ring is small. Here H is the magnetic field induced by an external magnet, $S = \pi R^2$ is the ring area and λ is the penetration depth of the magnetic field. We consider a ring whose critical temperature varies along the circumference $l = 2\pi R$, but is constant along the height h. In such a ring, the magnetic flux shifts the critical temperature of a section with a lowest T_c value only. When the superconducting state is closed in the ring, the current of the superconducting electrons must appear as a consequence of the relation (2) if Φ/Φ_0 is not an integer number. Therefore the lowest T_c value will be shifted periodically in the magnetic field as well as the homogeneous ring T_c.

2. Inhomogeneous superconducting rings as a thermo-electric machine of direct current

It is obvious that the current value must be constant along the circumference, because the capacitance is small. The value of the superconductor current must be constant if the current of the normal electrons is absent. Therefore the velocity of the superconducting electrons cannot be a constant value along the circumference of an inhomogeneous ring if the superconducting pair density is not constant. (I consider the ring with constant section area along the circumference.) Let us consider a ring consisting of two sections l_a and l_b $(l_a + l_b = l = 2\pi R)$ with different values of the critical temperature $T_{ca} > T_{cb}$. According to the relation (2) the superconducting current along the ring circumference, I_s, must appear below T_{cb} if Φ/Φ_0 is not an integer number. Then

$$I_s = I_{sa} = s_a j_{sa} = s_a 2 e n_{sa} v_{sa} = I_{sb} = s_b j_{sb} = s_b 2 e n_{sb} v_{sb} \qquad (4)$$

if the normal current is absent. Here n_{sa} and n_{sb} are the densities of the superconducting pair in the sections l_a and l_b; v_{sa} and v_{sb} are the velocities of the superconducting pairs in the sections l_a and l_b and s_a and s_b are the areas of wall section of l_a and l_b. $s = s_a = s_b = wh$. $\int_l dl v_s = v_{sa} l_a + v_{sb} l_b$. Therefore according to (2) and (4)

$$v_{sa} = \frac{2e}{mc} \frac{n_{sb}}{(l_a n_{sb} + l_b n_{sa})}(\Phi_o n - \Phi); \; v_{sb} = \frac{2e}{mc} \frac{n_{sa}}{(l_a n_{sb} + l_b n_{sa})}(\Phi_o n - \Phi) \; (5)$$

$$I_s = \frac{s 4 e^2}{mc} \frac{n_{sa} n_{sb}}{(l_a n_{sb} + l_b n_{sa})}(\Phi_o n - \Phi) \qquad (5a)$$

According to Eq.(5a) a change of the superconducting pair density induces a change of the superconducting current if $\Phi_o n - \Phi \neq 0$. The change of the superconducting current induces the change of the magnetic flux $\Phi = H\pi R^2 + L(I_s + I_n)$ and, as a consequence, induces the voltage and the current of the normal electrons (the normal current, I_n). The total current $I = I_s + I_n$ must be equal in both sections, because the capacitance is small. But I_{sa} cannot be equal to I_{sb}. Then $I_{na} \neq I_{nb}$ and consequently, the potential difference dU/dl exists along the ring circumference. The electric field along the ring circumference $E(r)$ is equal to

$$E(r) = -\frac{dU}{dl} - \frac{1}{l}\frac{d\Phi}{dt} = -\frac{dU}{dl} - \frac{L}{l}\frac{dI}{dt} = \frac{\rho_n}{s}I_n \qquad (6)$$

where ρ_n is the normal resistivity.

Because the normal current exists the relations (5) becomes no valid. The velocity v_{sa} and the current I_{ca} cannot be equal zero even at $n_{sb} = 0$. The current decreases during the decay time of the normal current L/R_{nb}. $R_{bn} = \rho_{bn}l_b/s$ is the resistance of the section l_b in the normal state.

The voltage, the normal current, and superconducting current change periodically in time if the n_{sb} value changes periodically. But in addition a direct potential difference U_b can appear if the l_b section is switched from the normal state in the superconducting state and backwards i.e. some times (t_n) $n_{sb} = 0$ and some times (t_s) $n_{sb} \neq 0$. Let us consider two limiting cases: $t_n \ll L/R_{nb}$ and $t_n \gg L/R_{nb}$. $t_s \gg L/R_{nb}$ and $l_a \gg l_b$ in the both cases.

In the first case the total current I is approximately constant in time and because $t_s \gg L/R_{bn}$

$$I \simeq s2en_{sa} < v_{sa} > \simeq s\frac{4e^2}{mc}\frac{n_{sa} < n_{sb} >}{l_b n_{sa} + l_a < n_{sb} >}(\Phi_0 n - \Phi) \qquad (7)$$

Here $< n_{sb} >$ is the average value of n_{sb}. The average value of the resistivity of the l_b section, $\rho_b \simeq \rho_{bn}t_n/(t_s + t_n)$. Consequently,

$$U_b = R_b I \simeq \frac{l_b < n_{sb} >}{l_b n_{sa} + l_a < n_{sb} >}\frac{(\Phi_0 n - \Phi)}{\lambda_{La}^2}\rho_b \qquad (8)$$

where $\lambda_{La} = (mc/4e^2 n_{sa})^{1/2}$ is the London penetration depth. According to (8) $U_b \neq 0$ at $< n_{sb} > \neq 0$ and $\rho_b \neq 0$. Consequently, the direct potential difference can be observed if we change the temperature inside the region of the resistive transition of the section l_b (where $0 < \rho_b < \rho_{bn}$).

In the second case

$$U_b \simeq \frac{s < n_{sb} >}{l_b n_{sa} + l_a < n_{sb} >}\frac{(\Phi_0 n - \Phi)}{\lambda_{La}^2}Lf \qquad (9)$$

where f is the frequency of the switching from the normal into the super-conducting state.

Thus, the inhomogeneous superconducting ring can be used as a thermo-electric machine of direct current. The power of this machine is small. It decreases with increasing the ring radius. Let us evaluate the maximum power at $l_b n_{sa} \ll l_a < n_{sb} >$. This condition means that the temperature of l_b changes strongly enough. The power will be maximum at $t_n \simeq t_s \simeq L/R_{bn}$. The switching frequency $f = R_{bn}/2L$ in this case. The power of the ring cannot exceed

$$W = IU_b = \frac{s l_b \rho_{bn} \Phi_0^2}{2 l_a^2 \lambda_{La}^4} (n - \Phi/\Phi_0)^2 < \frac{s l_b \rho_{bn} \Phi_0^2}{8 l_a^2 \lambda_{La}^4} \tag{10}$$

For example, at $s = 0.1 \mu m$, $l_b = 0.1 \mu m$, $l_a = 1 \mu m$, $\lambda_{La} = 0.1 \mu m$, $\rho_{bn} = 100 \mu\Omega\ cm$ the power is smaller than $W = 10^{-4} Vt$. The power can be increased by increasing the ring height h .

Above we considered the case when the change of the temperature is strong enough. In the opposite case, when the temperature change is small, $l_b n_{sa} \gg l_a < n_{sb} >$ and therefore the current and the voltage are propor-tional to the $< n_{sb} > /n_{sa}$ value. Because n_{sb} must be equal zero sometimes, this means that the voltage is proportional to the amplitude of the temper-ature change. Consequently, the inhomogeneous ring is a classical thermal machine with a maximum efficiency in the Carnot cycle [4].

At a small l_b/l_a value and a enough big $|(n - \Phi/\Phi_0)|$ value the supercon-ducting transition of l_b is first order. In this case in order to switch the l_b section from the normal state in the superconducting state and backwards the temperature must be changed by a finite value, because a hysteresis of the superconducting transition exists.

3. First order superconducting phase transition

According to Eq.(5) the v_{sb} value decreases with the n_{sb} value increasing. Therefore the dependence of the energy of the superconducting state on the n_{sb} value can have a maximum in some temperature region at $T \simeq T_{cb}(\Phi)$. The presence of such a maximum means that the superconducting transition is a first order phase transition.

The existence of the maximum and the width of the temperature region where the maximum exists depends on the $n - \Phi/\Phi_0$ value and on the ring parameters: l_a, l_b, w, h and T_{ca}/T_{cb}. These dependencies can be reduced to two parameters, B_f and L_I, which are introduced below. It is obvious that the maximum can exist at only $n - \Phi/\Phi_0 \neq 0$. Therefore only this case is considered below.

The Ginsburg-Landau free energy of the ring can be written as

$$F_{GL} = s[l_a((\alpha_a + \frac{mv_{sa}^2}{2})n_{sa} + \frac{\beta_a}{2}n_{sa}^2) + l_b((\alpha_b + \frac{mv_{sb}^2}{2})n_{sb} + \frac{\beta_b}{2}n_{sb}^2)] + \frac{LI_s^2}{2} \quad (11)$$

Here $\alpha_a = \alpha_{a0}(T/T_{ca} - 1)$, β_a, $\alpha_b = \alpha_{b0}(T/T_{cb} - 1)$ and β_b are the coefficients of the Ginsburg-Landau theory. We do not consider the energy connected with the density gradient of the superconducting pair. It can be shown that this does not essentially influence the results obtained below.

The Ginsburg-Landau free energy Eq.(11) consists of $F_{GL,la}$ (the energy of the section l_a), $F_{GL,lb}$ (the energy of the section l_a) and F_L (the energy of the magnetic field induced by the superconducting current):

$$F_{GL} = F_{GL,la} + F_{GL,lb} + F_L \quad (12)$$

Substituting the relation (4) for the superconducting current and the relation (5) for the velocity of the superconducting electrons into the relation (11), we obtain

$$F_{GL,la} = sl_a(\alpha_a(\Phi, n_{sa}, n_{sb})n_{sa} + \frac{\beta_a}{2}n_{sa}^2) \quad (12a)$$

$$F_{GL,lb} = sl_b(\alpha_b(\Phi, n_{sa}, n_{sb})n_{sb} + \frac{\beta_b}{2}n_{sb}^2) \quad (12b)$$

$$F_L = \frac{2Ls^2e^2}{mc} \frac{(\Phi_0 n - \Phi)^2 n_{sa}^2 n_{sb}^2}{(l_a n_{sb} + l_b n_{sa})^2} \quad (12c)$$

Here

$$\alpha_a(\Phi, n_{sa}, n_{sb}) = \alpha_{a0}(\frac{T}{T_{ca}} - 1 + (2\pi\xi_a(0))^2 \frac{(n - \Phi/\Phi_0)^2 n_{sb}^2}{(l_a n_{sb} + l_b n_{sa})^2})$$

$$\alpha_b(\Phi, n_{sa}, n_{sb}) = \alpha_{b0}(\frac{T}{T_{cb}} - 1 + (2\pi\xi_b(0))^2 \frac{(n - \Phi/\Phi_0)^2 n_{sa}^2}{(l_a n_{sb} + l_b n_{sa})^2})$$

$\xi_a(0) = (\hbar^2/2m\alpha_{a0})^{1/2}$; $\xi_b(0) = (\hbar^2/2m\alpha_{b0})^{1/2}$ are the coherence lengths at T=0.

According to the mean field approximation the transition into the superconducting state of the section l_b occurs at $\alpha_b(\Phi, n_{sa}, n_{sb}) = 0$. Because $n_{sa} \neq 0$ at $T = T_{cb}$ the position of the superconducting transition of the l_b section depends on the magnetic flux value:

$$T_{cb}(\Phi) = T_{cb}[1 - (2\pi\xi_b(0))^2 \frac{(n - \Phi/\Phi_0)^2 n_{sa}^2}{(l_a n_{sb} + l_b n_{sa})^2}] \quad (12d)$$

At $l_a = 0$ the relation (12d) coincides with the relation (3) for a homogeneous ring. A similar result ought be expected at $l_b \gg l_a$. But at $l_b \ll l_a$ the $T_{cb}(\Phi)$ value depends strongly on the n_{sb} value. At $n_{sb} = 0$ $T_{cb}(\Phi) = T_{cb}[1 - (2\pi\xi_b(0)/l_b)^2(n - \Phi/\Phi_0)^2]$ whereas at $l_a n_{sb} \gg l_b n_{sa}$ $T_{cb}(\Phi) = T_{cb}[1 - (2\pi\xi_b(0)/l_a)^2(n - \Phi/\Phi_0)^2]$. Consequently a hysteresis of the superconducting transition ought be expected in a ring for $l_b \ll l_a$.

To estimate the dependence of the hysteresis value on the ring parameters, we transform the relation (12) using the relations for the thermodynamic critical field $H_c = \Phi_0/2^{3/2}\pi\lambda_L\xi$; $\alpha^2/2\beta = H_c^2/8\pi$ and for the London penetration depth $\lambda_L = (cm/4e^2n_s)^{1/2}$. We consider a ring with $l_a \gg \xi_a(T) = \xi_a(0)(1 - T/T_{ca})^{0.5}$. $n_{sa} \simeq -\alpha_a/\beta_a$ in this case. Then

$$F_{GL} = F_{GLa} + Fn'_{sb}(\tau + \frac{1}{(n'_{sb}+1)^2} + n'_{sb}(B + \frac{1}{(n'_{sb}+1)^2}(2+L_I))) \quad (13)$$

Here $n'_{sb} = l_a n_{sb}/l_b n_{sa}$;

$$F_{GLa} = -sl_a\frac{H_{ca}^2}{8\pi}(1 + \frac{(2\pi\xi_a(T))^4}{l_a^4}(\frac{(n - \Phi/\Phi_0)n'_{sb}}{(n'_{sb}+1)})^4)$$

Because $l_a \gg 2\pi\xi_a$, $F_{GLa} \simeq -sl_a H_{ca}^2/8\pi$.

$$F = \frac{s\xi_a(T)H_{ca}^2}{2}\frac{2\pi\xi_a(T)}{l_a}(n - \Phi/\Phi_0)^2$$

$$\tau = (\frac{T}{T_{cb}} - 1)(n - \Phi/\Phi_0)^{-2}\frac{l_b^2}{(2\pi\xi_b(0))^2}$$

$$B_f = 0.5\frac{\beta_b}{\beta_a}\frac{l_b}{l_a}\frac{l_b^2}{(2\pi\xi_b(0))^2}(n - \Phi/\Phi_0)^{-2}$$

$$L_I = 4\pi\frac{s}{\lambda_{La}^2}\frac{L}{l_a}$$

For $h > R$, $L = k4\pi R^2/h$ where $k = 1$ at $h \gg R$. Consequently, $L_I = 4\pi(l/l_a)(lw/\lambda_{La}^2(T))$ in this case. At $h, w \ll R$, $L \simeq 4l\ln(2R/w)$, therefore $L_I = 16\pi(l/l_a)(s/\lambda_{La}^2(T))\ln(2R/w)$ in this case.

The numerical calculations show that the $F_{GL}(n'_{sb})$ dependence Eq.(13) has a maximum at small enough values of B_f and L_I in some region of the τ values. The width of the τ region with the $F_{GL}(n'_{sb})$ maximum depends on the B_f value first of all. At $L_I \ll 2$ the maximum exists at $B_f < 0.4$. For example at $B_f = 0.2$ and $L_I \ll 2$ the maximum takes place at $-1.02 < \tau < -0.89$. This means that the transition into the superconducting state of the section l_b occurs at $\tau \simeq -1.02$, (that is at $T_{cs} = T_{cb}(1 - 1.02(n - $

$\Phi/\Phi_0)^2(2\pi\xi_b(0)/l_b)^2)$ and the transition in the normal state occurs at $\tau \simeq -0.89$, (that is at $T_{cn} = T_{cb}(1 - 0.89(n - \Phi/\Phi_0)^2(2\pi\xi_b(0)/l_b)^2)$ if thermal fluctuations are not taken into account.

The inequality $\mathbf{L}_I \ll 2$ is valid for a tube (when $h > R$) with $2\pi lw \ll \lambda_{La}^2(T)$ and for a ring (when $h < R$) with $8\pi hw \ll \lambda_{La}^2(T)$. The hysteresis value increases with decreasing B_f value and decreases with increasing B_f value. The B_f value is proportional to $(n - \Phi/\Phi_0)^{-2}$. Consequently, the hysteresis value depends on the magnetic field value. Because the hysteresis is absent at $B_f > 0.4$, it can be observed in the regions of the magnetic field values, where Φ/Φ_0 differs essentially from an integer number. The width of these regions depends on the $0.5(\beta_b/\beta_a)(l_b^3/(2\pi\xi_b(0))^2 l_a)$ value (see above the relation for B_f). Since $(n - \Phi/\Phi_0)^2 < 0.25$ and $\beta_b \simeq \beta_a$ in the real case, the hysteresis can be observed in the ring with $l_b^3 < 0.2(2\pi\xi_b(0))^2 l_a)$. For example in the ring with $l_b = 2\pi\xi_b(0)$ and $l_a = 10l_b$, the hysteresis can be observed at $|n - \Phi/\Phi_0| > 0.35$ (if $\beta_a = \beta_b$). At $|n - \Phi/\Phi_0| = 0.5$ $B_f = 0.2$ and the hysteresis is equal to $T_{cn} - T_{cs} \simeq 0.03 T_{cb}$ in this ring.

4. Transformation of thermal fluctuation energy into electric energy

The hysteresis of the superconducting transition can be observed if the maximum is high enough. The maximum height is determined by a parameter F, which is introduced below. The hysteresis will be observed if the maximum height is much greater than the energy of the thermal fluctuation, $k_B T$. In the opposite case the thermal fluctuation switches the l_b section from the normal state into the superconducting one and backwards at $T \simeq T_{cb}(\Phi)$.

Above we have used the mean field approximation which is valid when the thermal fluctuation is small. In our case the mean field approximation is valid if the height of the $F_{GL} - F_{GLa}$ maximum, $F_{GL,max}$, is much greater than $k_B T$. This height depends on the F, τ, B_f and L_I values: $F_{GL,max} = FH(\tau, B_f, L_I)$. The F parameter is determined above. The $H(\tau, B_f, L_I)$ dependence can be calculated numerically from the relation (8). To estimate the validity of the mean field approximation we ought to know the maximum value of the $H(\tau)$ dependence: $H_{max}(B_f, L_I)$. We can use the mean field approximation if $FH_{max}(B_f, L_I) \gg k_B T$. This is possible if the height of the ring is large enough, namely

$$h \gg \xi_a(0)\frac{1}{\pi H_{max}(B_f, L_I)}\frac{l_a}{w}\frac{Gi^{1/2}}{T_{ca}/T_{cb} - 1}(n - \frac{\Phi}{\Phi_0})^{-2}$$

Here $Gi = (k_B T_{ca}/\xi_a(0)^3 H_{ca}^2)^2$ is the Ginsburg number for a three- dimensional superconductor. We have used the relation for the F parameter (see above). For conventional superconductors $Gi = 10^{-11} - 10^{-5}$. $H_{max} \simeq 10^{-2}$ for typical B_f and L_I values. For example in the ring with $B_f = 0.2$ and $L_I \ll 2$ the $H(\tau)$ dependence has a maximum $H_{max}(B_f = 0.2, L_I \ll 2) = 0.024$ at $\tau \simeq -0.94$. Consequently, the value of h cannot be very large. As an example, for a ring with parameter value $B_f = 0.2$, $L_I \ll 2$, $l_a/w = 20$, $T_{ca}/T_{cb} - 1 = 0.2$, and fabricated from an extremely dirty superconductor with $Gi = 10^{-5}$, the mean field approximation is valid at $h \gg 20\xi_a(0)$ if $|n - \Phi/\Phi_0| \simeq 0.5$.

If the mean field approximation is not valid, we must take into account the thermal fluctuations which decrease the value of the hysteresis. The probability of the transition from normal into superconducting state and that of the transition from superconducting into normal state are large when the maximum value of $F_{GL} - F_{GLa}$ is not much more than $k_B T$. Therefore the hysteresis cannot be observed at $FH_{max}(B_f, L_I) < k_B T$. This inequality can be valid for a ring made by lithography and etching methods from a thin superconducting film, where h is the film thickness in such a ring.

As a consequence of thermal fluctuations, the density $n_{sb}(r, t)$ changes with time. We can consider $n_{sb}(r, t)$ as a function of the time only if $h, w, l_b \simeq$ or $< \xi_b(T)$. At $T \simeq T_{cb}(\Phi)$ (at the resistive transition) l_b is switched by the fluctuations from the normal state in the superconducting state and backwards i.e. some times ($\simeq t_n$) $n_{sb} = 0$ and some times ($\simeq t_s$) $n_{sb} \neq 0$. Consequently, according to the relations (8) and (9) the direct potential difference can appear in the region of the resistive transition of the section l_b. Thus, the energy of the thermal fluctuations can be transformed into electric energy of direct current in the inhomogeneous ring at the temperature of the resistive transition of the section with the lowest critical temperature.

In order to evaluate the power of this transformation one ought to take into account that as a consequence of the thermal fluctuation the Ginsburg-Landau free energy F_{GL} changes in time with an amplitude $k_B T$. According to the relations (7), (9) and (12c) $U_b I/f \simeq F_L$. Because F_L is a part of F_{GL} (see (12)) the power can not exceed $k_B T f$. The maximum value of the switching frequency f is determined by the characteristic relaxation time of the superconducting fluctuation τ_{GL}: $f_{max} \simeq 1/\tau_{GL}$. In the linear approximation region [5]

$$\tau_{GL} = \frac{\hbar}{8k_B(T - T_c)} \tag{14}$$

The width of the resistive transition of the section l_b can be estimated by the value $T_{cb}Gi_b$. $Gi_b = (k_B T/H_c^2(0)l_b s)^{1/2}$ is the Ginsburg number of the section l_b. Consequently the power value cannot be larger than

$$W = \frac{8Gi_b}{\hbar}(k_B T_{cb})^2 \tag{15}$$

and the U_b value cannot exceed

$$U_{b,max} = \left(\frac{8R_b Gi_b}{\hbar}\right)^{1/2} k_B T_{cb} \tag{16}$$

The $U_{b,max}$ value is large enough to be measured experimentally. Even at $T_c = 1\ K$ and for real values $R_b = 10\Omega$ and $Gi_b = 0.05$, the maximum voltage is equal to $U_{b,max} \simeq 3\mu V$. In a ring made of a high-Tc superconductor, $U_{b,max}$ can exceed $100\mu V$. One ought to expect that the real U_b value will be appreciably smaller than $U_{b,max}$. This voltage can be determined by the periodical dependence on the magnetic field value (see the relations (8) and (9)).

Transformation of thermal fluctuation energy into electric energy does not contradict the second thermodynamic law, because the effect is valid within the thermal fluctuation regime [6].

Acknowledgements

I thank for financial support the International Association for the Promotion of Co-operation with Scientists from the New Independent States (Project INTAS-96-0452) and the National Scientific Council on Superconductivity of SSTP "ADPCM" (Project 95040).

References

1. M.Tinkham, *Introduction to Superconductivity*. (McGraw-Hill Book Company, New York, 1975).
2. W.A.Little and R.D.Parks, Phys. Rev. Lett **9**, 9 (1962) ; Phys. Rev. **133**, A97 (1964).
3. M.Tinkham, Phys. Rev. **129**, 2413 (1963).
4. C.Kittel, *Thermal Physics*. (John Wiley and Sons, Inc., New York, 1970).
5. W.J.Skocpol and M.Tinkham, Rep.Prog.Phys. **38**, 1049 (1975).
6. L.D.Landau and E.M.Lifshitz, *Statistical Physics* Part 1, (Pergamon Press, New York, 1989).

PHASE TRANSITION RELATED WITH

'SPIN DENSITY LATTICE' IN $TbNiSn$

HIDENOBU HORI
Department of Material Science,
Japan Advanced Institute of Science and Technology
(JAIST), Tatsunokuchi, Ishikawa 923-1292, Japan

AND

KOICHI KINDO
The Research Center for Materials Science
at Extreme Conditions, Osaka University,
Toyonaka, Osaka 560-8531, Japan

Abstract. A semi-empirical analysis method on the magnetoresistance in magnetically ordered rare-earth metal is proposed and is applied to the data of TbNiSn. Assuming the formation of the lattice like periodic distribution of the conduction electron spin around the $4f$-spins, rich physical properties in conduction electrons around the Fermi level can be deduced. As an example, extraordinary large change in the magneto resistance with the multi-step magnetic transition on TbNiSn was analyzed. The temperature dependent magneto resistance is explained well by use of the magnetization data. The results requires that many "incoherent electrons" appear in the magnetic phase above 2.1K. We also conclude that there is a change in the effective mass due to the modulation of the band structure.

1. Introduction

Transport experiment on highly correlated spin conductors are expected to give us rich and basic information on the region near the Fermi level. Indeed s-f exchange interaction couples the conduction electron with the localized $4f$-spins and plays the main role for magnetic ordering. In fact, some rare-earth magnetic metals show drastic change in the magneto-resistance

M. Ausloos and S. Kruchinin (eds.), Symmetry and Pairing in Superconductors, 383–391.
© 1999 *Kluwer Academic Publishers. Printed in the Netherlands.*

because of the magnetic phase transitions [1]. The analysis, however, has not been well performed and the microscopic information on the electronic state has not been well obtained, so far. The purpose of this work is to present a semi-empirical method in order to analyze the conductivity of magnetically ordered crystals based on a simple and clear model.

To discuss this method in detail, the magnetoresistance is analyzed for TbNiSn which is antiferromagnetic(AF) at low temperature and shows one interesting drastic change in the magnetoresistance around the field induced phase transition temperature at 2.1K. The isomorphous rare-earth inter-metallic compounds of equiatomic ternary compounds RNiSn-type crystals(R= rare-earth ions), which have the orthorhombic (Pnma) structure [2,3], allow us to systematically investigate the transport properties in these similar spin systems [1]. Ni^{2+} ion in these crystal is non-magnetic and the magnetic moment originated from the R^{3+} ion. In the case of TbNiSn, the magnetic moment with J=6, g=3/2 and $g[J(J + 1)]^{1/2}$ of 9.7 was experimentally obtained and the value is consistent with the moment of Tb^{3+} [4]. As it is well known, when the rare-earth metal ion in these isomorphous crystals is changed, a variation in the energy depth from the Fermi energy is expected. CeNiSn, for example, has the 4f-magnetic level close to the Fermi energy and indicates a remarkably strong spin correlation effect : the so-called Kondo semiconductor problem [5].

The magnetic measurements of TbNiSn, which is antiferromagnet below T_N of 18.5K, have been investigated in the temperature range between 1.5 K and 200 K[2,3]. As was reported in our previous work, the temperature dependence of successive phase transitions has been observed for single crystals of TbNiSn [6].

2. Experimental Results in TbNiSn

As seen in Fig.1, the temperature dependence of the magnetic susceptibility and resistivity [6] shows four successive phase transitions and the b-axis is most characteristic direction in this crystal. In a preliminary result on neutron experiment, it has been reported that the change in spin structure was not observed between both phases at zero-field. Moreover, another new phase transition with small kinks around 2.1K on the resistance and the susceptibility have been also observed by our group, as is shown in the insets of Fig.1.

According to the experimental results, it is remarkable that the difference in the specific heat [7], spin structure [8] and magnetization are quite slight between both phases across the transition temperature at 2.1K. However, only magnetoresistance shows quite a large change in the phase transition. The results show that the most important origin of the phase

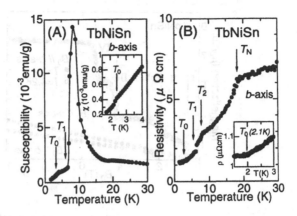

Figure 1. Temperature dependent phase transition in TbNiSn : (A) magnetic susceptibility, (B) resistance along the *b*-axis.

transition is in the scattering mechanism near the Fermi level. Because of good correspondence between magnetoresistance and field induced phase transition found in Fig.2 and 3, the data of the magnetoresistance should be analyzed in comparison with the magnetization data to obtain the relation with the magnetic phase.

Figure 2 shows the anisotropic magnetization process in applied magnetic fields along the *b*-axis at 1.4 K and 4.2 K. To investigate the characteristics of the field induced phase transition, our attention is focused on the data with respect to the *b*-axis. On the whole, both magnetization processes are the same and each phase is defined as is indicated in Fig.2. A good correspondence between the abrupt change in magnetoresistance and the step magnetization shows that the variation in the magnetoresistance originated from the 4*f*-spin structure, although the 4*f*-level is fairly deep from the Fermi energy ϵ_F.

The transition fields H_{c1}, H_{c2}, H_{c3} and H_{c4} at 1.4 K are given by 0.6T, 1.8T, 4.3T and 5.3T, respectively. The magnetization at each step in 1.4K are estimated to be 0, $(1/6)M_s$, $(1/3)M_s$, $(2/3)M_s$ and M_s for the phases I:(0-H_{c1}), II:(H_{c1}-H_{c2}), III:(H_{c2}-H_{c3}), IV:(H_{c3}-H_{c4}) and V(H_{c4}()), respectively. The observed saturation moment M_s of 8.1 μ_B at 1.4K is roughly consistent with the value of $g\mu_B[J(J+1)]^{1/2}$ of 9.72μ_B and $gJ_z = 9.0$, but a slight moment contraction is observed.

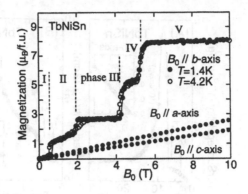

Figure 2. Anisotropic magnetization process at 1.4 and 4.2K

Taking account of these effects, the observed magnetization M per one Tb^{3+} is given by use of M_z of 4f-spins and the parameter A as,

$$M = AM_z \qquad (1)$$

The value of A represents the shielding effect. As was discussed in ref.[6], the contraction of the moment mainly arises from the magnetic shielding effect.

As is seen in Fig.3, the extraordinary increase of the magneto-resistance is found in phase II and IV. Moreover, it should be noticed that the magnitude of the resistance does not always increase with increasing magnetization. Such a phenomenon has not been observed in previous work until now.

The problems in the experimental results are summarized as follows: 1) why does the localized moment of a deep 4f-level sensitively affect the conduction electrons; the 4f-level being of the order of an eV away from the Fermi level ϵ_F. 2) how can we explain the relation between the magnetoresistance and magnetization data in which an increase in the magnetization does not always lead to an increase of the magnetoresistance. 3) what kind of phase transition appears at 2.1K. In the following a quite simple model is introduced in order to analyze these problems : we assume that the localized 4f-spin on each Tb^{3+} induces an AF spin polarization of the conduction electrons near ϵ_F through the s-f exchange interaction. The induced spin

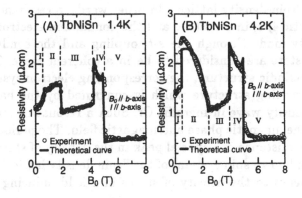

Figure 3. Resistance in magnetic fields. The solid curves are theoretical ones.

moments in the conduction electrons interfer with each other and a lattice like spatial distribution of the conduction electron spins is made. Its structure and period follow the spin structure of Tb^{3+}, which is called here a "spin density lattice". The spatial period of the magnetic phase does not always match that of the real lattice. In some phases, the contradiction in the periodicity can induce a large spin fluctuation due to such a disagreement. The spin-fluctuation on the conduction electrons affects the resistance through the spin dependent relaxation time. When the periodic property of the "spin density lattice" has some commensurate periodicity with the real lattice, it can be assumed that it induces a variation of the conduction band structure near ϵ_F. Such a variation affects the resistivity through a change in the effective mass m^*. Taking account the discussion by Miwa [9] on the so-called "magnetic super zone effect"[10], this model is considered to be reasonable.

3. Analysis

In order to analyze the data in Fig.3 within the single particle theory [11], the following model is considered: the conduction electrons near ϵ_F are separated into two parts in the ground state: 1) the first one is the usual electron states of a metal or teh liquid Fermion state and 2) another one is the electron system with a special wave vector which reflects the periodicity of the "spin density lattice". In other words, a resonance state between the magnetic 4f-spin structure and the conduction electron system is self-consistently made through the s-f coupling and the conduction electrons in such a state are considered to be in a coherent state. To make such a coherent periodic structure, the corresponding electron system should have a kind of nesting structure which is confirmed by the band structure in the spin density wave in metallic Cr. Such a resonating state strongly depends on the magnetic phase and magnetic field. The periodicity properties require the existence of a special peak in the density of state near ϵ_F. Moreover, up and down spin systems of the "spin density lattice" may contribute different peaks to the density of states in a field inducing a ferrimagnetic phase.

The coherent state electrons cannot make a large contribution on the resistance, because the electrons in the state have the same periodic property with the "spin density lattice". The incoherently fluctuated electrons mainly affect the resistance. The incoherent electron sees a 4f-spin moment during the scattering which induces the resistance. The scatterer of the isolated spin has the effective moment M, see eq.(1) which are reflected in the properties of the magnetically ordered system. In this model, the formula in ref.[11] is applicable by rewriting the magnetic parameters which characterizes the magnetic order. The magnetoresistance formula for magnetic impurity was already presented using the expectation values of the localized spin and the Zeeman energy normalized with temperature, $g\mu_B H/k_B T$ [11].

The total resistance R is given by the sum of the magnetic part $R(T,H_0)$ and another part $R_0(T)$ originating from non-magnetic scattering:

$$R = R_0(T) + R(T, H_0). \qquad (2)$$

The resistivity R_0 in low temperature depends on the number of thermally excited incoherent electrons in each phase. Therefore, R_0 is considered to be a measure of incoherence in each phase. R_0 is also phase dependent.

In the case of the magnetic ordered crystal at low temperature, the magnetoresistance formula based on the model can be commonly represented for all phases with following the change in magnetic parameters: the parameters are given by the observed values of $\langle M_z^2 \rangle$ ($\cong \langle M_z \rangle^2$), etc. which include the information of magnetically ordered state. The formula in ref.[11] is based on s-f Hamiltonian for Bloch electron and the Boltzmann transport equation. The formula rewritten in terms of the ordered state parameters is given by

$$R(T, H_0) \cong \rho(m^*)[\{\langle M_z^2 \rangle + F(g\mu_B H^*/k_B T)\langle \delta M_\perp^2 \rangle\}$$
$$-\langle \Delta M_z \rangle^2/\{1 + F(g\mu_B H/k_B T)\langle \delta M_\perp^2 \rangle/\langle M_z^2 \rangle\}]. \qquad (3)$$

The parameters of magnetic moment are defined as,

$$\langle M_z^2 \rangle = M^2 \cong \langle M_+^2 + M_-^2 \rangle / 2, \tag{4}$$

$$\langle \Delta M_z \rangle^2 = \langle \vec{M}_+ + \vec{M}_- \rangle^2 / 2, \tag{5}$$

$$\langle \delta M_\perp^2 \rangle \cong \langle M^2 - (\alpha \langle \Delta M_z \rangle)^2 \rangle, \tag{6}$$

$$F(x) = x / \{1 - \exp(-x)\}, \tag{7}$$

$$\rho(m^*) = (\rho_0 m^* / \epsilon_F)(J/N)^2, \tag{8}$$

where r_0, J, N and S are the numerical constant, exchange parameter, number of electrons and magnitude of electron spin, respectively. H* is the effective field on the electron which includes the applied field and the exchange scattering of the electrons. The most simple form of magnetic anisotropy in the band is given by the form of exchange field on each conduction electron. The fitting parameter α in eq.(6) represents another anisotropy of the localized 4f-spin. The term of $\langle M_z^2 \rangle$ in eq.(3) arises from the spin-conserved scattering and the term of $\langle \delta M^2 \rangle$ arises from the spin-reversal scattering process.

In the case of TbNiSn, the anisotropy effect in the conduction band is ignored, because it can be assumed that the large anisotropic moment with 4f-spin makes the main contribution to the anisotropy of the conductivity. The field dependence of the resistances at 1.4K and 4.2K are calculated by use of the magnetization data. The fitting parameter $\rho(m^*)$ or $r_1/2$ proportional to $m^* J/N$ includes the information of the conduction band near ϵ_F. The values of the fitting parameters are listed in Table I. As is seen in Fig.3, the theoretical solid curves at 1.4K and 4.2K given by solid curves are in good agreement with the experimental data. In particular, the parameter α is determined to be 1.15 at the saturation field, consistently with the value of A in eq.(1). As is seen in ref.[6], the remarkably large value of α in phase II at 4.2K is considered as follows: phase II with the magnetization $(1/6)M_s$ has the large magnetic cell and many Brillouin zone boundaries intersect the Fermi surface. The difference in R_0 on phases I and II means that the number of incoherent electrons at 4.2K is much larger than at 1.4K. The difference in ρ suggests that the temperature variation of the conduction band near ϵ_F arises from the super zone effect. It is also remarkable that the determined value of R_0 in Phases II means that the electron number excited from the "spin density lattice" state is 6.7 times larger than the value in Phase I. These results lead us to the physical conclusion that the phase above 2.1K has many more incoherent electrons than the phase below 2.1K ruled by the "magnetic lattice" electrons.

TABLE 1. Fitting parameters : α is approximately isotropic except in phase II.

		Phase				
		I	II	III	IV	V
$\rho^{1/2}(\times 10^{-3}\Omega^{1/2}cm^{1/2}\mu_B^{-1})$	1.4K	0.080	0.045	0.033	0.069	0.067
	4.2K	0.11	0.12	0.094	0.11	0.11
α	1.4K	1.1	1.5	1.1	1.0	1.2
	4.2K	1.4	3.5	1.1	1.1	1.2
$R_0(\times 10^{-6}\Omega cm)$	1.4K	0.13	0.3	0.99	0.13	0.13
	4.2K	0.74	1.3	0.74	0.74	0.74

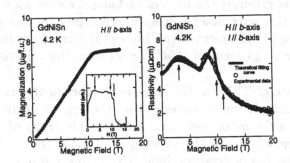

Figure 4. The magnetization and magnetoresistance data in GdNiSn.

4. Discussion

As in our previous work [13], we can also make a fit of the data for GdNiSn though it has quite different characteristics (Fig.4) from those of TbNiSn. The physical validity and the consistency of the resulting parameters will be published in the near future. The fitting curve is consistent with the experimental one. This method gives a good result in the low temperature region because of low excitation numbers.

The most important problem in this method is the proof of the existence of the "spin density lattice" based on the exact solution. If the existence is theoretically confirmed and the structure and properties of the "magnetic lattice" are known, although the work is difficult, the various phenomena which arise from various correlated spin structure in deep $4f$-level can be physically understood. This method may also be effective in analyzing the strongly correlated spin systems, although the strong coupling makes the broadening and overlapping of the peaks in the density of states. These are widely opened as an interesting problem in the near future.

References

[1]   Much work has been done and one example presented in ICM'94 is :J.M.Moreira, M.Salgueiro da Silva, V.S.Amaral, J.B.Sousa, S.B.Palmer, J. Magn. & Magn. Materials 140-144 (1995) 747.

[2]   A.E.Dwight, J. Less-Common Met. 93 (1983) 411.

[3]   C.Routsi, J.K.Yakinthos and H.Gamari-Seale, J. Magn. & Magn. Materials 117 (1992) 79

[4]   M.Kurisu, H.Hori, M.Furusawa, M.Miyake, A.Andoh, I.Oguro, K.Kindo, T.Takeuchi and A.Yamagishi, Physica B 201 (1994) 107.

[5]   T.Takabatake, Y.Nakazawa and M.Ishikawa, Jpn. J. Appl. Phys. 26 Suppl. 26-3 (1987) 547.

[6]   H.Hori, A.Oki, M.Kurisu, M.Furusawa, Y.Andoh, K.Kindo, Phys. Letters A224 (1997) 293.

[7]   T.Suzuki, M.Kitamura, T.Fujita, Y.Andoh M.Kurisu: Abstract of the Meeting of Phys. Soc. of Japan 51 (1993) 124.(in Japanese)

[8]   Y.Andoh, M.Kuris et al., in preparation.

[9]   H.Miwa, Prog. Theor. Phys. 29 (1963) 477.

[10]   R.J. Elliott and F.A. Wedgwood, Proc. Roy. Soc. 81 (1963) 846.

[11]   K.Yosida, Phys.Rev. 107 (1957) 396.

[12]   A.Oki, H.Hori, N.Nunomura, M.Kurisu, M.Furusawa, S.Yamada, A.Andoh, S.Mitsudo, M.Motokawa and K.Kindo, J. Magn.& Magn. Materials 177-181 (1998) 1089.

quote from "THE INNOCENTS ABROAD" (1869)
by MARK TWAIN

J. A. CLAYHOLD
Physics and Astronomy Dept., 306 Kinard Laboratory,
Clemson, SC, 29634 USA

... We anchored here at Yalta two or three days ago. To me the place was a vision of the Sierras. The tall, gray mountains that back it, their sides bristling with pines–cloven with ravines–here and there a hoary rock towering into view–long, straight streaks sweeping down from the summit to the sea, marking the passage of some avalanche of former times–all these were as like what one sees in the Sierras as if the one were a portrait of the other. The little village of Yalta nestles at the foot of an amphitheater which slopes backward and upward to the wall of hills, and looks as if it might have sunk quietly down to its present position from a higher elevation. This depression is covered with the great parks and gardens of noblemen, and through the mass of green foliage the bright colors of their palaces bud out here and there like flowers. It is a beautiful spot.

M. Ausloos and S. Kruchinin (eds.), Symmetry and Pairing in Superconductors, 393.
© 1999 *Kluwer Academic Publishers. Printed in the Netherlands.*

THEORY AND EXPERIMENTAL EVIDENCE FOR PAIR FQHE STATE CAUSED BY ZERO-POINT HARMONIC OSCILLATION IN BLOCK-LAYER POTENTIAL

MASANORI SUGAHARA, SYUHEI MITANI, HONG FEI LU,

YASUHIKO KUMAGAI, HISAYOSHI KANEDA,

NOBUO HANEJI, AND

NOBUYUKI YOSHIKAWA

Faculty of Engineering, Yokohama National University,
Hodogaya, Yokohama, 240-8501, Japan

Abstract. The zero point oscillation state of hole carriers confined in block layer potential of $La_{2-x}Sr_xCuO_4$ is similar to the zero point state in a Landau gauge, when the effective "magnetic field" is greater than $10^3 T$. It is shown that a strong effective field and space charge form layered fractional quantum Hall effect (FQHE) state for c-axis oriented $La_{2-x}Sr_xCuO_4$ films with localization character, for which there is an anomalous dielectric polarization with negative dielectric constant is expected to appear. Experimental study of the dielectric properties of c-axis oriented $La_{2-x}Sr_xCuO_4$ film with oxygen deficiency is made by "ac method" and "dc method". The negativeness of the electric field in $La_{2-x}Sr_xCuO_4$ film and of the film capacitance are observed with hysteretic polarization.

1. INTRODUCTION

Polycrystalline $La_{2-x}Sr_xCuO_4$ samples showed a systematically cyclic-like Sr-doping-level (x) dependence of small current normal resistivity [1]. For $c-$axis oriented epitaxially grown $La_{2-x}Sr_xCuO_4$ oxygen deficient thin films, we found a "negative" dielectric constant which was enhanced at doping levels $x \approx 1/2n$ [2] with hysteretic polarization [3]. A relationship with the fractional quantum Hall effect (FQHE) [4, 5] was pointed out. In this report we discuss the relationship between the doping level dependent anomalies

395

M. Ausloos and S. Kruchinin (eds.), Symmetry and Pairing in Superconductors, 395–408.
© *1999 Kluwer Academic Publishers. Printed in the Netherlands.*

and pair FQHE state based on $//c$ zero-point oscillation of the $2D$ carriers in CuO_2 layers.

2. CARRIER STATE AND DIELECTRIC ANOMALY OF $La_{2-x}Sr_xCuO_4$ FILM

2.1. DOPING-LEVEL-DEPENDENT DIELECTRIC PROPERTY OF $La_{2-x}Sr_xCuO_4$ FILM

We found [2] that the capacitance device (see the inset of Fig. 1) with c-axis-oriented $La_{2-x}Sr_xCuO_4$ layer shows anomalous doping level (x) dependence as shown in Fig. 1, where the ordinate gives the capacitance ratio C_t/C_{STO}, and abscissa x. C_t has the multi-layer structure $Pd/La_{2-x}Sr_xCuO_4$ (300nm) /(100) $SrTiO_3$ (1mm)/Pd, and C_{STO} has the structure $Pd/(100)SrTiO_3$ (1mm)/Pd.

In order to observe the reproductibility of the anomalous x dependence, $La_{2-x}Sr_xCuO_4$ must have a localization character. The latter is introduced by a (constant) deoxidization process during the cooling stage of the epitaxial film, fabricated by sputtering under full oxidation condition. It is known that the slight existence of localization suppresses the normal conductivity at low temperature and superconductivity in $La_{2-x}Sr_xCuO_4$. However, a gradual increase of the dielectric anomaly is found with localization increase.

In Fig. 2, the filling-factor (x) dependence of the FQHE-quasiparticle-excitation energy [4] E_s (order s) per unit quasiparticle charge q is shown: (a) for the FQHE state formed in the Fermi-particle system, and (b) for the FQHE state in the Bose-particle system. E_s is given on the basis of the level of (ground or quasiparticle) Laughlin state [5] for a given filling factor. The ordinates are given in units of $|q_0|/(4\pi\epsilon l)$ (q_0 is the particle charge, l the magnetic length). Comparing Figs.1 and 2, we notice a relationship between the doping-level dependence of the dielectric characteristic and the filling-factor dependence of the FQHE excitation energy of the boson system. We note the following:

1. In high temperature superconductors, we empirically know that a charge-localization property is incompatible with superconductivity (with definite phase and indefinite charge number).
2. On the other hand, it is known that the existence of moderate localization strength enhances [5] the observation of FQHE due to FQHE quasiparticle charge pinning.
3. There is a possibility as seen below that the carrier state of the $La_{2-x}Sr_x$ CuO_4 film with localization character can be explained by a FQHE model for boson systems when identifying the doping level (carrier-

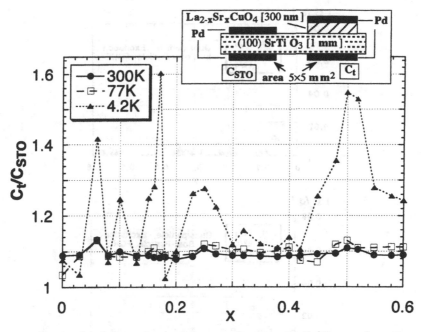

Figure 1. Doping-level dependence of the capacitance ratio C_t/C_{STO}, where the capacitance C_t has the multi-layer $Pd/La_{2-x}Sr_xCuO_4$ (300nm) /(100) $SrTiO_3$ (1mmm)/Pd and C_{STO} has the $Pd/(100)SrTiO_3$(1mm)/Pd structure.

doping rate per Cu site) with the filling factor (carrier-supply rate per Larmor site).

4. The large excitation energy at $x = 1/(2n)$ $(n = 1, 2, 3, ...)$ in Fig. 2 (b) means that there is a large stability of the FQHE state, - when one can expect a strong dielectric anomaly as in the normal state of $La_{2-x}Sr_xCuO_4$ with localization (see below).

2.2. STRONG EFFECTIVE "MAGNETIC FIELD" AND CARRIER PAIRING

At first we point out that the "single-particle" ground-state wave function of the $2D$ carriers in the strong-coupling system has a form similar to the Landau-gauge wave function in a strong "magnetic field". In $La_{2-x}Sr_xCuO_4$ crystal, the $//c$ normal-carrier motion is known to be made by tunneling between CuO_2 layers separated by $d \approx 0.66nm$. This suggests that the $//c$

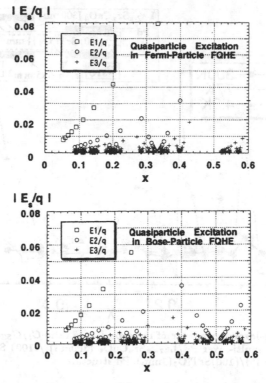

Figure 2. Filling factor x dependence of the quasi-particle excitation energy[4] E_s (order s) per unit quasiparticle charge q; (a) FQHE state of a Fermi-particle system; (b) FQHE state of a Bose-particle system.

wave-function size of the carriers is $r(//c) \leq 0.66nm$, and the related zero-point kinetic energy $\Delta E_0(//c) \geq 0.1eV$. According to the Pippard theory, the $//ab$ wave-function size $\Delta r(//ab)$ is larger than the observed $//ab$ coherence length $\xi(//ab) \approx 3.7nm$, giving the zero-point energy $\Delta E_0(//ab) \leq 0.0028eV$. Based on the large energy inequality $\Delta E_0(//c) \gg \Delta E_0(//ab)$ and the assumption of a parabolic shape of the block-layer carrier-confinement potential, we employ as a first approximation, the following Hamiltonian and zero-point wave function in order to consider the zero-point "single-particle" state in the strong coupling system:

$$H_{\zeta 0} = -(\hbar^2/2m)d^2/dx^2 + (m\omega_b^2/2)(x - \zeta d)^2$$

for

$$\zeta d - d/2 < x < \zeta d + d/2,$$

with

$$\Psi_{\zeta 0} = (\sqrt{\pi}l_{H0})^{-1/2}exp[-(x - \zeta d)^2/2l_{H0}^2]$$

where $\omega_b (\hbar\omega_b/2 \geq 0.1eV)$ is the oscillation frequency, $l_{HO} = \sqrt{\hbar/m\omega_b} \leq 0.66nm$ is the "magnetic length" and ζ is the CuO_2 layer number. The wave function $\Psi_{\zeta 0}$ equals the (zero current) zero-point wave function in the "Landau gauge" $\vec{A}_0(x, \zeta) = (0, B_0(x - \zeta d), 0)$ in a "magnetic field" $B_0 = (0, 0, B_0 \sim 10^3 T)$, whose intensity may be strong enough to cause FQHE even at room temperature on a 2D xy plane. We must note, however, that $A_0(x, \zeta)$ changes its functional form from CuO_2 layer to layer through its ζ dependence, and that only by the presence of the periodic block-layer potential one cannot expect the appearance of FQHE. Indeed this presupposes the existence of a "unique gauge" over the "single particle" states of carriers on a 2D plane.

We must note concerning the effective "magnetic field" B_0 that it affects only the orbital part, and that it is ineffective for the spin part of the carrier wave function. Therefore singlet pairing is allowed in a strong "field" B_0 contrary to real FQHE for a 2D electron gas where the presence of Zeeman energy provides unidirectional spin alignment. Since the zero-point energy $\hbar\omega_b/2 = \hbar^2/2ml_{HO}^2 \propto 1/m$ is lower in the singlet-pairing state than in the unpaired state, pairing of the 2D carriers in $La_{2-x}Sr_xCuO_4$ is plausible even at room temperature for which $kT \approx 0.026eV \ll 0.1eV$.

Next we show that the presence of a $\rho_0 = $ const. space charge works to materialize the "unique gauge". Supposing the existence of a $//y$ current, we rewrite the "single-particle" carrier wave function in the range ($\zeta d - d/2 < x < \zeta d + d/2$) and its Hamiltonian including an electrostatic potential $\phi(x) = -(q_0\rho_0/2\epsilon)x^2$ due to ρ_0 :

$$\Psi_\zeta(x, y) = \chi_\zeta(x)exp(ik_{\zeta y}y) \tag{1}$$

and

$$H_\zeta = (\hbar^2/2m)(-d^2/dx^2 + k_{\zeta y}^2) + (m\omega_b^2/2)(x - \zeta d)^2 - (q_0\rho_0/2\epsilon)x^2 \tag{2}$$

where ϵ is the dielectric constant, and q_0 is the carrier charge. After a simple mathematical manipulation, we can rewrite Eq. (2) as

$$H_\zeta = (\hbar^2/2m)[-(d^2/dx'^2 + l'^{-4}(x' \mp l'^2 k_{\zeta y})^2] \tag{3}$$

where $\omega_0 = \sqrt{q_0\rho_0/\epsilon m}$, $k_{\zeta y} = \pm m\omega_0\zeta d/\hbar\sqrt{1 - \omega_0^2/\omega_b^2}$, $x' = (\omega_b/\omega_0)x$, $m' = (\omega_0/\omega_b)^2 m$, $\omega_b' = \omega_b\sqrt{1 - \omega_0^2/\omega_b^2}$ and $l'^2 = \hbar/m'\omega_b'$. Equation (3) denotes that carriers are found to be in a "unique gauge" $\vec{A}_0'(x') = (0, B_0'x', 0)$ ($B_0' = m'\omega_b'/q_0$) which is independent of ζ.

The following must be noticed concerning the appearance of FQHE state in c-axis oriented $La_{2-x}Sr_xCuO_4$ films:

1. Since a FQHE quasiparticle is known to have fractional excess charge,[5] and since electrostatic shielding length in FQHE system increases due to $E \perp J$ (E electric field, J current density) property, a quasiparticle Laughlin state provides the ground state with $\pm\rho_0$ constant-charge,

2. The seemingly unnatural "single-particle" wave-number property $k_{\zeta y} \propto \zeta$ may be feasible with the help of the strong coupling character and the strong effective "magnetic field" where the FQHE configuration[5] is stabilized,

3. Equation (3) shows that the $//c$ confinement potential V_ζ suffers deformation by the presence of the electrostatic potential $\phi(x) = \mp(q_0\rho_0/2\epsilon) x^2$ made by a space charge $\pm\rho_0$, i.e.

$$V_\zeta(x,0) \propto [x - \zeta d]^2 \rightarrow V_\zeta(x, \pm\rho_0) \propto [x - \zeta d/(1 \mp \omega_0^2/\omega_b^2)]^2 \quad (4)$$

for $\zeta d - d/2 < x < \zeta d + d/2$.

Expression (4) leads to the existence of a limiting ζ value, i.e. $\zeta_{max} = \gamma\omega_b^2/\omega_0^2$ (with $\gamma \sim 1$). For $|\zeta| < \zeta_{max}$, we can safely disregard the V_ζ deformation. Since a FQHE state may spread within the range where the $//c$ carrier confinement is practically identical on each CuO_2 layer, there exists a $//x$ limiting thickness $\eta \approx \zeta_{max} d$ for one FQHE (layer) region.

2.3. "CHARGE DOMAIN" AND POLARIZATION PROPERTY

Based on the above discussion, we introduce a model for the carriers of a c-axis oriented $La_{2-x}Sr_xCuO_4$ film crystal with localization character, for which the lamination of $\pm\rho_0$ charged quasiparticle Laughlin states of the hole-pairs are established in the ground state. The lowering of the electrostatic energy of "ground charge" $\pm\rho_0$ necessitates the formation of a "charge domain" structure composed of thin charged dipole layers, where all the electric force lines starting from $+\rho_0$ terminate at the closest $-\rho_0$. Based on the concept of "charge domains" and their "inversion" dynamics in polarization process, i.e. by analogy to the magnetic domain dynamics in ferromagnets, we can expect to find a hysteretic polarization behavior in $La_{2-x}Sr_xCuO_4$ film with localization character, where FQHE quasiparticle fixing by irregular potential provides the "pinning force".

In Fig. 3 (a) is schematically shown the structure of a dipole layer of "charge domain" with thickness $2\eta = 2\zeta_{max}d$ and potential difference $\phi_d = \rho_0\eta^2/\epsilon$. One must notice that the "ground electric field" due to the gradient of $\phi(x)$ does not provide the gradient of the chemical potential, lacking the ability to cause macroscopic current of FQHE particles. This is seen from the fact that the modified Hamiltonian Eq. (3) in the presence of $\phi(x)$ gives an energy eigenvalue which is independent of the coordinates. Suppose a c-axis oriented $La_{2-x}Sr_xCuO_4$ film with localization character in zero exter-

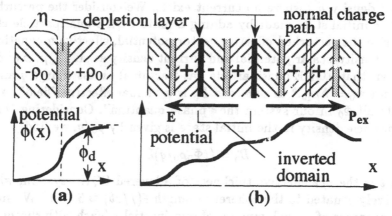

Figure 3. (a) Structure of a "charge domain". (b) Array of "charge domains" with one inverted "domain" when finite macroscopic electrostatic potential difference exists across the film.

nally applied $//c$ potential difference. The layered domains in the film must develop an array like $(-\rho_0/+\rho_0)(+\rho_0/-\rho_0)(-\rho_0/+\rho_0)\cdots$ with no macroscopic field $\langle E \rangle$ and macroscopic polarization $\langle P \rangle$. In Fig. 3 (b) is shown the case when a finite $//c$ macroscopic potential difference exists, where finite $\langle E \rangle$ and $\langle P \rangle$ appear with some charge domain inversion. One must notice that the senses of $\langle E \rangle$ and $\langle P \rangle$ are reverse to each other suggesting that the macroscopic dielectric constant is negative. The increase of an externally imposed macroscopic field (or polarization $\langle P \rangle$) strengthens the domain alignment as $(-\rho_0/+\rho_0)(-\rho_0/+\rho_0)(-\rho_0/+\rho_0)\cdots$. Since this configuration has lower electrostatic energy than $(-\rho_0/+\rho_0)(+\rho_0/-\rho_0)(-\rho_0/+\rho_0)\cdots$, we may expect the self-alignment of domain polarization just as in ferromagnets. An ordinary c-axis oriented $La_{2-x}Sr_xCuO_4$ film is composed of many $//ab$ flat crystals each of which has different polarization characteristics. Therefore the averaging $\langle\rangle$ concerning $\langle E \rangle$ and $\langle P \rangle$ must be taken not only $//c$ but also $//ab$.

2.4. NORMAL CHARGE REPULSION IN FQHE REGION

Here we discuss the "normal" charge repulsion in 2D (xy surface) for the FQHE state made of a q_0-charged boson system with q-charged quasiparticles. The discussion is made supposing an ideal state such as seen in c-axis oriented $La_{2-x}Sr_x CuO_4$ films with moderate localization character at $T \cong 0$. In the initial state under no normal-charge influence the FQHE system is supposed to have a FQHE-particle density $|\Psi_0|^2 = n_0$, effective vector potential A_0', effective magnetic field $B_0' = (0, 0, B_0')$ and "ground charge" density ρ_0, where no current exists. We consider the perturbation from the initial state made by adding a finite FQHE quasiparticle charge, and write the perturbed effective vector potential, effective magnetic field, charge density, scalar potential and current density as $\tilde{A}$, $\tilde{B}$, $\tilde{\rho}$, $\tilde{\phi}$ and $\tilde{J}$. It is known that FQHE quasiparticle has fractional excess charge q and flux quantum $\Phi_0 = h/q_0$.[5] In our case B_0' is a virtual effective "field", and we should call Φ_0 in our system the "phase quantum". Considering that the quasiparticle density in the initial state is given by ρ_0/q, we find

$$B_0' = (t\Phi_0\rho_0/q)k ,$$

where t is the $//z$ effective thickness of a layered FQHE system, which is tentatively equated to the coherence length $\xi(//ab) \approx 3.7 nm$. We suppose the appearance of a small amount of quasiparticles (each with charge q and phase quantum Φ_0) with density $|\Psi|^2 - n_0$, where Ψ is the macroscopic wave function of the FQHE system.[7] The excess charge density

$$\tilde{\rho} = q(|\Psi|^2 - n_0)$$

produces an electrostatic field with FQHE-particles acceleration, and the magnetic field

$$\nabla \times \tilde{A} = \tilde{B} = (t\Phi_0/q)\tilde{\rho}k \qquad (5)$$

The current is given by the usual relation

$$\tilde{J} = -(q_0/2m)[\Psi^*(i\hbar\nabla + q_0 \tilde{A})\Psi + \Psi(-i\hbar\nabla + q_0 \tilde{A}\Psi^*)].$$

In this case the FQHE wave function is robust[5] to perturbation, whence we put $\nabla\Psi = 0$ and find

$$\tilde{J} = (q_0^2|\Psi|^2/m)\tilde{A}. \qquad (6)$$

The excess electric field $\tilde{E} = (\partial\tilde{\phi}/\partial x, \partial\tilde{\phi}/\partial y, 0)$ made by $\tilde{\rho}$ forms the FQHE relationship with current

$$\tilde{J} = (q/\Phi_0 t)(-\partial\tilde{\phi}/\partial y, \partial\tilde{\phi}/\partial x, 0) = -(q/\Phi_0 t)\nabla \times (\tilde{\phi}k) \qquad (7)$$

keeping $\tilde{E}\cdot\tilde{J} = 0$. From Eqs. (5)-(7) and Maxwell equation $\nabla\times\tilde{B} = \mu_0\tilde{J}$ (μ_0, vacuum permeability), we find

$$\tilde{\rho} - (\mu_0 q^2/t^2\Phi_0^2)\tilde{\phi} \qquad (8)$$

$$\nabla^2\tilde{\phi} = -\tilde{\rho}/\epsilon_s \qquad (9)$$

where $\epsilon_s = mq^2/q_0^2\Phi_0^2 t^2 n_0$, and

$$\nabla^2 \tilde{\phi} = -\tilde{\phi}/\lambda_s^2 \tag{10}$$

where $\lambda_s^2 = m/\mu_0 q_0^2 n_0$.

These equations (8)-(10) show the following:

1. based on Eq. (8) which is an analog of the London equation, we know the existence of an intrinsic negative capacitance

$$C_{in} \sim \lambda_s \rho/\phi = -\epsilon_s/\lambda_s < 0$$

per unit $//ab$ area per charge domain in the $//c$ polarization,

2. equation (9) denotes that a FQHE region has the dielectric property with dielectric constant ϵ_s for the field $\tilde{E}$, where $\tilde{E} \cdot \tilde{J} = 0$,

3. equation (10) denotes that the pertubation electric field induced by the excess charge is shielded at a distance $\lambda_s \sim 10^{-7} m$,

4. because of the charge repulsion property analogous to the flux repulsion property in the Meissner effect, externally applied "normal" charge can get into a FQHE region only through the "normal charge paths" which are formed at the charge-domain inversion as shown in Fig. 3(b). There the FQHE property is weak with no $//c$ potential gradient.

2.5. POTENTIAL IRREGULARITY AND QUASIPARTICLE EXCITATION ENERGY

As is known for ordinary 2D electron gas FQHE, a strong potential irregularity destroys FQHE, but moderate irregularity enhances FQHE observation due to quasiparticle localization.[5] The optimum irregularity density for stabilization of our FQHE may be $10^{2\sim3}$ times larger than for ordinary FQHE considering "magnetic length" difference. In comparison with ordinary FQHE, our FQHE in a 3D sample has extra geometrical degrees of freedom. Therefore the clear observation of FQHE phenomena may be difficult except in special cases for which some particle localization conditions are satisfied : (a) small sample with 2D-like geometrical feature such as thin film crystal, (b) crystalline sample having inhomogeneity composition with micro-structures such that the possibility of a 2D boundary condition exists.

3. EXPERIMENT

The capacitance measurements have been made by two methods: (a) the capacitance - meter method ("ac method") where we study the differential capacitance $C_{ac} = dQ/dV$, and (b) the step-voltage-response method ("dc method") where we study the large amplitude capacitance $C_{dc} = [Q(V) - Q(0)]/V$ measuring the stored charge for step-voltage input.[3] The "ac method" is fine for the study of the small amplitude polarization at constant dc bias voltage, and "dc method" for hysteresis characteristics. Using these methods the capacitance C_t of Pd/La$_{2-x}$Sr$_x$ CuO$_4$/(100)SrTiO$_3$(1mm)/Pd structure and C_{STO} of Pd/(100)SrTiO$_3$(1mm)/Pd structure (see the inset of Fig. (2) are measured and compared. The epitaxially grown La$_{2-x}$Sr$_x$CuO$_4$ films were made by sputtering on a 1 mm thick (100)SrTiO$_3$ substrate at $700°C$ in fully oxidization condition. The (001) crystallization is confirmed by XRD and RHEED. Since carrier localization is important[6] in this experiment as noticed above, we intentionally introduce irregularity in the La$_{2-x}$Sr$_x$CuO$_4$ film by "deoxidization" in the cooling stage, i.e. pumping out oxygen at $T \leq T_d$. We call a La$_{2-x}$Sr$_x$CuO$_4$ film with $T_d = 700°C$ a "type 700", and for $T_d = 500°C$ a "type 500". All La$_{2-x}$Sr$_x$CuO$_4$ films have semiconductive $R - T$ characteristics below 100K. The doping-level (x) dependence of C_t/C_{STO} of Fig. 1 is obtained using "type 700" sample with 300 nm thick La$_{2-x}$Sr$_x$CuO$_4$ film by the "ac method" (without dc bias voltage). A C_t/C_{STO} increase is observed at $x = 1/2n$ (n = integer), suggesting a strong relationship with FQHE.[2]

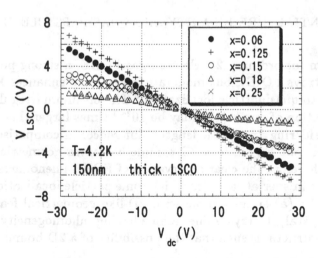

Figure 4. Bias voltage (V_{dc}) dependence of the "internal voltage" of a La$_{2-x}$Sr$_x$CuO$_4$ film measured by the "ac method".

In Figs. 4 and 5, the dc-bias-voltage (V_{dc}) dependence of the dielectric property is shown as obtained by the "ac method" for a 150nm thick La$_{2-x}$Sr$_x$CuO$_4$ film of "type 500". It is seen that the //c "internal voltage" V_{LSCO} of La$_{2-x}$Sr$_x$CuO$_4$ layer is negative, i.e. the electric field in La$_{2-x}$Sr$_x$CuO$_4$ has a reverse sense to that of the polarization in the SrTiO$_3$ substrate. |V_{LSCO}| is found to increase monotonically with the increase of the polarization with some saturation tendency. In Fig. 5 the $La_{2-x}Sr_xCuO_4$ "capacitance" C_{LSCO} is seen to be negative. Because of the inverse proportionality $C_{LSCO} \propto V_{LSCO}^{-1}$ samples with a larger $|V_{LSCO}|$ value in Fig. 4 gives a smaller $|C_{LSCO}|$ value in Fig. 5.

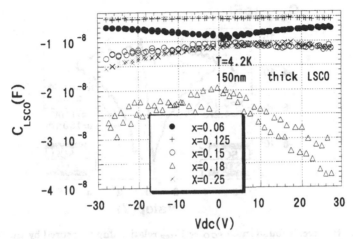

Figure 5. dc-bias-voltage(V_{dc}) dependence of the "capacitance" of a $La_{2-x}Sr_xCuO_4$ film measured by the "ac method".

Considering the ferroelectricity of the $SrTiO_3$ substrate, the values of V_{LSCO} and C_{LSCO} are determined as follows under the condition of the same charge storage in the substrate layers in C_t and C_{STO}:

1. from the experimentally observed relationships of $C_t \sim V_{dc}$ and $C_{STO} \sim V_{dc}$, we obtain $C_t V_{dc}(\rightarrow Q_1) \sim V_{dc}(\rightarrow V_{STO} + V_{LSCO})$ and $C_{STO}V_{dc}(\rightarrow Q_2) \sim V_{dc}(\rightarrow V_{STO})$,
2. setting $Q_1 = Q_2 = Q$, we get $V_{STO}(Q)$ and $V_{LSCO}(Q)$;
3. then we find the "capacitances" $C_{STO}(Q) = Q/V_{STO}(Q)$ and $C_{LSCO}(Q) = Q/V_{LSCO}$.

Next we show in Fig. 6 the typical "hysteresis" found in $C_{LSCO} \sim V_{step}$ relationship of a 150 nm thick $La_{2-x}Sr_xCuO_4$ film measured by the "dc method", and in Fig.7 "dips" found in the $La_{2-x}Sr_xCuO_4$-thickness dependence of C_t/C_{STO} ($V_{dc} = 0$) by the "ac method" on "type 700" samples.

4. COMPARISON BETWEEN THE FQHE MODEL AND EXPERIMENT

1. The polarization mechanisms discussed in Secs. 2.3 and/or 2.4 suggest the existence of negative "internal voltage" and the negativity of "ca-

pacitance" of the FQHE region. They are actually found in the experiment on the dc-bias-voltage dependence of C_{LSCO} of $La_{2-x}Sr_xCuO_4$ film with localization character as seen in Figs. 4 and 5.

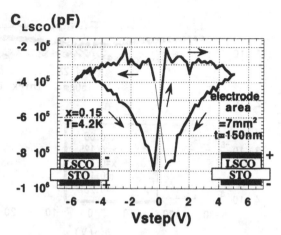

Figure 6. Hysteresis found in $C_{LSCO} \sim V_{step}$ relationship measured by the "dc method".

2. The hysteresis in the step-voltage (V_{step}) dependence of the "capacitance" C_{LSCO} measured by the "dc method" in Fig. 6 is well explained by "charge domain" pinning in the polarization process originating from the localization of FQHE quasiparticles by disorder.

3. the peaks and dips found in the dependence of C_T/C_{STO} on $La_{2-x}Sr_x$ CuO_4 film thickness shown in Fig. 7 may be a direct evidence of the existence of "charge domain" structure. When the film thickness satisfies certain geometrical quantum conditions, the dielectric response may be enhanced (or suppressed). This leads to a regular change of the dielectric anomaly at special thickness values of a $La_{2-x}Sr_xCuO_4$ layered system.

5. CONCLUSION

The carrier state in $La_{2-x}Sr_xCuO_4$ has been discussed considering the $//c$ zero-point-oscillation of 2D carriers confined by block-layer potential. The carrier behavior is found to be described by a FQHE model with anomalous dielectric characteristics.

The dielectric anomaly of c-axis oriented $La_{2-x}Sr_xCuO_4$ films with localization character has been experimentally studied by the "ac method"

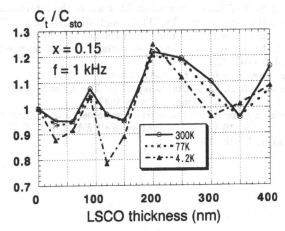

Figure 7. La$_{2-x}$Sr$_x$CuO$_4$ thickness dependence of C_T/C_{STO} measured by the "ac method".

measuring the differential capacitance $C_{ac} = dQ/dV$ and the "dc method" measuring the large amplitude capacitance $C_{dc} = [Q(V) - Q(0)]/V$. In the "ac method" we confirm that the sense of the "internal" electric field in La$_{2-x}$Sr$_x$CuO$_4$ film is reverse to that of the polarization in the SrTiO$_3$ substrate. The observed "capacitance" C_{LSCO} of La$_{2-x}$Sr$_x$CuO$_4$ film is also negative. Besides, the dependence of the capacitance ratio C_t/C_{STO} on La$_{2-x}$Sr$_x$CuO$_4$ film thickness has dip. In the "dc method" we observe a hysteresis in the $C_{LSCO} \sim V_{step}$ relationship. These experiments are found to be consistent with the FQHE model based on the $//c$ zero-point oscillation. Thus,

1. the x dependence of the FQHE excitation energy well explains the x dependence of C_t/C_{STO} in Ref. 2.
2. the formation of "charge-domain" structure and polarization based on the charge domain inversion and/or on charge repulsion explain well the negative capacitance and hysteresis.
3. the dip found in the film-thickness dependence C_t/C_{STO} is interpreted as the direct evidence of the existence of charge domains.

References

1. M.Sugahara and J.F.Jiang, Appl.Phys.Lett. **63**, 255 (1993); Physica B **194-196**, 2177 (1994).
2. M.Sugahara, S.B. Wu, X.Y. Han, H.F. Lu, H. Kaneda, N. Haneji and N. Yoshikawa, Physica C **293**, 216 (1997); M. Sugahara, S.B. Wu, X.Y. Han, H.F. Lu, H.Q. Yin,

408

N. Haneji and N. Yoshikawa, Extended Abstract of the International Workshop on Superconductivity cosponsored by ISTEC and MRS, Hawaii, 1997, 377.

3. M. Sugahara, H.F. Lu, H.Q. Yin, Y. Kumagai, M. Miyata, N. Haneji and N. Yoshikawa, to be published in the Proceedings of 10th International symposium on Superconductivity, Gifu, 1997.

4. B.I.Halperin, Phys.Rev.Lett. **52**, 1583 (1984).

5. See, for example, R.B. Laughlin, Chap. 7 in *The Quantum Hall Effect*, (ed. R.E. Prange and S.M. Girvin, Springer, Berlin 1987).

6. In samples which show superconductivity we observe no dielectric anomaly..

7. S.M. Girvin, Chap. 10 in *The Quantum Hall Effect*, (ed. R.E. Prange and S.M. Girvin, Springer, Berlin 1987).

INDEX

Abrikosov state 131, 302
Andreev scattering 272
anisotropy 289
ARPES 164, 219, 234, 265

Bi-based compounds 27, 127, 152
156, 168, 195, 319, 363

charge density wave 121, 187
clusters 347
coherence length 32, 48, 138
conductivity 33, 293, 363
correlations 111

diffusivity,
- thermal 226
dimensionality
- 1D 213, 330
- 2D 83, 199, 212,
doping 155, 178, 182, 227, 298
- oxygen 291
d-wave, d-pairing 71, 105, 129, 173,
221, 345

EPR 347

Fermi surface 64, 102, 123, 222, 361
film 35
fluctuation 36, 109, 219, 295, 319, 359,
368, 381
flux 141, 337

gap
- anisotropy 75, 259, 261, 274
- equation 113, 213
- pseudo- 42, 83
- symmetry 23, 121
Ginzburg-Landau 71, 248, 255, 378
grain boundary 167, 338
Green function 46, 112, 189, 233

heat capacity 163, 227
Hg-based compounds 127, 195, 363-5
Hubbard model 44, 85, 111, 212, 246,
260, 361

interlayer
- coupling 328, 360
- tunneling 167, 175, 211, 262
I-V characteristics 12, 145, 194-8

Jahn-Teller 174
Josephson junction 166